Analytical Instrumentation

Analytical Instrumentation A Guide to Laboratory, Portable and Miniaturized Instruments

First Edition

GILLIAN MCMAHON

School of Chemical Sciences
Dublin City University
Ireland

John Wiley & Sons, Ltd

Telephone (+44) 1243 779777

Email (for orders and customer service enquiries): cs-books@wiley.co.uk
Visit our Home Page on www.wileyeurope.com or www.wiley.com

Other Wiley Editorial Offices

John Wiley & Sons Inc., 111 River Street, Hoboken, NJ 07030, USA

Jossey-Bass, 989 Market Street, San Francisco, CA 94103-1741, USA

Wiley-VCH Verlag GmbH, Boschstr. 12, D-69469 Weinheim, Germany

John Wiley & Sons Australia Ltd, 42 McDougall Street, Milton, Queensland 4064, Australia

John Wiley & Sons (Asia) Pte Ltd, 2 Clementi Loop #02-01, Jin Xing Distripark, Singapore 129809

John Wiley & Sons Ltd, 6045 Freemont Blvd, Mississauga, Ontario L5R 4J3, Canada

Wiley also publishes its books in a variety of electronic formats. Some content that appears in print may not be available in electronic books.

Anniversary Logo Design: Richard J. Pacifico

British Library Cataloguing in Publication Data

A catalogue record for this book is available from the British Library

ISBN 978-0470-027950

Typeset in 10/12 pt Times by Thomson Digital
Printed and bound in Great Britain by Antony Rowe Ltd, Chippenham, Wiltshire, England

This book is printed on acid-free paper responsibly manufactured from sustainable forestry in which at least two trees are planted for each one used for paper production.

For Sophie and Charlie

Contents

Foreword

This book has arisen from a series of lectures developed by Dr Gillan McMahon and delivered to students on the taught postgraduate module on instrumentation at Dublin City University. Gillian was previously herself a student in DCU and since graduating, she has developed her analytical background initially in industry in the pharmachem arena, and more recently, as a very successful academic teacher and researcher. She gained a wealth of experience over a broad range of analytical techniques in the pharmachem industry, working with the Geotest Chemical Company (USA), Newport Pharmaceuticals (Ireland), Bristol-Myers Squibb (Ireland) and Zeneca Pharmaceuticals (UK). This experience applied not just to the use of techniques and methods, but also to data tracking and compliance, which is a critical aspect for this sector. While with BMS, she was engaged in training of staff in advanced analytical techniques and compliance at other sites in Italy and Puerto Rico prior to production campaigns.

Her academic career as an analytical scientist is equally impressive. She completed her PhD research at the Lombardi Cancer Centre, Georgetown University, USA and currently is a lecturer at Dublin City University, where she teaches on the two national flagship analytical courses (Analytical Science BSc and Instrumental Analysis MSc). In addition to her impressive research publications, and activities in professional bodies like the Royal Society of Chemistry and Institute of Chemistry of Ireland, Gillian has also won significant external research funding, and has been the recipient of numerous individual awards for dissemination.

Gillian therefore brings a rare, but vitally important mix of experience to this text. Analytical science is a complex discipline, ranging from instrumentation, electronics, optics, through data processing and statistics, to the fundamental science of molecular recognition and transduction. Analytical techniques are employed in an every-increasing range of applications. Along with synthetic chemistry, it provides the cornerstone of the pharmceutical industry. Without analytical information and new methods, the human genome project would never have been realised, and high throughput bioanalytical instruments are now helping to unravel the secrets of human genetic disposition to disease.

Analytical instruments are routinely used to monitor the status of our environment and the quality of our food, and to enable individuals to track personal health indicators. And of course, where would forensic science be without analytical instruments? Devising a text to teach the principles and practice of analytical science to students with a wide diversity of educational backgrounds requires a balance between depth and breadth, and above all, a systematic, consistent approach. In this text, Gillian has met this formidable challenge, and the result is a clearly written and structured text that reveals the basis of the key instrumental methods, and the importance they play in many aspects of modern life. The clarity of the explanations will appeal to both undergraduate and postgraduate students, as well as scientists in industry and will help guide them in a practical way towards particular specialisms they may find interesting as they move through their career.

The text breaks new ground in that it takes the reader all the way from large, lab-based instruments through to on-line and in-line instruments for industry, to portable and hand-held equipment and finally to micro-scale lab-on-a-chip devices. This offers an alternative approach for teaching modern instrumentation. It covers a wide range of modern instrumental methods in a practical and relevant way, including techniques not traditionally covered in analytical instrumentation texts, such as the imaging techniques which are becoming ubiquitous in modern analytical laboratories. Gillian's background in compliance comes through in the section covering on-line and in-line instruments wherein she covers not only the sensing and analytical techniques used in process analysis, but also the new FDA-driven phenomenon of process analytical technology (PAT).

Always appealing to students is the ability to make the technology and science relevant. Gillian excels in this respect, linking analytical platforms to numerous specific examples of applications ranging across healthcare, the environment and the pharmaceutical industry.

In conclusion, this is an exciting new resource for analytical science education that, I have no doubt, will prove to be popular with students and educators alike. I will certainly have a copy on my shelf!

Dermot Diamond
BSc, PhD (QUB), PGCE, MICI, MRSC, C.Chem.
Science Foundation Ireland Principle Investigator
in the 'Adaptive Information Cluster'

Preface

The idea for this book on analytical instrumentation came after I was given the task of writing and delivering a new lecture course entitled 'Instrumentation'. The course comprised a module in a taught postgraduate Masters in Instrumental Analysis. When I examined the overall course content, the background of the students and the aim of the qualification, I realised that I wanted it to be more than an explanation of the theory and practice of standard analytical instrumental techniques – something that is already a formidable task on its own. I felt strongly that the course should mirror recent trends in instrumentation such as the development of portable and point-of-care instruments, use of field devices, the significant integration of analytical equipment into industrial processes and the area of miniaturisation. And since the course is pitched mainly at professional scientists working in industry, the emphasis, I felt, should lean towards the practical rather than the theoretical side of such knowledge.

While preparing the module lectures, I found myself reading across many disciplines, from chemistry to engineering, learning about a range of technologies from biology to physics and browsing many different aisles in the library, from medicine to regulation. And yet, there was no one textbook that I could find to help me teach the course as I felt it should be taught. And so the seed was sown and the rest is history....

I have tried to take a logical approach in the book by moving from the discussion of large instruments at the beginning of the book to small instruments at the end of the book. This also means that the book moves from traditional equipment through modern technology to instruments only described in the literature, and at the same time from commercially available equipment to devices only at the research and development stage.

Chapter 1 is a short introduction to analytical instrumentation and the analytical process in general. I also explain a little about my approach.

Section I covers the more conventional equipment available for analytical scientists. I have used a unified means of illustrating the composition of instruments over the five chapters in this section. This system describes each piece of equipment in terms of five modules – source, sample, discriminator, detector and output device. I believe this system allows for easily comparing and contrasting of instruments across the various categories, as opposed to other texts where different instrument types are represented by different schematic styles. Chapter 2 in this section describes the spectroscopic techniques of visible and ultraviolet spectrophotometry, near infrared, mid-infrared and Raman spectrometry, fluorescence and phosphorescence, nuclear magnetic resonance, mass spectrometry and, finally, a section on atomic spectrometric techniques. I have used the aspirin molecule as an example all the way through this section so that the spectral data obtained from each technique for a simple organic compound can be compared and contrasted easily. Chapter 3 discusses separation techniques such as the well-known gas and liquid chromatographies, capillary electrophoresis and supercritical fluid chromatography. The latter part of Chapter 3 is devoted to hyphenated (hybrid) techniques since these are so important in today's laboratory where complex mixtures often need to be separated prior to identification and quantitation and where these demands can be met in one run. I also explain some of the challenges that have been overcome in coupling instruments together effectively. Chapter 4 outlines the imaging methods that are becoming so much more prevalent in analytical science where single-atom resolution is now possible. Not only are these appliances useful as stand-alone instruments but often they are linked to spectral devices to enable spectral imaging, an even more powerful tool. Chapter 5 describes the electrochemical methods of potentiometry, voltammetry and conductivity measurement. Chapter 6 briefly covers thermoanalytical and diffraction methods.

Section II moves into the realm of smaller instruments with a discussion of why there is a drive to make devices more portable, the use of portable instruments in the laboratory (with plenty of commercially-available examples) and uses of portable devices in medical and environmental applications. Special emphasis is placed on point-of-care meters for blood glucose testing and coagulation monitoring as their technologies are based on simple, rugged chemical tests. Portable instruments in environmental monitoring have made field testing a reality.

Section III discusses process analytical instrumentation, which is a big growth area in science, especially in the petrochemical, food and beverage and pharmaceutical industries. Manufacturers have had to shift the analytical emphasis of their equipment from sensitive to rugged and analytical scientists have had to think like and work with engineers in order to install on-line and in-line assays. After discussing in-process sampling and in-process analysis, a number of examples are given of instruments that are being used in process analytics applications.

Section IV then tackles the most recent trend in analytical instrumentation, which is miniaturisation and the drive to create lab-on-a-chip devices. In this section, I discuss the development of chip-based technologies and the challenges associated with this such as pumping fluids on the microscale, fitting components onto a chip, detection strategies and how processes such as mixing are so different in the microworld when compared to the macroworld.

As a final note, it is clear that analytical instrumentation is developing at a very fast pace and getting smaller, smarter and faster every year. I hope that by reading some or all of this book that the reader will have learned something new and found the journey interesting along the way.

Gillian McMahon

Acknowledgements

I wish to express my gratitude to my colleagues in the School of Chemical Sciences in Dublin City University, Ireland, who gave their time and expertise to review chapters of this book for me. Their feedback kept me on track and their suggestions have helped shape the text into its current form. My sincere thanks to: Prof. Han Vos, Prof. Dermot Diamond, Prof. Conor Long, Prof. Brett Paull, Dr Fiona Regan, Dr Mary Pryce, Dr Dermot Brougham, Dr Peter Kenny and Dr Tia Keyes

I would also like to thank my brother Graham McMahon, MD, Assistant Professor of Medicine, Harvard Medical School, for his very helpful comments.

Acknowledgements

Acronyms and Abbreviations

2-D	two-dimensional
3-D	three-dimensional
AAS	atomic absorption spectrometry
AC	alternating current
ACN	acetonitrile
ADC	analogue-to-digital converter
AES	atomic emission spectrometry
AFM	atomic force microscopy
AFS	atomic fluorescence spectrometry
APCI	atmospheric pressure chemical ionisation
API	atmospheric pressure ionisation
ASV	adsorptive stripping voltammetry
ATR	attenuated total reflectance
CCD	charge-coupled device
CE	capillary electrophoresis
CEC	capillary electrochromatography
CFA	continuous flow analysis
CFM	chemical force microscopy
CGE	capillary gel electrophoresis
CGMS	continuous glucose monitoring system
CI	chemical ionisation
CIT	cylindrical ion trap
CID	charge-injection device
CME	chemically modified electrode
COSY	correlation spectrometry
CT	computerised tomography
CZE	capillary zone electrophoresis
DAD	diode array detector
Dart	direct analysis in real time
DC	direct current
DEPT	distortionless enhancement by polarisation transfer
Desi	desorption electrospray ionisation
DME	dropping mercury electrode
DNA	deoxyribonucleic acid
DPP	differential pulsed polarography

DPV	differential pulsed voltammetry
DSC	differential scanning calorimetry
DTA	differential thermal analysis
DTG	derivative thermogravimetry
DTS	digital transform spectrometer
ECD	electron capture detector
EDX	energy-dispersive X-ray
EI	electron impact
EOF	electro-osmotic flow
ESED	environmental secondary electron detector
ESI	electrospray ionisation
ETAAS	electrothermal atomic absorption spectrometry
FAAS	flame atomic absorption spectrometry
FAB	fast atom bombardment
FANSOM	fluorescence apertureless near-field scanning microscope
FBRM	focused beam reflectance measurement
FDA	Food and Drug Administration
FIA	flow injection analysis
FID	flame ionisation detector
FID	free induction decay
FPD	flame photometric detector
FRET	fluorescence resonance transfer
FT	Fourier transform
FT–ICR	Fourier transform–ion cyclotron resonance
FTIR	Fourier transform infrared
GC	gas chromatography
GFAAS	graphite furnace atomic absorption spectrometry
GLP	good laboratory practice
GMP	good manufacturing practice
GO	glucose oxidase
GPS	global positioning system
GSM	global system for mobile communications
hCG	human chorionic gonadotrophin
HCL	hollow cathode lamp
HGAAS	hydride generation atomic absorption spectrometry
HIV	human immunodeficiency virus
HPLC	high performance liquid chromatography
IC	internal conversion
IC	integrated circuit
IC	ion chromatography
ICP	inductively coupled plasma
IMS	ion mobility spectrometry
INR	international normalised ratio
IR	infrared
ISC	inter-system crossing

ISE	ion-selective electrode
ISFET	ion-selective field effect transistors
IT	ion-trap
LC	liquid chromatography
LCD	liquid crystal display
LED	light emitting diode
LIF	laser-induced fluorescence
LIMS	laboratory integrated management system
LOC	lab-on-a-chip
LOD	limit of detection
LOQ	limit of quantitation
LOV	lab-on-valve
LTQ	linear trap quadrupole
m/z	mass-to-charge ratio
MALDI	matrix-assisted laser desorption ionisation
MDSC	modulated differential scanning calorimetry
MECC	micellar electrokinetic capillary chromatography
MEKC	micellar electrokinetic chromatography
MEMS	micro electro mechanical systems
MIP	molecular-imprinted polymer
ML	mercury liberation
MRI	magnetic resonance imaging
MRM	multiple reaction monitoring
MS	mass spectrometry
MS^2	tandem mass spectrometry
MST	micro system technology
NACE	non-aqueous capillary electrophoresis
Nd:YAG	neodymium yttrium aluminium garnet
NFOM	near-field optical microscopy
NIR	near infrared
NIRA	near infrared reflectance analysis
NMR	nuclear magnetic resonance
NPD	nitrogen phosphorus detector
OES	optical emission spectrometry
ORP	oxidation reduction potential
PAH	polycyclic aromatic hydrocarbons
PAT	process analytical technology
PC	personal computer
PCB	polychlorinated biphenyls
PCR	polymerase chain reaction
PDA	photodiode array
PDMS	polydimethylsiloxane
PID	photoionisation detector
PIOP	paramagnetic iron oxide particles
PMMA	poly(methylmethacrylate)

PMT	photomultiplier tube
POC	point-of-care
ppb	parts per billion
ppm	parts per million
ppt	parts per trillion
PSA	prostate specific antigen
PT	prothrombin time
PTT	partial thromboplastin time
PVC	polyvinyl chloride
PVM	particle video microscopy
RCP	reducing compound photometer
RF	radiofrequency
RFIC	reagent-free ion chromatography
RI	refractive index
RNA	ribonucleic acid
RR	resonance Raman
RSD	relative standard deviation
SAW	surface acoustic wave
SCE	saturated calomel electrode
SEC	size exclusion chromatography
SELDI	surface enhanced laser desorption ionisation
SEM	scanning electron microscopy
SERS	surface enhanced Raman scattering
SFC	supercritical fluid chromatography
SFE	supercritical fluid extraction
SIA	sequential injection analysis
SIC	single ion current
SIM	single ion monitoring
SLED	superluminescent light emitting diode
SMDE	static mercury drop electrode
S/N	signal-to-noise
SNOM	scanning near-field optical microscopy
SOP	side-on-plasma
SNP	single nucleotide polymorphism
SPE	solid phase extraction
SPM	scanning probe microscopy
STEM	scanning transmission electron microscopy
STM	scanning tunnelling microscopy
TAS	total analysis system
μTAS	micro total analysis systems
TCD	thermal conductivity detector
TDS	total dissolved solids
TEM	transmission electron microscopy
TG	thermogravimetry
TGA	thermogravimetric analysis
THF	tetrahydrofuran

TIC	total ion chromatogram
TISAB	total ionic strength adjustment buffer
TMS	tetramethylsilane
TOF	time-of-flight
UV	ultraviolet
Vis	visible
VOC	volatile organic compounds
XRD	X-ray diffraction

Note

Where relevant it is usual for an acronym to be used for both the technique and the instrument, e.g. MS for Mass Spectrometry and Mass Spectrometer. The context of use should always be considered.

1

Introduction

Scientific data and results have to be accurate, precise and reliable and are subject to ever-increasing scrutiny by regulators in industry, the environment and medicine, in validation and also in research and development. Therefore, the choice of instrument to be used in particular circumstances is an important decision. Hence, analytical scientists today need a good working knowledge of the available techniques and equipment so that they can get the most out of analytical instruments and devices. Instrumentation is developing at such a rapid pace – getting smarter, smaller and faster – that it is difficult to keep up to date. This book attempts to bring together the key laboratory-based analytical techniques, hyphenated, field and portable instruments, process instrumentation in industry and trends in instrumentation such as miniaturisation. This should enable any analytical scientist to critically evaluate equipment, design suitable instrumentation for particular applications and use the most appropriate devices to solve problems and obtain results.

1.1 The Analytical Scientist

Analytical science is all around us. It pervades many disciplines such as chemistry, biology, physics, geology and engineering. It encompasses different types of analysis, such as chemical, physical, surface, materials, biomedical and environmental. Hence analytical scientists are found in all types of industrial and academic positions, from food and beverages to forensics to toxicology to pharmaceuticals to research.

A good analytical scientist must have a sound knowledge of experimental techniques in the laboratory as well as a strong theoretical knowledge of the fundamental science behind them – the analytical principles. This enables critical thinking and an understanding behind the techniques used, something that distinguishes the analytical scientist from other scientists. It is necessary to know how the equipment works, its range of applications and limitations and whether it is the best choice for the task at hand. The next step is to select the most appropriate technique for the job and analytical method development,

Analytical Instrumentation: A Guide to Laboratory, Portable and Miniaturized Instruments G. McMahon

which involves optimising the conditions for the analyte of interest. Analytical scientists also need to be able to critically evaluate a problem and decide on the best course of action taking into account time (people and sample turnaround), cost, availability of people and instruments, accuracy and knowledge of any knock-on effects or consequences that the result will have, and this is the analytical procedure. Hence, the analytical scientist must be proficient in all aspects of the process, from analytical principles to analytical methods and the final analytical procedure.

1.2 The Analytical Process

The analytical process is the science of taking measurements in an analytical and logical way. In practice, identifying or quantifying an analyte in a complex sample becomes an exercise in problem solving. To be efficient and effective, an analytical scientist must know the tools that are available to tackle a wide variety of different challenges. Armed also with a fundamental understanding of analytical methods, a scientist faced with a difficult analytical problem can apply the most suitable technique(s). This fundamental knowledge also makes it easier to identify when a particular problem cannot be solved by traditional methods, and gives an analyst the knowledge that is needed to develop creative approaches, hybrid instrumentation or new analytical methods. The analytical process is a logical sequence of steps that may take the form of the flowchart shown in Figure 1.1.

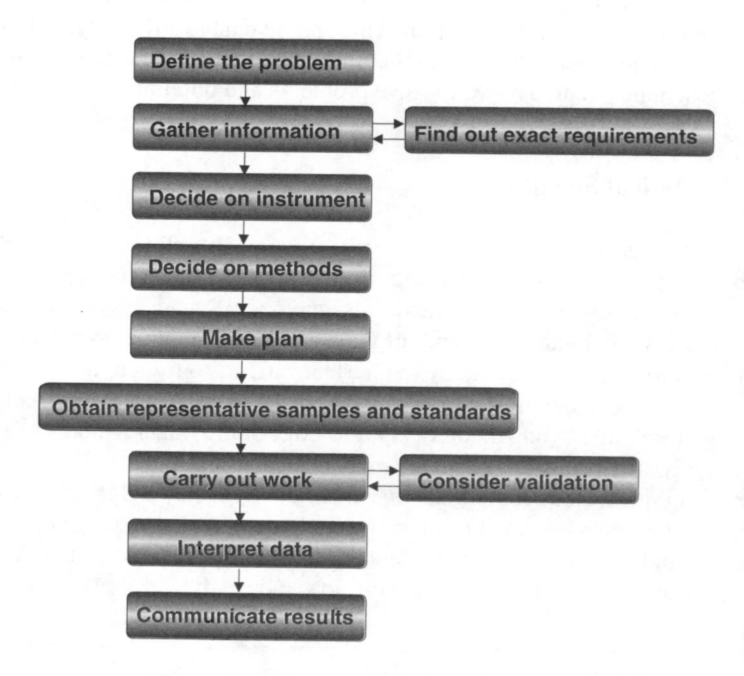

Figure 1.1 *Steps in the analytical process.*

When presented with a problem to solve, the analyst is likely to ask some of the following questions – What is it I am looking for? How much of it is likely to be there? Am I carrying out qualitative or quantitative analysis? What analytical technique should I use? How long will it take? How much will the assay(s) cost? The way an analysis is to be performed depends on experience, time, cost and the instrumentation available. Analysts must be able to evaluate available instruments in an open minded and critical way. Most assays used to be performed using classical methods of analysis such as gravimetry but a move towards instrumental methods began in the 1960s. Instrumental analysis is based on the measurement of a physical property of the sample and while they are generally more sensitive and selective than the older methods, they are sometimes less precise. Modern instruments are usually rapid, automated and capable of measuring more than one analyte at a time. Most instrumental methods of analysis are relative. Hence the equipment must be calibrated and the instrumental methods used on them must be validated to prove that they work reliably.

In answering the question 'what analytical technique should I use?', it must be selective for the compound of interest over the required concentration range, it must exhibit acceptable accuracy, precision and levels of sensitivity, it should be reliable, robust and easy to use, the frequency of measurement and speed of analysis must be suitable and the cost per sample should not be prohibitive. Looking at the big picture, the outlay is normally large when investing in instrumentation but in terms of the saving of time due to less labour-intensive analytical steps, in the long run such equipment usually works out to be econom.

1.3 Analytical Instrumentation

An instrument is a device that enables analytical measurements to be carried out automatically and objectively. Analytical instruments help analysts to work out composition, characterise samples, separate mixtures and yield useful results. Historically, instruments are often broken down into four component parts:

- Signal generator, e.g. radiation source
- Input transducer (detector)
- Electronic signal modifier, e.g. filter or amplifier
- Output transducer, e.g. computer.

As an example, for a UV–Vis spectrophotometer, the signal generator is the radiation source, the input transducer is the photodetector, the electronic signal modifier is a current-to-voltage converter and the output transducer is the digital display.

However, when comparing and contrasting instruments, especially across a multitude of disciplines, this and similar means of describing instruments can fall short. Another approach has been developed and reported by G. Rayson,[1] which allows a more unified description of instruments. This proposes that analytical instruments are comprised of five distinct modules:

- Source
- Sample

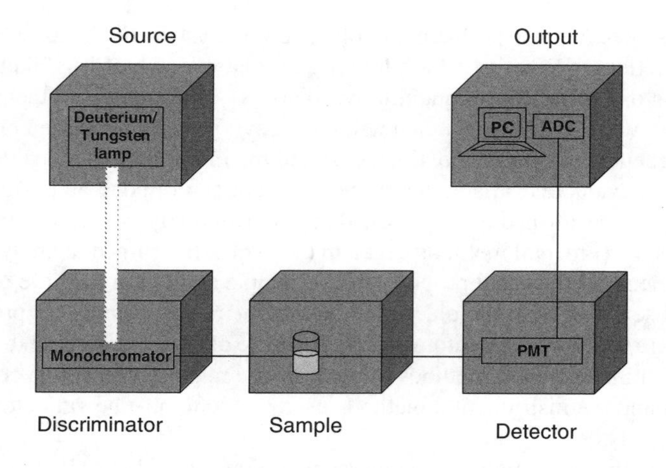

Figure 1.2 *Modules of an analytical instrument.*

- Discriminator
- Detector
- Output device

So, using this approach, for a UV–Vis spectrophotometer, the source is a deuterium/tungsten lamp, the sample is the cuvette or flowcell, the discriminator is the monochromator, the detector is a photomultiplier tube and the output device is a computer with an analogue-to-digital converter (Figure 1.2).

This means of describing instruments has been employed in this book, especially in the first section where many of the laboratory analytical instruments are discussed.

The demands on analytical instruments today are greater than ever before due to more challenging limits of sensitivity, smaller sample sizes, a wider range of applications and the growing list of new compounds that must be detected. It is fortunate, therefore, that modern instruments are improving all the time due to the availability of new technologies supporting their development. These include fibre optics, chemometrics, lasers, smaller components and more powerful computers.

1.4 Choosing the Right Instrument

To make a decision about which instrument is best to use for the job at hand, the analyst needs to know about the different types available.

Spectroscopic instruments are normally based on a compound's interaction with radiation, which yields information about its identity, quantity and/or its structure.

Separation instruments are usually based on chromatographic or electrophoretic separation of a mixture of compounds such that each can be identified and quantified. They are particularly powerful for complex samples. The detectors used in conjunction with separation techniques allow further identity information to be obtained and often structure to be elucidated.

Imaging instruments are based on close examination of the surface of a compound or material, which can allow identification, structural elucidation and an understanding of what is happening on a very small scale.

Electrochemical instruments are normally based on the changes in electrical energy that occur when a chemical reaction takes place, for example ion-selective electrodes (potentiometry) and voltammetric techniques. These can be measured in different ways and can give various qualitative and quantitative information about the reactants or products. In the case of conductivity measurements, changes in ionic content are monitored and, although nonspecific, can give useful data.

Thermoanalytical instruments allow the study of chemical and physical changes that occur with temperature, allowing the characterisation of materials and an understanding of their thermal events.

Diffraction instrumentation allows the structure of a compound at the atomic level to be understood.

As well as the general working knowledge of what instruments are capable of doing, the analyst also needs to understand the problem, the plan for solving it and the instruments that are available. If a number of techniques can be used, the analyst will need to know what the priorities are – is it sensitivity, is speed the most important factor or does expense play a role? This will help in deciding which equipment will best serve the purpose. Finally, experience is also a big factor. If an analyst is very familiar with a piece of equipment and with using it to analyse a variety of samples, it makes a big difference to the confidence in the results obtained. While, if newly trained in a technique, results may be given more tentatively. Hence, in choosing the right instrument, there are two sets of performance criteria to be considered.

The performance criteria affecting quality of the result include:

- Accuracy
- Precision (repeatability and reproducibility)
- Sensitivity (LOD and LOQ)
- Selectivity
- Linearity
- Dynamic range
- Stability.

The performance criteria for the economics include:

- Cost of purchase, installation and maintenance
- Analysis time
- Safety aspects
- Running costs – supplies, gases, consumables
- Training
- Sample throughput.

Overall, analytical instrumentation is as important as ever to the analytical scientist and to their analytical approach to problem solving. The range of equipment available today is enormous with manufacturers vying to make equipment smaller, better, more sensitive or less expensive than competitors in the market. Hence there is a need for analysts to have a strong understanding of the workings and capabilities of analytical instruments and devices so that wise decisions are made when purchasing or choosing to use one instrument over another.

Reference

1. Rayson, G. (2004) A unifying description of modern analytical instrumentation within a course on instrumental methods of analysis. *J Chem Ed*, **81 (12)**, 1767–71.

Section I

Laboratory Analytical Instrumentation

2

Spectrometric Instruments

2.1 Molecular Spectrometry

At room temperature, most compounds are in their lowest energy or ground state. Upon interaction with the appropriate type of electromagnetic radiation (Figure 2.1), characteristic electronic, vibrational and rotational transitions can occur. Excited states thus formed usually decay back to the ground state very quickly, either by emitting the energy they absorbed with the same or lower frequency or by 'radiationless' relaxation through heat loss. Infrared radiation causes the vibrations in molecules to increase in amplitude, absorption of visible and ultraviolet radiation cause electrons to move to higher electronic orbitals while X-rays actually break bonds and ionise molecules. Molecular spectra can be obtained by measuring the radiation absorbed or emitted by gases, liquids or solids and yield much analytical information about a molecule. These phenomena are exploited by spectrometric instruments.

2.1.1 Ultraviolet, Visible and Near Infrared

Principles

Many molecules absorb ultraviolet (UV), visible (Vis) or near infrared (NIR) radiation. In terms of the electromagnetic spectrum, UV radiation covers the region from 190–350 nm, visible radiation covers the region 350–800 nm and NIR radiation covers the region 800–2500 nm (and maybe a little higher). Absorption of UV and/or Vis radiation corresponds to the excitation of outer electrons in the molecule. Typically, radiation with a specific intensity is passed through a liquid sample, often in a quartz cuvette. When the radiation emerges on the other side of the cuvette, it is reduced in intensity owing to losses from a) reflection off the cuvette windows, b) scattering and c) absorption by the sample itself. Often, a reference solution which has no analyte is also analysed to account for the losses due to reflection and scattering; thereby the intensity attenuation due to absorption alone can be worked out by simple subtraction. In organic molecules, this absorption is restricted

Analytical Instrumentation: A Guide to Laboratory, Portable and Miniaturized Instruments G. McMahon

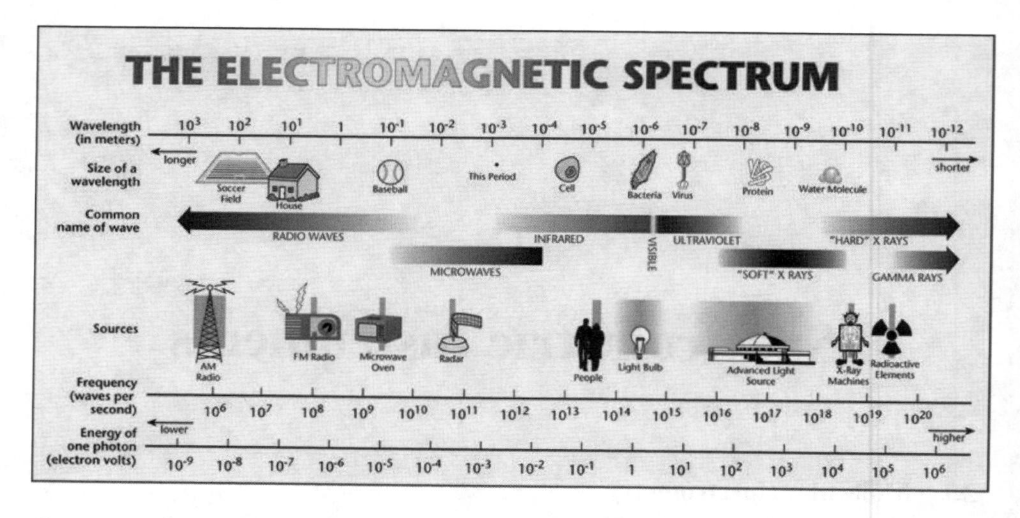

Figure 2.1 *The electromagnetic spectrum (Courtesy of the Advanced Light Source, Berkeley Laboratory).*

to certain functional groups (chromophores) that contain electrons of low excitation energy. Aromatic molecules, for example, mostly absorb UV in the 200–300 nm region.

An absorption spectrum is usually a plot of absorbance versus wavelength and is normally continuous and broad with little fine structure (Figure 2.2). The broad spectrum is due to the fact that the higher energy radiation involved means that vibrational and rotational transitions co-occur as well as electronic transitions; all of these are superimposed on each other resulting in broad bands rather than sharp peaks. In UV–Vis absorption spectrometry, concentration of the species is related to absorbance by the Beer–Lambert Law (Equation (2.1)):

$$A_\lambda = \varepsilon_\lambda c l \qquad (2.1)$$

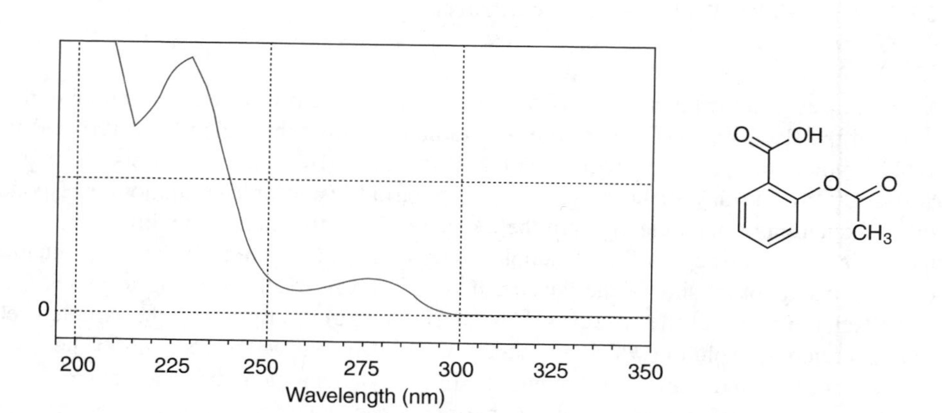

Figure 2.2 *UV absorption spectrum of aspirin in aqueous solution (The benzoic acid group is known to absorb at 230 and 270 nm in aqueous solution and the carbonyl and benzene groups also absorb at ≤200 nm).*

where A_λ = absorbance at a particular wavelength (λ), ε_λ = extinction coefficient at a particular wavelength (λ), c = concentration and l = pathlength. During most experiments, ε and l remain constant, so absorbance is proportional to concentration, a relationship that is exploited for quantitative analysis.

Absorption of NIR radiation corresponds to certain vibrations of the molecule and is due to overtones and combinations of parent absorption bands in the mid-infrared (IR) region. Generally, these absorptions are weaker than the parent absorptions in the IR but the decrease in intensities is not the same for all molecules. It can be seen as a complementary technique to conventional IR, exploiting a different region of the electromagnetic spectrum. For example, water has less absorption in the NIR compared with mid IR, so NIR spectra of aqueous samples are often sharper. A NIR absorption spectrum is usually a plot of absorbance versus wavelength and has more fine structure than a UV or UV–Vis spectrum. NIR absorbance also follows the Beer–Lambert Law, so can be used as a quantitative technique.

Instrument

A spectrophotometer can be either of single beam or double beam design. In the older single beam instrument, all of the light passes through the sample cell and hence the sample must be replaced by a blank/reference sample to account for any matrix effects. In a double-beam instrument, the light is split into two beams before it reaches the sample. One beam is used as a reference beam while the other beam passes through the sample. Some double-beam instruments have two detectors, so the sample and reference beam can be measured at the same time. In other designs, the source irradiation is alternately passed through the sample and the reference samples to compensate for any changes in the intensity of the source or response of the detector. A modular schematic of a simple single beam UV–Vis–NIR spectrophotometer is shown in Figure 2.3.

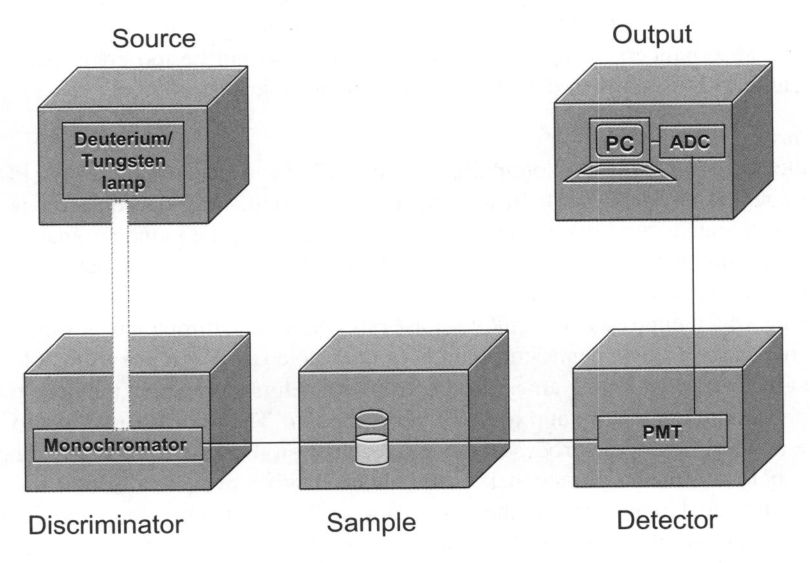

Figure 2.3 Schematic diagram of a single beam UV–Vis–NIR spectrophotometer.

Source

Deuterium lamps are commonly used as a UV radiation source in the range 200–400 nm and tungsten incandescent lamps as sources for the visible and NIR regions covering the range 400–2500 nm. For the NIR work, the source is operated at 2500–3000 K, which results in more intense radiation.

Discriminator

A monochromator is usually used as a wavelength selector. Monochromators are composed of a dispersing medium to 'separate' the wavelengths of the polychromatic radiation from the source, slits to select the narrow band of wavelengths of interest and lenses or mirrors to focus the chosen radiation.

The dispersing medium can be a diffraction grating, a prism or an optical filter. Those based on a grating are most effective at producing spectra with reduced stray light. The more finely separated the ruled lines on the grating are, the higher the resolution. However, especially for NIR, interferometers are becoming more common in Fourier Transform (FT) instruments. FT is more effective at longer wavelengths such as in IR and NIR but can be used for UV–Vis also. The Michelson interferometer is described in more detail in the section on IR.

Sample

The sample holder must be transparent in the wavelength region being measured. Quartz cuvettes are normally used for UV–Vis and NIR measurements. For UV–Vis absorbance, cuvettes are usually 1 cm in pathlength in laboratory based instruments, though shorter pathlengths can be employed. For NIR, longer pathlengths of 5–10 cm are used in the short wavelength NIR (750–1100 nm) and shorter path lengths of 0.1–2 cm are used for the long wavelength NIR (1100–2500 nm). Fibre optic cables can be used over longer distances. Flow-through, cylindrical, micro and thermal cells can also be used. All cuvettes and cells should be handled carefully to avoid leaving fingerprints. The sample compartment must be able to prevent stray light and dust from entering because this will adversely affect the absorbance readings. The sample should also not be too concentrated as the Beer–Lambert Law starts to deviate at high absorbance levels.

Detector

The detector is typically a photomultiplier tube (PMT), a photodiode array (PDA) or a charge-coupled device (CCD). In a monochannel system, only one detector is used. It measures the intensity of one resolution element at a time as the monochromator is slowly scanned through the spectrum. The multichannel system uses an array detector where all intensities are measured simultaneously. This gives rise to two advantages – multichannel advantage, which improves the signal-to-noise ratio (S/N), and throughput advantage, which allows the use of a single deuterium source for the whole UV–Vis range from 200–780 nm.

The greatest recent improvement in spectrophotometers has been in the detector. PMTs are monochannel detectors and are still very popular. They consist of a photosensitive surface and a series of electrodes (dynodes), each at an increased potential compared to the one before. When a photon strikes the photosensitive surface, a primary electron is emitted and accelerates towards the first dynode. This electron impacts the dynode and causes the release of a number of secondary electrons, which hit the next electrode and so on, until the signal is amplified many times over. Even extremely small signals can

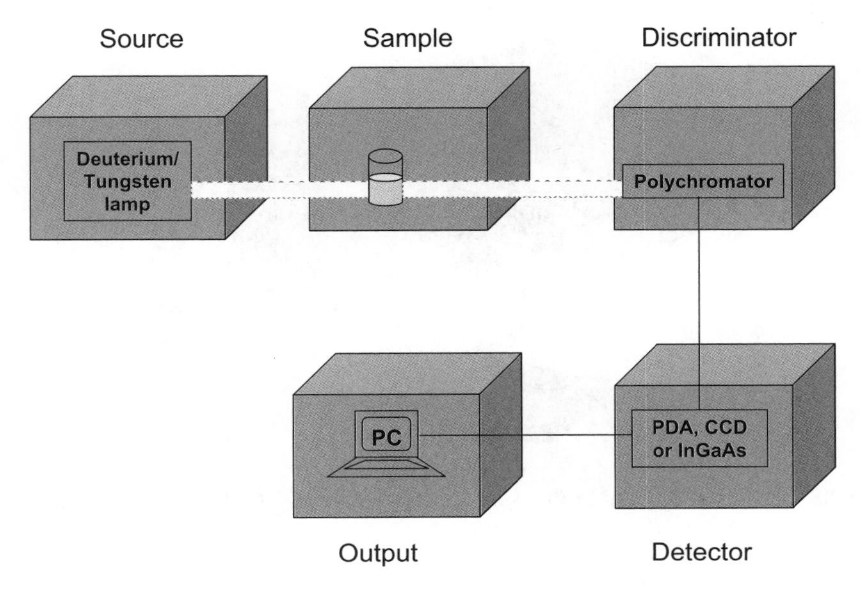

Figure 2.4 *Schematic diagram of a UV–Vis–NIR spectrophotometer with a multichannel detector.*

now be detected. However, multichannel detectors are increasing in popularity. These consist of arrays of diodes such as are found in PDAs, CCDs and charge-injection devices (CIDs). They have the advantage over PMTs of being able to measure many wavelengths simultaneously. Hence, these instruments have a different configuration to that shown in Figure 2.3, with no monochromator before the sample and, instead, a polychromator placed after the sample and before the detector (Figure 2.4).

These multichannel detectors work by having hundreds of silicon photodiodes positioned side by side on a single collision crystal or chip. Each photodiode has an associated storage capacitor that collects and integrates the photocurrent generated when the photons strike the photodiode. Periodically, they are discharged and the current read. A spectrum can be recorded if radiation dispersed into its different wavelengths falls on the surface of the diode array.

A CCD is also based on semi-conductor technology. It is a two-dimensional array which stores photo-generated charge. The electrons in each element are transferred out for reading until the array has been fully read. For NIR, the detectors described above can cover the shorter wavelength NIR spectrum but for the longer wavelength NIR spectrum lead sulfide or indium/gallium/arsenic (InGaAs) detectors are used. The InGaAs detector is about 100 times more sensitive than the mid IR region detectors and therefore, with NIR measurements, there are very low noise levels. It can also be array-based.

Output

The PC collects the data, converts it from transmission to absorbance and displays the spectrum. The PC can often carry out baseline subtraction and smoothing and filtering tasks as well as qualitative and quantitative analysis. It may have other capabilities, such

Figure 2.5 Agilent 8453E UV Visible Spectroscopy System (© Copyright 2006 Agilent Technologies, Inc. reproduced with permission).

as the ability to compare a spectrum to those in a spectral library and to carry out peak purity checks.

Information Obtained

As spectra for molecules with similar absorbing groups tend to be very alike, UV and UV–Vis spectrometry are not the most useful identification (qualitative) techniques. But as the absorbance is directly proportional to the concentration of the absorbing species according to the Beer–Lambert Law, UV–Vis is a very useful quantitative technique. It has wide applicability and good sensitivity (10^{-6} M). An example of a UV–Vis spectrophotometer is shown in Figure 2.5. UV and UV–Vis instruments are often used as flow-through detectors for separation techniques and are be discussed further in that section. NIR is used mainly as a nondestructive quantitative technique. Even though the intensities of the bands are weak, sources are often more intense and detectors more sensitive so noise levels are reduced.

Developments/Specialist Techniques

Even though NIR spectra have low absorption which can result in low sensitivity, the signal-to-noise ratio is high because of the use of intense radiation sources and sensitive detectors. Also, NIR radiation travels very well over long distances in fibre optic probes, so this technique lends itself well to portable instruments and field work. It is now possible to buy all-in-one UV–Vis–NIR instruments (Figure 2.6) but it is still worth weighing up the needs of the laboratory carefully before deciding whether to buy separate UV–Vis and NIR equipment or the combined instrument[1]. Shimadzu has recently launched a UV–Vis–NIR spectrophotometer with a range from 185–3300 nm that has three different detectors built in – a PMT, an InGaAs detector and a lead sulfide detector.

Figure 2.6 A combination UV–Vis–NIR spectrophotometer (Reproduced by kind permission of Jasco Corporation).

Applications

There are a huge number of applications for UV, Vis and NIR instruments. UV–Vis is routinely used for the determination of solutions of transition metals, which are often coloured, and highly conjugated organic compounds. For example, determination of iron by forming a coloured complex with 1,10-phenanthroline can be detected by visible spectrophotometry[2,3]. The analysis of nitrate nitrogen in water[4,5], phosphate in water[6] and soil[7] and lead on the surfaces of leaves[8] can also be determined colorimetrically. Many pharmaceuticals, dyes and other organic compounds can be detected easily by UV due to their strong chromophores. Moisture, fat, sugars, fibre, protein and oil can be determined in foodstuffs such as soy bean[9], corn, rice, milk[10], meat and cheese by NIR.

2.1.2 Infrared and Raman

Principles

Infrared (IR) and Raman are vibrational spectroscopy techniques. They are extremely useful for providing structural information about molecules in terms of their functional groups, the orientation of those groups and information on isomers. They can be used to examine most kinds of sample and are nondestructive. They can also be used to provide quantitative information. In this book, IR refers to the mid IR region, which covers the range 4000–400 cm^{-1} (2500–25,000 nm). Raman radiation spans the range 4000 down to about zero cm^{-1}. IR and Raman spectroscopies are similar insofar as they both produce spectra because of vibrational transitions within a molecule and use the same region of the electromagnetic spectrum. They differ in how observation and measurement are achieved, since IR is an absorption (transmission) method and Raman is a scattering method.

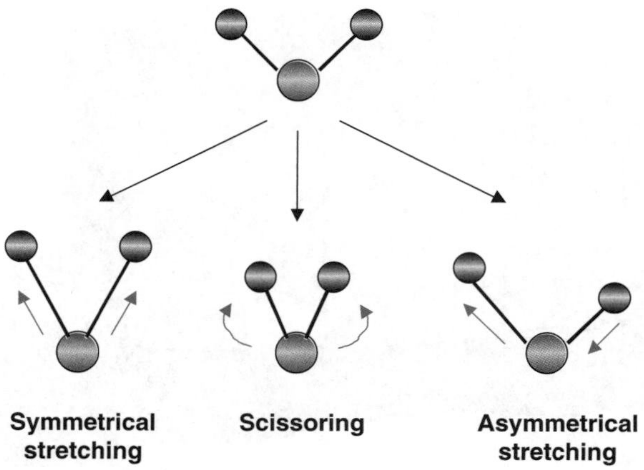

Figure 2.7 Some of the possible vibrations for a simple molecule upon absorption of infrared radiation.

Many molecules absorb IR radiation, which corresponds to the vibrational and rotational transitions of the molecules. For this absorption to occur, there must be a change in polarity of the molecule. IR radiation is too low in energy to excite electronic transitions. There are a number of vibrations and rotations that the molecule can undergo (a few of these are shown in Figure 2.7) which all result in absorption of IR radiation.

In a similar fashion to a UV–Vis spectrum, an IR spectrum is a plot of transmittance versus wavelength. It is normally a complex series of sharp peaks corresponding to the vibrations of structural groups within the molecule. The IR spectrum for aspirin is shown in Figure 2.8 and it can be seen by comparing this with Figure 2.2 that IR yields much more useful qualitative data than the corresponding UV or UV–Vis spectrum. For quantitative

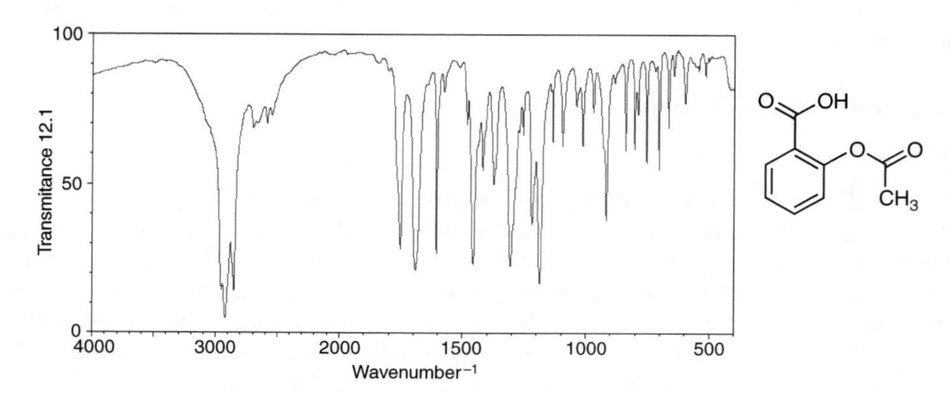

Figure 2.8 IR transmission spectrum of aspirin (The transmission peaks at approximately 3000 cm^{-1} (broad band), 2900 cm^{-1} (sharp bands), 2600–2700 cm^{-1} and 1700 cm^{-1} are due to vibrations of the OH, aromatic CH, aliphatic CH and the C=O bonds in the molecule respectively. The stretches in the region between 1500 cm^{-1} and 500 cm^{-1} are more difficult to assign as this is the 'fingerprint region').

work, IR measurements can deviate from the Beer–Lambert law due to some scattered radiation and the use of relatively wide slits. Hence, a ratio method is often used where a peak that is apart from those being used for quantitative measurement is chosen and is employed as an internal standard. This strategy serves to minimise relative errors, such as those due to differences in sample size. However, under controlled experimental conditions, IR can comply with the Beer–Lambert Law directly for quantitative measurements.

Conventional IR spectrometers are known as dispersive instruments but have now been largely replaced by Fourier Transform infrared (FTIR) spectrometers. Rather than a grating monochromator, an FTIR instrument uses an interferometer to obtain a spectrum. The advantages are greater signal-to-noise ratio, speed and simultaneous measurement of all wavelengths. This gain in speed due to the simultaneous acquisition of data is sometimes called the Felgett Advantage and the gain in sensitivity due to the use of certain detectors is called the Jacquinot Advantage. There is a third advantage to be gained by using FTIR over dispersive IR and that is the Connes Advantage, whereby a helium–neon (HeNe) laser is used as an internal calibration standard which renders these instruments self-calibrating.

The Raman effect arises when incident light distorts the electron density in the molecule, which subsequently scatters the light. Most of this scattered light is at the same wavelength as the incident light and is called Rayleigh scatter. However, a very small proportion of the light is scattered at a different wavelength. This *inelastically* scattered light is called Raman scatter or Raman effect. For this to occur, there must be a change in polarisability of the molecule. It results from the molecule changing its molecular vibrational motions and is a very weak signal. The energy difference between the incident light and the Raman scattered light is equal to the energy involved in getting the molecule to vibrate. This energy difference is called the Raman shift. These shifts are small and are known as Stokes and anti-Stokes shifts, which correspond to shifts to lower and higher frequencies respectively. Several different Raman shifted signals will often be observed, each being associated with different vibrational or rotational motions of molecules in the sample. The Raman signals observed are particular to the molecule under examination. A plot of Raman intensity versus Raman shift is a Raman spectrum. An example of a Raman spectrum for aspirin is shown in Figure 2.9.

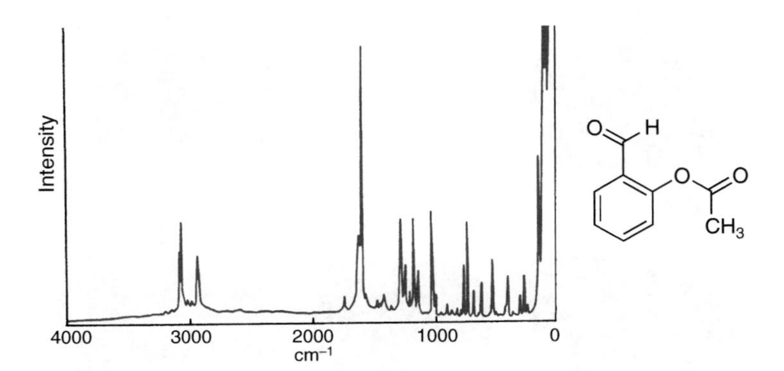

Figure 2.9 *Raman spectrum of aspirin (with no sample preparation required) (The Raman signals at approximately 3100 cm^{-1}and 2900 cm^{-1} are due to CH vibrations and the signal at 1700 cm^{-1}is due to C=O vibrations in the molecule. It can be seen by comparing this spectrum to Figure 2.8 that the infrared and Raman spectra are complementary to each other).*

Raman signals can be obscured by fluorescence so care must be taken when choosing the energy source. Raman is primarily used as a non-contact quantitative technique. Raman and IR are complementary techniques in that molecules that vibrate/rotate in this region of the spectrum will generally give a spectrum in both. However, polar functional groups with low symmetry generally give strong IR signals while molecules with polarisable functional groups with high symmetry generally give strong Raman signals. Hence, strong infrared absorptions appear usually as weak Raman ones and vice versa. Raman is not as sensitive to the environment of the molecule as IR. Also, Raman is relatively insensitive to water whereas the opposite is true of IR.

Instrument

The components for an FTIR instrument and an FT–Raman instrument are shown, respectively, in Figures 2.10 and 2.11.

Source

There are a variety of black-body heat sources used in IR, such as the Nernst glower, the Globar (silicon carbide rod) and even synchotron (in research applications). Tunable lasers are also now coming into use. For FT–Raman spectrometers, lasers in the NIR region (to eliminate fluorescence) such as neodymium yttrium aluminium garnet (Nd:YAG) at 1060 nm are used with high power output or lasers in the visible region, e.g. argon at 488 nm or helium–neon at 632.8 nm, can be employed. Since the intensity of the Raman lines varies with the fourth power of the exciting frequency, the 488 nm argon line provides scattering that is nearly three times as strong as that produced by the helium–neon laser with the same input power. Visible region lasers are usually operated at 100 mW, while Nd:YAG lasers can be used up to 350 mW without causing photodecomposition of organic samples. UV sources for Raman are now available.

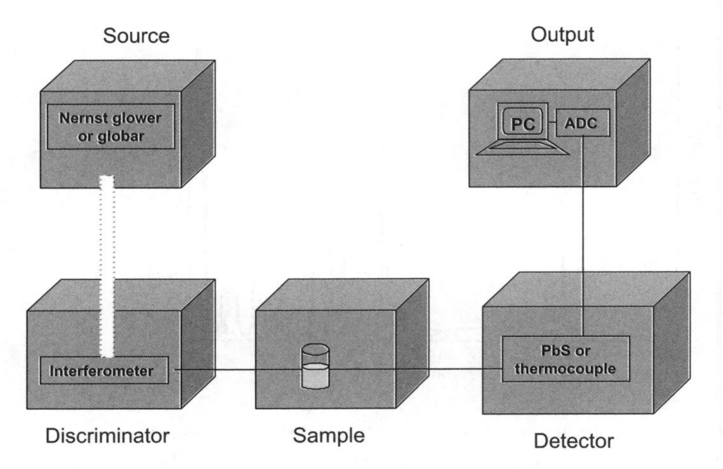

Figure 2.10 *Schematic diagram of an FTIR spectrometer.*

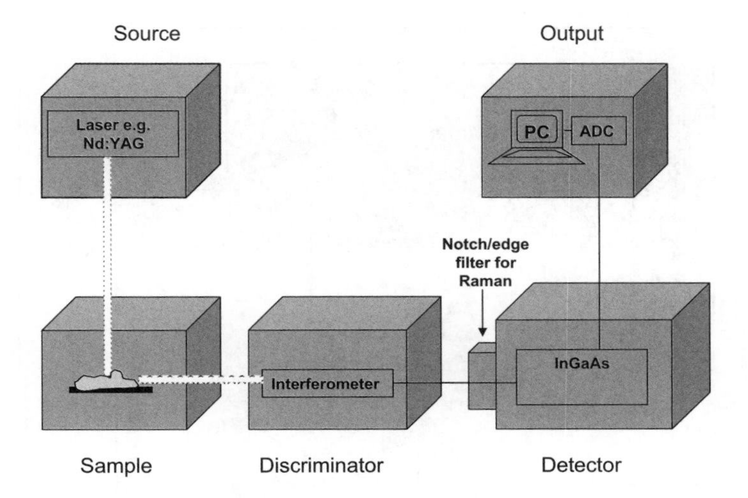

Figure 2.11 *Schematic diagram of an FT–Raman spectrometer.*

Discriminator

Most IR and many Raman instruments today use an interferometer, where the spectral encoding takes place, instead of a monochromator. The Michelson interferometer is a multiplex system with a simple design – a fixed mirror, a moving mirror and an optical beam splitter. It is illustrated in Figure 2.12 and its use within an FTIR instrument is shown in Figure 2.13. The source radiation hits the beam splitter from where some of the light is reflected to the moving mirror and some is transmitted to the fixed mirror. The mirrors

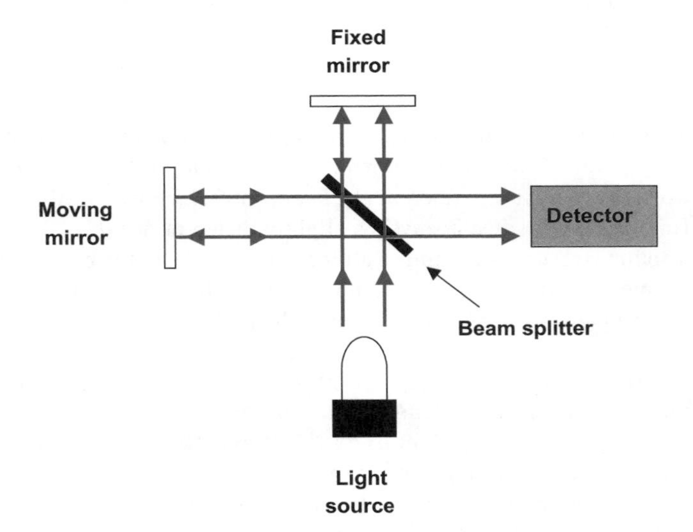

Figure 2.12 *Schematic diagram of a Michelson interferometer.*

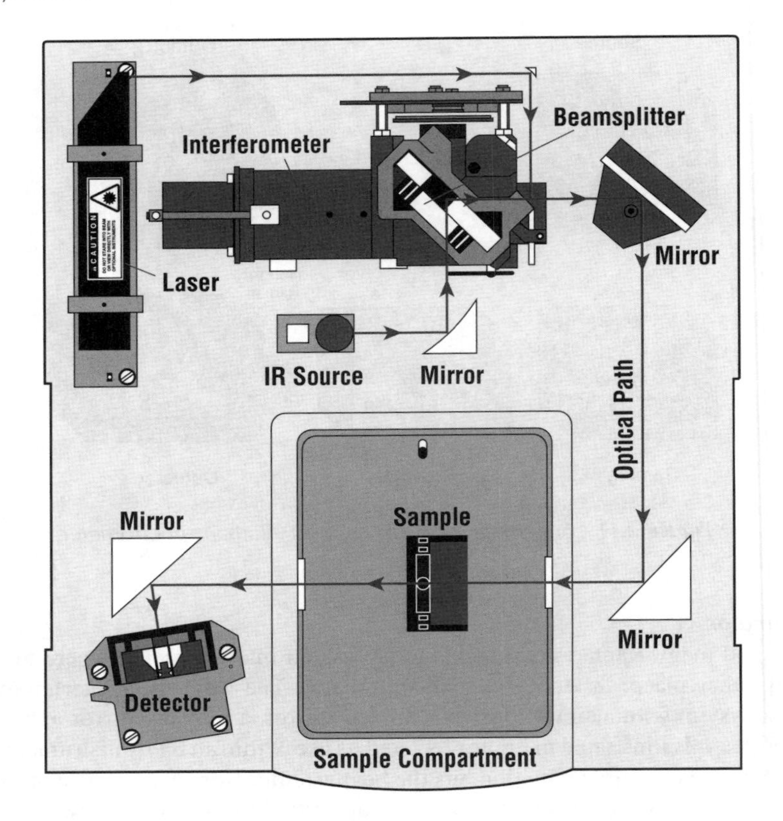

Figure 2.13 *Schematic diagram of the inside of an FTIR spectrometer showing the position of the Michelson interferometer (Reproduced by kind permission of Thermo Electron Corp.).*

reflect the light back to the beam splitter, some of which recombines and goes on to the detector. The key point of the moving mirror is to generate a difference in the optical paths of the two paths of light separated by the beam splitter; consequently, one is slightly out of phase from the other since it travels a slightly different distance. The recombined light produces an interference spectrum of all the wavelengths in the beam before passing through the sample. In other words, the sample sees all the wavelengths simultaneously and the interference pattern changes with time as the mirror is continuously scanned at a linear velocity.

The result of the sample absorbing radiation is a spectrum in the time domain called an interferogram. Fourier transformation (FT) converts this very complex signal to the frequency domain. Combined FTIR and Raman spectrometers based on the Michelson interferometer are commercially available. Depending on the spectral regions of interest required, it can be operated with a single calcium fluoride (CaF_2) beam splitter or a combination of materials with an automatic changer.

As with IR, both dispersive and FT–Raman instruments exist. The main advantage of FT–Raman is the ability to change from using visible to NIR excitation with an associated reduction in broad-band fluorescence. Secondly, an FTIR instrument can be configured to carry out FT–Raman experiments without significant effort. However, Raman instruments that do not employ an interferometer are still commonly used owing to their high radiation throughput and lower cost.

Sample

For IR analysis, the beam is transmitted through (or reflected off the surface of) the sample, depending on the type of analysis being performed. Liquids are sampled in cells, which must be free of interferences, or as a thin film between two IR windows, e.g. sodium chloride (NaCl) or potassium bromide (KBr) for non-aqueous media which have energy cutoffs at 650 and 350 cm^{-1} respectively, or calcium fluoride for aqueous samples. Some solids can be dissolved in suitable solvents or cast onto films but they usually need to be ground and dispersed and compressed into a transmitting medium, e.g. potassium bromide disk or pellet. Typically 1 mg of sample is mixed with 200 mg of dry potassium bromide. Other solids that are hygroscopic or polar can be ground into a mineral oil such as nujol to give a mull (dispersion). Hard materials can be compressed between two parallel diamond faces to produce a thin film as diamond transmits most of the mid IR.

A number of special sampling techniques for solids exist. One that is used to obtain IR spectral information on the surface of a sample is called diffuse reflectance. Here reflected radiation instead of transmitted radiation is measured. Another technique is attenuated total reflectance (ATR), in which the sample is brought into close contact with the surface of a prism made of a material with a high refractive index, e.g. sapphire. A light beam approaching the interface from the optically denser medium at a large enough angle of incidence is totally reflected. However, the beam does penetrate a small distance into the optically thinner medium (the sample). If the sample absorbs IR radiation, an IR spectrum can be obtained. By changing the angle of incidence, depth profiling can be achieved. The film (~200 nm) is made thin enough to allow light to pass through to the adjoining aqueous phase. A diagram of such a cell is shown in Figure 2.14. ATR can also be exploited in other types of spectrometry and is useful in probes as well as in cells.

The pathlength for IR samples is in the range 2–3 mm. Gases are easiest to sample in a longer pathlength cell, typically 10 cm in length. Fibre optic cables can be used where the sample is remote from the spectrometer.

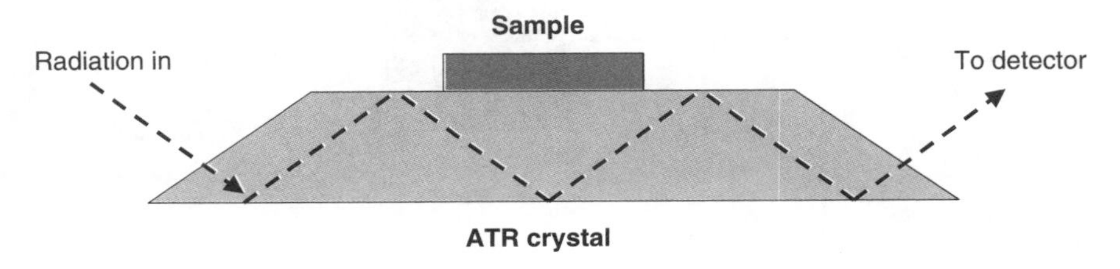

Figure 2.14 *Diagram of an attenuated total reflectance (ATR) cell.*

For Raman analysis, sample preparation is much easier than with IR. In fact, the source light is simply focussed onto the solid or liquid sample directly. If a cuvette is used, quartz or glass windows can be used. If a slide or surface is used, a background spectrum should be taken to remove the possibility of any interfering peaks. Glass tubes are often used and since water is a weak Raman scatterer, aqueous samples can be easily analysed. Reflectance measurements, as distinct from transmissive measurements above, can also be made and are useful for studying films on metal surfaces or samples on diamond surfaces. Measurements should also ideally take place in the dark to remove ambient light interferences.

Detector

The most common detectors in IR are thermal, i.e. thermocouples, thermistors and bolometers. A thermocouple is based on the use of two different conductors connected by a junction. When a temperature difference is experienced at the junction, a potential difference can be measured. A series of thermocouples together is called a thermopile. Thermistors and bolometers are based on a change in resistance with temperature. They have a faster response time than thermocouples. With a Fourier Transform IR (FTIR), where rapid response and improved sensitivity is key, lead sulfide and InGaAs detectors are used as for NIR. Some arrays are also used.

The most common detectors in Raman instruments are PDAs and CCDs but for FT–Raman, single channel detectors are used, e.g. InGaAs. An extra requirement for the FT–Raman instrument is a notch or edge filter; it is included to reject scattered laser light at the strong Rayleigh line, which could otherwise obscure the FT–Raman spectrum.

Output

With an FT instrument, the main function of the PC is to carry out the Fourier Transformation of the interferogram, i.e. conversion of the information from the time domain to the frequency domain (see Figure 2.15). However, the PC also carries out both qualitative and quantitative analysis. Library searching, spectral matching, chemometrics and other software are readily available. IR measurements can deviate from the Beer–Lambert Law and so it is not as easy to use as a quantitative technique. Raman spectral databases are also available but have less coverage of compounds than IR. Raman can be used as quantitative technique as well using scattering as the basis of signal generation, which can be difficult to standardise.

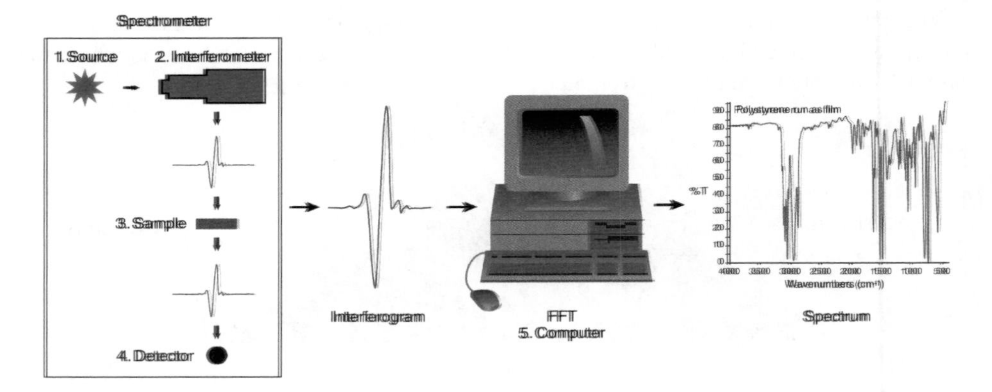

Figure 2.15 *Schematic diagram showing conversion of an interferogram to the required FTIR spectrum (Reproduced by kind permission of Thermo Electron Corp.).*

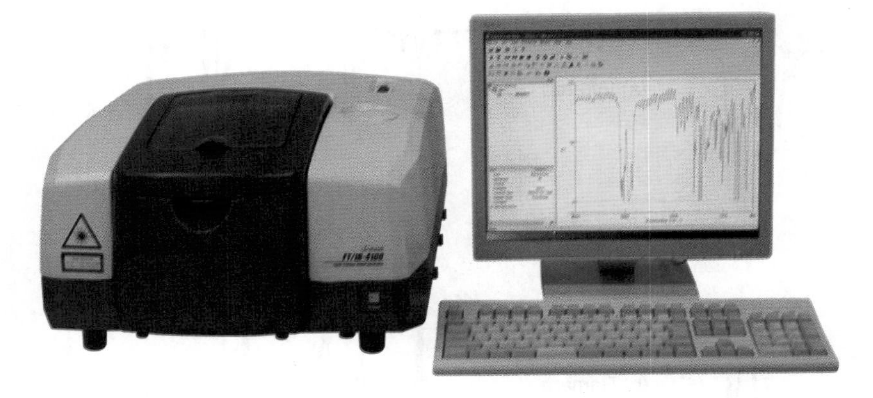

Figure 2.16 An FTIR spectrometer (Reproduced by kind permission of Jasco Corporation).

Information Obtained

The real strength of IR is its ability to identify the structural groups in a molecule, e.g. olefin, carbonyl, and so IR absorption spectroscopy is a powerful identification technique. In particular, the 'fingerprint' region below 1500 cm^{-1} is very dependent on the molecule's environment and it may be possible to identify a molecule by comparing its transmission bands in this region with spectra from an IR library. Mid IR can also determine the quality or consistency of a sample and the amount of components in a mixture. Examples of an FTIR and an FT–Raman instrument are shown in Figures 2.16 and 2.17, respectively.

The main advantage of Raman as a technique is its ability to obtain spectra through glass or plastic, which is very useful in the food and pharmaceutical industries where Raman spectra are routinely used for quality control purposes after a product has been packaged. Raman is relatively unaffected by water and so is an ideal technique for aqueous samples.

Figure 2.17 An FT–Raman spectrometer (Reproduced by kind permission of Thermo Electron Corp.).

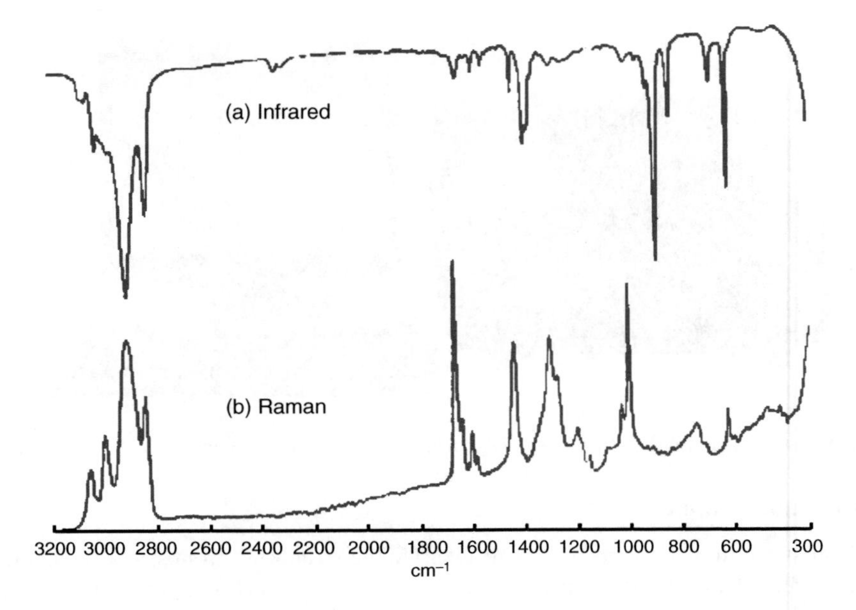

Figure 2.18 *The infrared and Raman spectra of styrene/butadiene rubber illustrating the complementary nature of the two techniques (From 'What is Raman Spectroscopy?', Hendra, P., Int J Vib Spect, [www.ijvs.com]* **1** *(5), pp. 6–16 (1998)[11]. Copyright John Wiley & Sons Limited. Reproduced with permission).*

Raman and IR spectra are often complementary for a given compound. A comparison of the IR and Raman spectra for a sample of styrene/butadiene rubber is shown in Figure 2.18.

Developments/Specialist Techniques

When dedicated quantitative work is required in IR, simple filter instruments (also called photometers) are often used. These employ a narrow bandpass filter with a wavelength that corresponds to the absorption of the component being measured. This type of absorption is the more popular method for gas analysis, e.g. carbon monoxide, nitrogen dioxde and methane. It is also possible to have variable bandpass filters for use in scanning instruments. These are rugged and reliable dedicated analysers and commonly used in on-line systems.

In Raman spectroscopy, only a tiny proportion of the scattered light will be Raman (maybe one out of a million) so the signals are weak. However, signal intensities can be greatly enhanced by exploiting the resonance Raman (RR) effect. This is of particular interest for molecules containing chromophores. RR is achieved by using an exciting wavelength that matches the electronic transitions of the chromophore and hence UV laser sources are commonly employed. When RR happens, the vibrations associated with the absorbing chromophore are enhanced by a few orders of magnitude (up to 10^6). Thus Raman becomes a highly specific probe for certain functional groups or chemical sites.

Surface Enhanced Raman Spectroscopy (SERS) is another specialist technique where the Raman signal is greatly enhanced by adsorbing the sample onto a roughened metal surface. It is limited in application to surfaces or interfaces and most studies to date have focused on electrode surfaces. Intensity may be amplified upto 10^{12} on copper, gold and silver surfaces, allowing detection down to a single molecule.

Remote sampling in IR and Raman is a big development area – from measurements of atmospheric air to sample probes attached by optical fibres to the rest of the instrument at some distance away. Another growing area is that of IR microscopes and spectral imaging (Section 4.5), which will be discussed at the end of the section on imaging techniques.

Applications

Infrared spectrometry can be used for testing emissions. For example, IR has been reported for determining carbon monoxide and nitrogen oxide[12] and other trace gases[13,14], remote sensing of volcanic gases and trace analysis of halocarbons in the atmosphere, among many other applications. IR is also employed in the beverage industry for monitoring alcohol, sugar and water content in drinks[15], and sugars, fibres and acidity in juices. In the food industry, IR spectrometry is very useful for determining protein, oil, ash, moisture and particle size in flour[16], for following fermentation and microbiological reactions[17] and for the study of microorganisms in food products.

Raman spectrometry is often used for investigating biological samples since water does not interfere significantly, e.g. transformational changes in proteins and lipids can be readily followed[18]. Raman spectrometry for inorganic compounds, especially metal complexes, is also employed as it is a well understood area[19].

Raman spectrometry is used extensively too in areas that are less well known such as space research, deep ocean investigation and is a useful technique in medical research and healthcare[20].

2.1.3 Luminescence

Principles

Luminescence is a phenomenon where light is emitted by a substance and that substance appears to 'glow'. It occurs because an electron has returned to the electronic ground state from an excited state and loses its excess energy as a photon of light (Figure 2.19). Luminescence techniques are based on the measurement of this emitted radiation which is characteristic of the molecule under study. Compared to absorption techniques, these methods are more selective, sensitive and can have much wider linear dynamic ranges but, unfortunately, the number of compounds which naturally undergo luminescence is small.

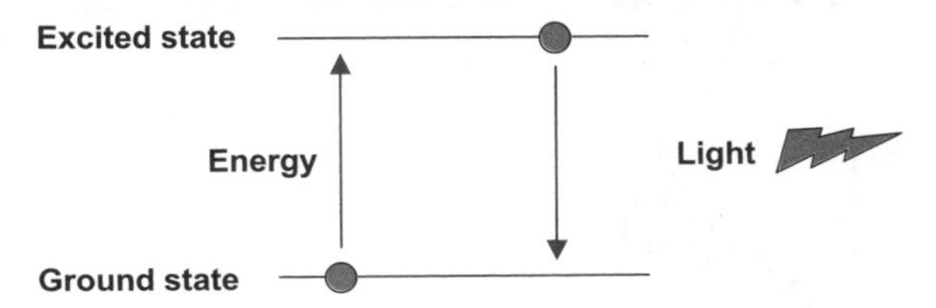

Figure 2.19 *A simplified energy diagram showing a molecule absorbing radiation, its electron being promoted to an excited state and that electron relaxing back down to the ground state by emission of a characteristic photon of light.*

Radiation is first absorbed by the molecule. Electrons in the molecule use this energy to attain a higher energy state. The molecule is now 'excited'. There are two possible transitions for the electron depending on the spin quantum numbers in the excited state. If the spins are opposite, the state is a singlet one and if they are in parallel, the state is a triplet one. After becoming excited, the electrons need to relax back down to the ground state. This emitted radiation, which is slightly lower in energy than the absorbed radiation, is luminescence.

There are many types of luminescence:

- Photoluminescence
  - Fluorescence
  - Phosphorescence
- Radioluminescence
- Bioluminescence
- Chemiluminescence.

The most important of these analytically are the photoluminescence modes of fluorescence and phosphorescence. Fluorescence occurs when UV radiation provides the energy needed to excite electrons in the molecule to the excited singlet state and subsequently, after some radiationless decay, emit photons almost instantaneously ($\sim 10^{-9}$ s) in order to return to the ground state. What occurs in the case of a fluorescing molecule is depicted schematically in Figure 2.20.

Various possible vibrational levels are superimposed on the two electronic levels (excited singlet state and ground state). When this molecule is excited, there are five possible transitions for the electrons, of which the excitation to the V' = 1 is most probable. This transition corresponds to absorption of a photon at 254 nm, which is therefore the λ_{max}

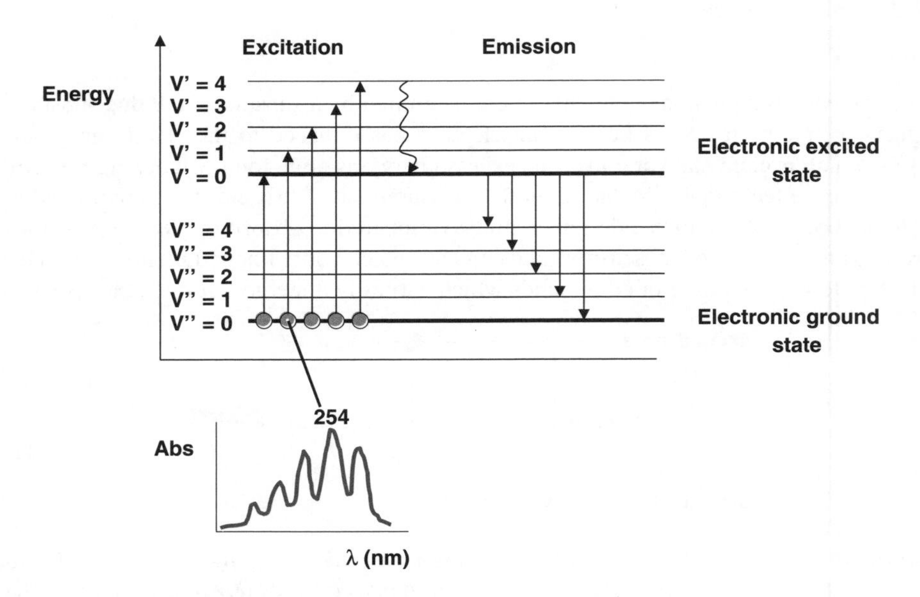

Figure 2.20 Energy diagram for fluorescence in a molecule.

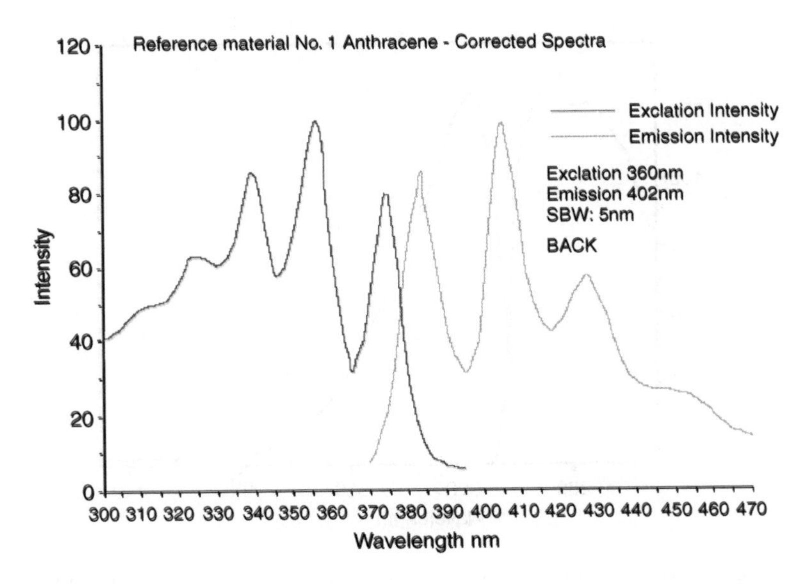

Figure 2.21 *The excitation and emission spectra for anthracene in alcohol (Reproduced by permission of Starna UK).*

for the excitation spectrum. The emission spectrum occurs at lower energy and longer wavelength due to some radiationless losses of energy mainly due to vibrations internally in the molecule. Each of the excitation and emission spectra has five peaks corresponding to the five possible transitions. The excitation and emission spectra are also approximate mirror images of each other, shifted by the small loss of energy above (Figure 2.21). Stokes shift is the term given to describe the difference between the excitation maximum and the emission maximum.

Only about 10–15 % of organic compounds are able to fluoresce naturally and these tend to have certain features in common such as a rigid and planar structure, e.g. polyaromatics, heterocyclics and uranyl compounds (Figure 2.22). However, most molecules can be derivatised if fluorescence is the desired technique. Fluorescence does not have wide applicability but has excellent sensitivity as the signal is measured against a zero background. It can also be a very selective technique when the compound of interest fluoresces and other components in the sample do not or fluoresce at different wavelengths.

The UV absorption spectrum and fluorescence excitation spectrum of a molecule often occur at similar wavelengths. The same phenomenon is responsible for both spectra – absorption of a photon and promotion of an electron from the ground state into a higher

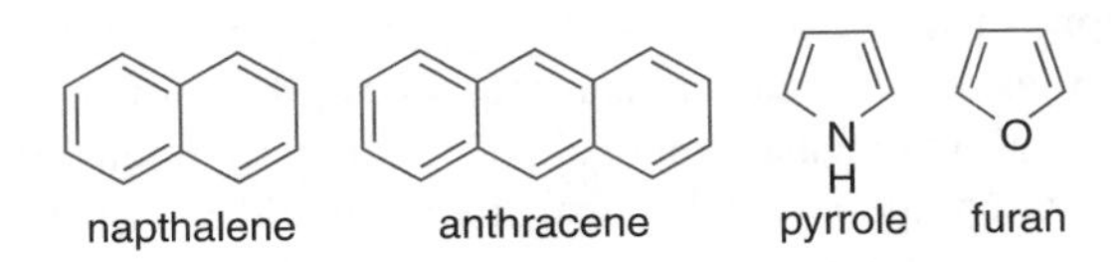

Figure 2.22 *Some examples of compounds that readily fluoresce due to their rigid, planar structures.*

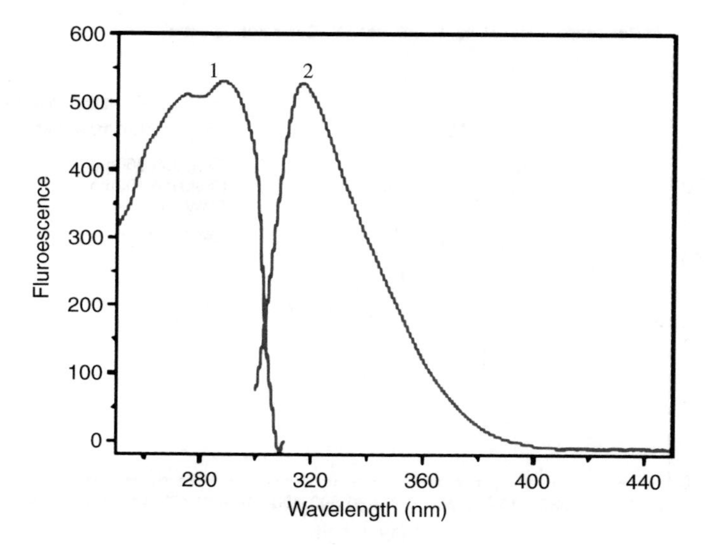

Figure 2.23 *Excitation (1) and emission (2) spectra of aspirin in the solid phase (Reprinted from Analytica Chimica Acta. Moreira, A.B. et al., 'Solid-phase fluorescence spectroscopy for the determination of acetylsalicylic acid in powdered pharmaceutical samples',* **523***(1), 49–52*[21]*. Copyright 2004, with permission from Elsevier).*

energy level. The fluorescence excitation and emission spectra for aspirin are shown in Figure 2.23. Spectrum (1) can be compared with the UV absorption spectrum for aspirin given in Figure 2.2.

Phosphorescence begins in the same way as fluorescence. Radiation provides the energy needed to excite the molecule and the electrons undergo a small amount of internal, radiationless loss of energy. However, instead of emitting fluorescence at this point, there is inter-system crossing (ISC) to the triplet excited state, from where the electrons undergo more radiationless energy loss and finally emit photons of lower energy as phosphorescence as they return to the ground state. This is shown diagrammatically in Figure 2.24.

The emission spectrum of a phosphorescent molecule occurs at even longer wavelength than fluorescence. If quenching is prevented, e.g. by keeping the sample as cold or as solid as possible to minimise collisions between molecules, and by ensuring the sample is oxygen-free, phosphorescence can be seen. Phosphorescence is a longer-lasting luminescence than fluorescence and can sometimes be observed even after the excitation source has been removed, lasting for seconds or even minutes. The number of compounds that naturally phosphoresce is also small.

Phosphorescence can be observed without interference from fluorescence by a process called time resolution. Instruments for measuring phosphorescence are very similar to those used for fluorescence but a mechanism that allows the sample to be irradiated and then, after a time delay, allows measurement of phosphorescent intensity (phosphoroscope) is required as an extra component. The instrument should also have the capability of keeping samples at very low temperatures. Another type of long-lived photoluminescence is time-delayed fluorescence, where the electrons in the molecule obtain enough energy to be excited from a 'special' excited state to the normal excited state and then fluoresce.

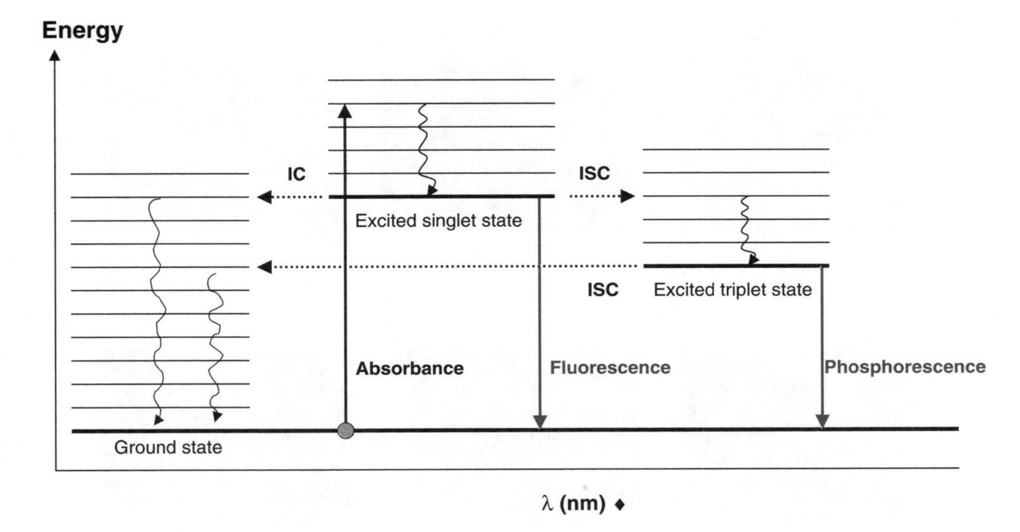

Figure 2.24 *Energy diagram for both fluorescence and phosphorescence in a molecule (Following absorption of radiation, an electron is promoted into the excited singlet state. Following radiationless loss of energy, the electron moves by inter-system crossing (ISC) to the excited triplet state, from where it can phosphoresce. As can be seen from the diagram, internal conversion (IC) and other ISC transitions are possible between states).*

Instrument

A schematic diagram of a fluorescence instrument is given in Figure 2.25.

Source
Generally, a UV source is required, e.g. xenon arc lamps, halogen lamps or lasers. The output of a xenon arc lamp is continuous over a broad wavelength range from UV to NIR. The nitrogen laser (337.1 nm) is a good example of a laser that is used for fluorescence. Tunable lasers can provide multiwavelength excitation.

Discriminator
Filters, monochromators or polychromators can be used. The excitation wavelength selector (discriminator 1) is used to select only the excitation wavelength and to filter out other wavelengths. The emission wavelength selector (discriminator 2) allows passage of only the emission (fluorescence) wavelength. Filters provide better detection limits but do not provide wavelength scanning abilities, which are often required if spectra need to be acquired. Instruments that use filters are called fluorometers. Those which use monochromators are spectrofluorometers; these allow scanning of the wavelengths and hence determination of the optimal wavelengths of both excitation and emission. Lasers generally provide monochromatic light, so an excitation wavelength selector (discriminator 1) is not needed in this case.

Sample
Cuvettes are normally 1 cm in path length and made of quartz (UV region) or glass (Vis region) for laboratory instruments. At least two adjacent faces will be transparent. Disposable

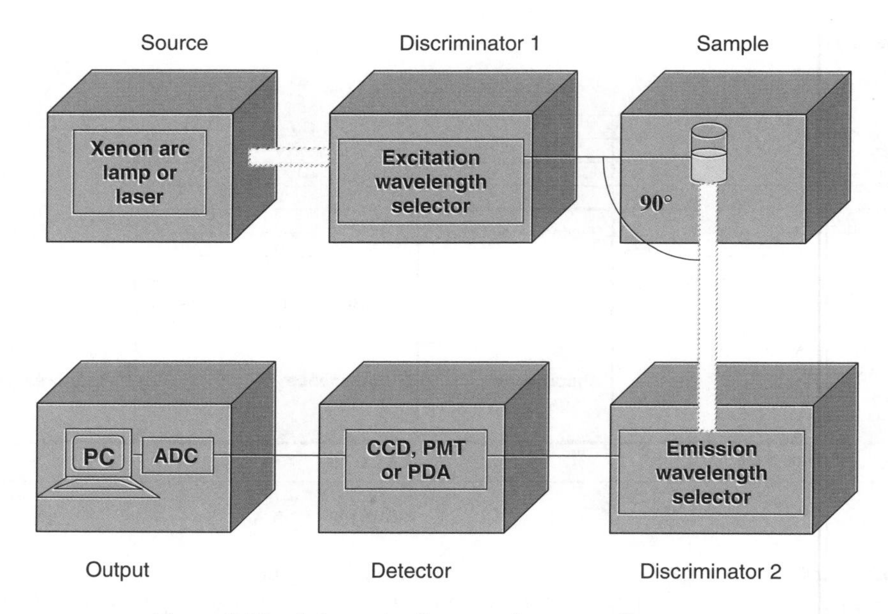

Figure 2.25 *Schematic diagram of a spectrofluorometer.*

plastic cuvettes are sometimes used for work on aqueous solutions in the visible region of the spectrum. Optical fibres can be used to transport light to and from luminescent samples in relatively inaccessible locations, such as in the field, process streams and *in vivo*. For phosphorescence measurements at room temperature, conventional fluorescence cuvettes work well, though steps may be taken to eliminate oxygen by controlling the environment. If low temperature control of the phosphorescent sample is required, the samples may be contained in a quartz cell surrounded by liquid nitrogen in a Dewar flask. Molecular luminescence measurements can be strongly affected by parameters such as solvent, temperature, pH and ionic strength so these need to be kept constant from sample to sample.

Detector

In fluorescent detection, the sample absorbs light and then re-emits it in all directions. A 90° arrangement is usually used in the instrument because this minimises any incident light that could interfere. Photomultiplier tubes (PMTs) are the most common detectors and are sensitive in the range 200–600 nm. Multichannel detectors such as PDAs, CCDs and CIDs can also be employed.

Output

The computer collects, stores, manipulates, analyses and displays the spectral data. It can carry out both qualitative and quantitative analysis as well as spectral searching and matching if the libraries are available.

Information Obtained

As spectra for molecules with similar fluorescing groups tend to be very alike, this is not a very useful identification (qualitative) technique. But fluorescence is an optical technique, and is subject to the Beer–Lambert Law, so for substances that do fluoresce, it is very useful for quantitative measurement. As already mentioned, fluorescence is a highly selective and sensitive technique due to very low background interference, so the limits of detection afforded are normally excellent.

Developments/Specialist Techniques

Generally, the source must be more intense than that required for UV–Vis absorption spectrometry as the magnitude of the emitted radiation is directly proportional to the power of the source. Hence, lasers are sometimes used as excitation sources, resulting in the technique of laser-induced fluorescence (LIF), which is often used in biological applications and allows some of the lowest detection limits possible, e.g. 10^{-11} M.

A technique called fluorescence resonance transfer (FRET) is where one excitation wavelength can be used to excite a number of moieties in the same molecule to fluoresce. This occurs due to internal transfer of energy between the donor moiety and an acceptor moiety. This technique has recently been used in a biological nanosensor to determine the neurotransmitter glutamate in living tissue[22].

Applications

Some cancer drugs are naturally fluorescent, e.g. anthracyclines. Hence, their levels can be measured in the blood of patients undergoing chemotherapy using fluorescence detection after separation by liquid chromatography[23,24]. This therapeutic drug monitoring can help to understand how the patient is metabolising the drugs. Many other drugs exhibit enough fluorescence to exploit for quantitative purposes[25,26]. Fluorescence was also used extensively on the human genome project. Native fluorescence has been used to study proteins[27] and fluorescent probes have been employed in the study of micelles, nucleic acids and other biological molecules[28]. Immunochemical methods for compounds of biological interest such as antibodies and antigens involving fluorescent or phosphorescent labels have largely replaced older radiolabelling methods[29].

2.1.4 Nuclear Magnetic Resonance

Principle

Nuclear magnetic resonance (NMR) is a spectroscopic technique used primarily to elucidate the structures of organic compounds, especially following synthesis or isolation of products. It is based on the fact that atomic nuclei that have an angular momentum and a magnetic moment have a special property called nuclear spin. For example, protons (^{1}H) and carbon-13 (^{13}C) have nuclear spin while oxygen (^{16}O) and carbon-12 (^{12}C) do not. The ^{1}H and ^{13}C nuclei, which due to their inherent spin are most commonly examined in NMR experiments, act like small magnets in a magnetic field as they align themselves parallel with or antiparallel to the applied magnetic field. The parallel orientation is lower in energy and therefore preferred.

Nuclear magnetic resonance exploits this interaction of spin with strong magnetic fields by using radiofrequency (RF) radiation to stimulate transitions between different nuclear spin states of samples in a magnetic field. When irradiated, the parallel, lower energy nuclei move to the higher energy spin state where they are now 'in resonance'.

This is akin to electrons being in the 'excited state' in UV spectroscopy. When the pulse of irradiation disappears, the nuclei relax to the lower energy spin state once more. This spin relaxation data can be collected as a free induction decay (FID) signal. This FID data is subjected to Fourier Transformation (FT) to yield a spectrum giving information on each type of nucleus in the molecule. However, because the nuclei are surrounded by other electrons and atoms (shielding effect), more than one signal may be observed for each type of nucleus. Three important spectral parameters are obtained in a ^{1}H-NMR spectrum: chemical shifts, coupling constants and intensities (integrals). These provide information on the environment and proximity of the structural groups, the molecular structure and the number of nuclei involved respectively.

The nuclides of most interest are protons (^{1}H) and carbon-13 (^{13}C) for organic molecules, though others such as phosphorous and silicon can be used. NMR spectroscopy is most useful as a qualitative tool for determining the structure and identity of molecules. It is rich in information content but can be poor in sensitivity. Most NMR instruments today are based on FT–NMR.

Instrument

A schematic of an NMR instrument is illustrated in Figure 2.26 and the components are shown in Figure 2.27.

Source

Radiofrequency (RF) transmitters generate frequencies of a few MHz to almost 1 GHz, which irradiate the sample molecules. If the energy difference between the relevant spin states is matched by the RF pulse, the nuclei will move to the higher spin state and be 'in resonance' with the magnetic field. In older instruments, either the frequency sweep

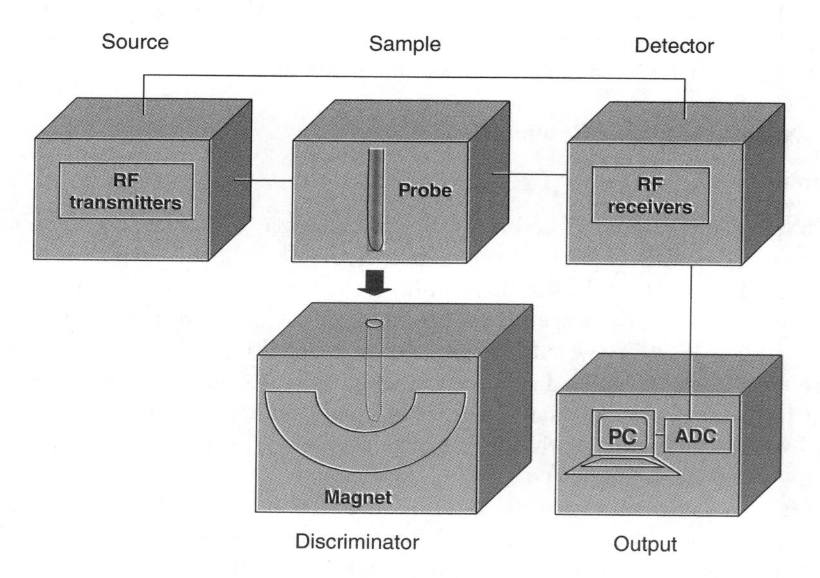

Figure 2.26 *Schematic diagram of an NMR spectrometer.*

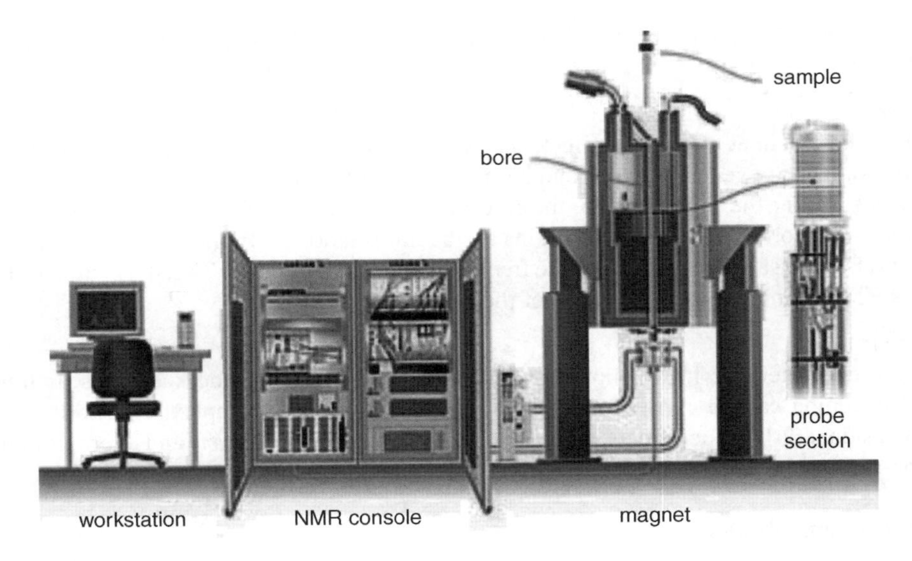

Figure 2.27 *Components of an NMR instrument (Reproduced with permission from Varian Inc.).*

or field sweep methods are used. In the frequency sweep method, the magnetic field is held constant while the frequency is continuously varied until the resonant condition is met for each nuclide. In the field sweep method, the frequency is held constant while the magnetic field is varied. More recently, pulsed NMR has taken over in which RF transmitters provide pulses of radiation to simultaneously excite all of the one type of nuclide, e.g. carbon-13 in the molecule.

Sample
The sample is placed in the probe which is placed in the centre of the bore. At the top of the probe are coils surrounded by electronics and tuning components. There are also now flow probes, solid state probes and cryogenic (cooled) probes. The cryogenic probes cool the electronics to reduce background noise, which can improve the S/N ratio by 3–4 times. Recently, capillary probes have been reported to give the same sensitivity as cryoprobes but using a lot less sample (active volume only 3 μl)[30]. The variable spectrometer frequency is proportional to field strength and certain pulses of RF radiation cause particular nuclei to become resonant with the magnetic field.

Discriminator
The magnetic field is the discriminator. The vast majority of magnets used today are super-conducting cryomagnets, which allow higher field strengths than older permanent magnets or electromagnets. These cryomagnets contain a large solenoid immersed in liquid helium at a temperature of 4.2 K or below. The coil resides inside a liquid helium Dewar flask at the centre of the magnet through which the bore passes. The bore is at room temperature. To reduce boil-off of the helium, the Dewar flask is surrounded by a second Dewar flask containing liquid nitrogen. Some magnets include a second superconducting coil outside the main solenoid to cancel out the effects of the field in

the surrounding room. Supercooled magnets can obtain field strengths of up to 21 T (900 MHz).

Detector
When the resonant condition is met, the NMR signal is collected at the RF receivers. NMR signals are generally weak and need to be amplified and processed prior to further analysis. Using the pulsed mode, the free induction decay (FID) spectrum in the time domain is recorded and while it contains all the information on frequencies, splitting and integrals, it must be converted into the frequency domain by Fourier Transformation (FT). This FT step enhances the S/N ratio of the signal.

Output
At the computer, the huge amount of information is processed and spectral searching and matching can be carried out. NMR spectra can be very complex, especially two-dimensional (2-D) experiments, and may require detailed data analysis and interpretation. Libraries can be very useful for assigning structure and identifying compounds.

Information Obtained

Chemically nonequivalent nuclei in a molecule are shielded differently and so give separate signals in an NMR spectrum. This leads to peaks positioned at a particular chemical shift (ppm) on the x-axis of an NMR spectrum. The internal standard marker for 0 ppm is given by tetramethylsilane (TMS). Furthermore, splitting of the peaks arises due to interactions between different nuclei, which provides information about the proximity of different atoms in a molecule, i.e. environment. Thirdly, the area under a signal can be integrated; this integral indicates how many nuclei are contributing to that signal (Figures 2.28 and 2.29).

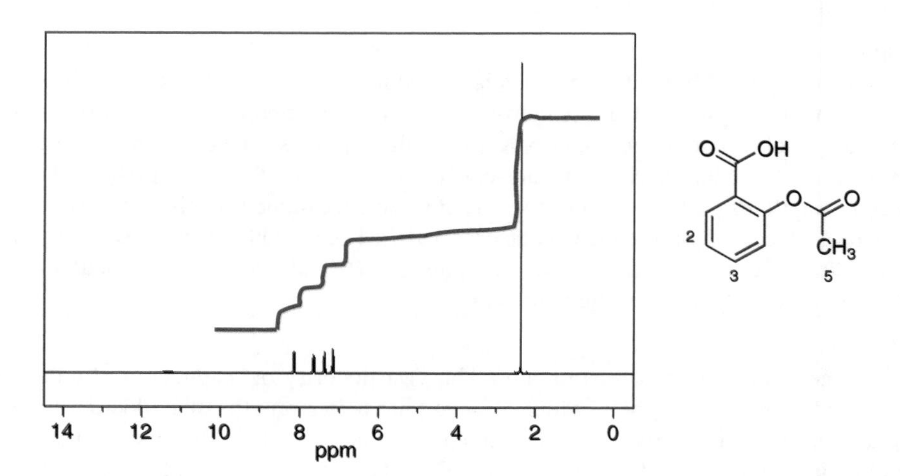

Figure 2.28 *^{1}H-NMR spectrum of aspirin (The broad NMR signal at 11.3 ppm corresponds to acidic proton number 6. The signals at 8.2, 7.7, 7.4 and 7.1 ppm correspond to the aromatic protons 1, 3, 2 and 4 respectively. The singlet signal at 2.3 ppm corresponds to proton 5. On integration, the ratio of the area under each peak is 1:1:1:1:3.).*

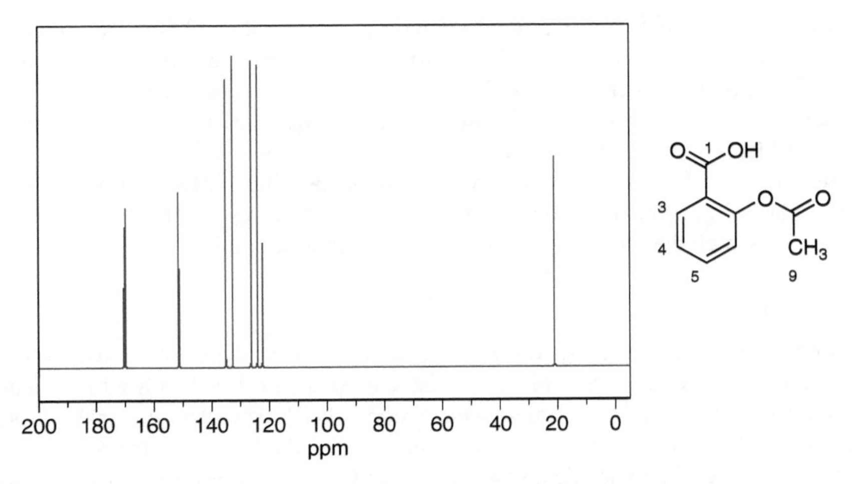

Figure 2.29 ^{13}C-NMR (decoupled) spectrum of aspirin (The NMR signal at 170 ppm (two peaks) corresponds to carboxyl carbons 1 and 8. The signals between 152 and 122ppm correspond to the aromatic carbons in the order 7, 4, 3, 5, 6 and 2. The peak at 21 ppm corresponds to the aliphatic carbon number 9.).

Developments/Specialist Techniques

In cases where assignment of all the signals in a spectrum is proving difficult, additional experiments can be carried out. These are spin decoupling, the DEPT experiment and two-dimensional experiments. There are two main types of spin decoupling experiments: heteronuclear decoupling, which simplifies the spectrum of one nuclide by eliminating the coupling to another, and homonuclear decoupling, which is a useful technique for analysing proton spectra. The DEPT (Distortionless Enhancement by Polarisation Transfer) technique uses a complex pulse sequence to enable the distinction to be made between the type and number of CH, CH_2 and CH_3 groups in a ^{13}C-NMR spectrum. Two-dimensional experiments allow the correlation of chemical shifts along two frequency axes and are called COSY (correlated spectroscopy) experiments. Homonuclear H,H–COSY involves correlating the chemical shifts from two ^{1}H experiments while heteronuclear H,C–COSY involves correlating the chemical shifts from both a ^{1}H and a ^{13}C experiment. By comparing the diagonal and cross peaks, signals can be more clearly assigned and some extra fine structure information gained in some cases.

New cryoprobes are coming on stream that are designed for multipurpose NMR measurements of four different nuclei – phosphorus, carbon, nitrogen and hydrogen. There is no need to change probe between the different experiments.

In medicine, NMR is employed as MRI (magnetic resonance imaging) and was introduced in the 1980s. The primary advantage of MRI is that it is a noninvasive technique and unlike CT (computerised tomography) scans, it does not use X-rays. Like proton NMR, MRI allows 3-D images to be made by applying RF waves to the water molecules present in soft tissue, which line up in an applied magnetic field. 3-D images of the soft tissue in the brain, eyes and spinal column, among others, can be produced. The magnets used in MRI instruments have a bore, or central opening, large enough for a human body to fit within the magnet. While in the magnetic field, the subject being examined can be irradiated with RF from multiple angles to produce a 3-D 'picture'. MRI can also be used to image blood vessels and provide insight into the chemical components of selected tissues. An MRI image can provide invaluable help in diagnosing disease. Further information on NMR and MRI can be found elsewhere[31].

Savukov *et al.* have reported a new way of measuring NMR in liquid samples[32]. Polarised laser light can be rotated by nuclear spin and the degree of rotation is related to the extent to which the nuclear spins are aligned with each other.

Applications

Nuclear Magnetic Resonance spectrometry is used for many types of analytical work but is key in the elucidation of structures of chemical compounds. When used in conjunction with mass spectrometry and infrared spectroscopy, the three techniques make it possible to determine the complete structures of novel compounds. Mass spectrometry is used to determine the size of a molecule and its molecular formula and infrared spectroscopy help identify the functional groups present in a molecule. NMR spectroscopy is used to determine the carbon–hydrogen framework of a molecule and works with even the most complex molecules. NMR is now being used to elucidate complicated protein structures[33, 34].

While mainly used as a qualitative technique for identification, NMR has been used for quantitative analysis. Examples include the determination of glyphosate in biological fluids[35], ions in serum[36] and fluoroquinolones in aqueous samples[37].

2.1.5 Mass Spectrometry

Principle

Mass spectrometry (MS) is based on generating ions in the gaseous state, separating them according to their mass-to-charge ratio (m/z) and detecting them. MS is therefore useful for identification as it can elucidate chemical and structural information about molecules from their molecular weights and distinctive fragmentation patterns. In fact, MS provides more information about the composition and structure of a compound from less sample than any other analytical technique. MS is also very important for the quantitative measurement of atoms or molecules. The sensitivity of MS for identifying molecules is in the 10^{-12}–10^{-15} molar range and for proteins, mass measurement accuracy can be extremely low. Not all compounds are amenable to being ionised, especially from liquids and solids, so the challenge with mass spectrometry is often at the front end of the instrument where the chemistry of the molecules and the ionisation chamber conditions play an important role.

Instrument

In general, the operation of a mass spectrometer involves creating ions in the gaseous phase, separating the ions in space or time based on their mass-to-charge ratio (m/z) and measuring the quantity of ions of each mass-to-charge ratio. A mass spectrometer consists of a sample introduction system, an ion source, a mass-selective analyser, an ion detector and a computer. Since mass spectrometers create and manipulate gaseous ions, they operate in a high-vacuum system. The magnetic-sector, quadrupole, and time-of-flight designs also require extraction and acceleration ion optics to transfer ions from the source region into the mass analyser. A MS instrument is schematically represented in Figure 2.30.

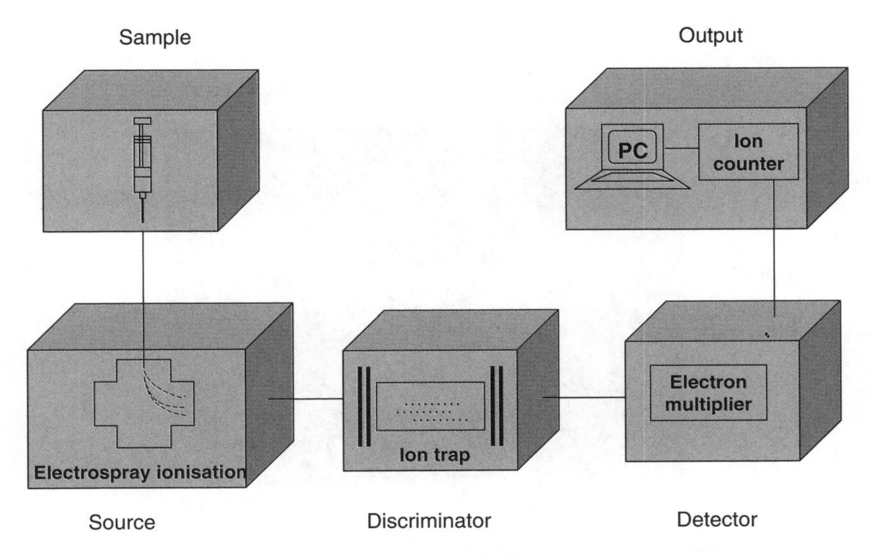

Figure 2.30 *Schematic diagram of a mass spectrometer.*

Sample
Obviously, the easiest sample to work with is a gas or volatile sample. However, there are now many ionisation and desorption techniques that can drive gas ions from a condensed liquid or solid phase. These allow even very large thermally labile molecules to be ionised and separated according to their m/z values. The sample can be introduced into a mass spectrometer in a number of ways. Two of the most common are by direct inlet probe and infusion by syringe at a set (slow) flow rate. Another common means of delivery is straight from another instrument, such as a high performance liquid chromatograph (HPLC) where a stream of liquid is infused from the outlet of the HPLC into the MS via the ion source chamber.

Source
The ion source is where the sample of interest is both ionised with a positive or negative charge and converted into the gas phase. There are a number of ion sources available:

- Electron impact
- Chemical ionisation
- Electrospray ionisation
- Atmospheric pressure chemical ionisation
- Fast atom bombardment
- Matrix assisted laser desorption ionisation
- Surface enhanced laser desorption ionisation.

Electron impact The gaseous sample enters the electron impact (EI) chamber via the inlet and is immediately bombarded by a beam of electrons. These electrons impact the sample molecules causing the loss of electrons, rendering the neutral molecules positively charged (Figure 2.31). With this positive charge, the ions are now attracted to the extraction plate, from where they pass on to the mass analyser. EI is considered to be a ‘hard’

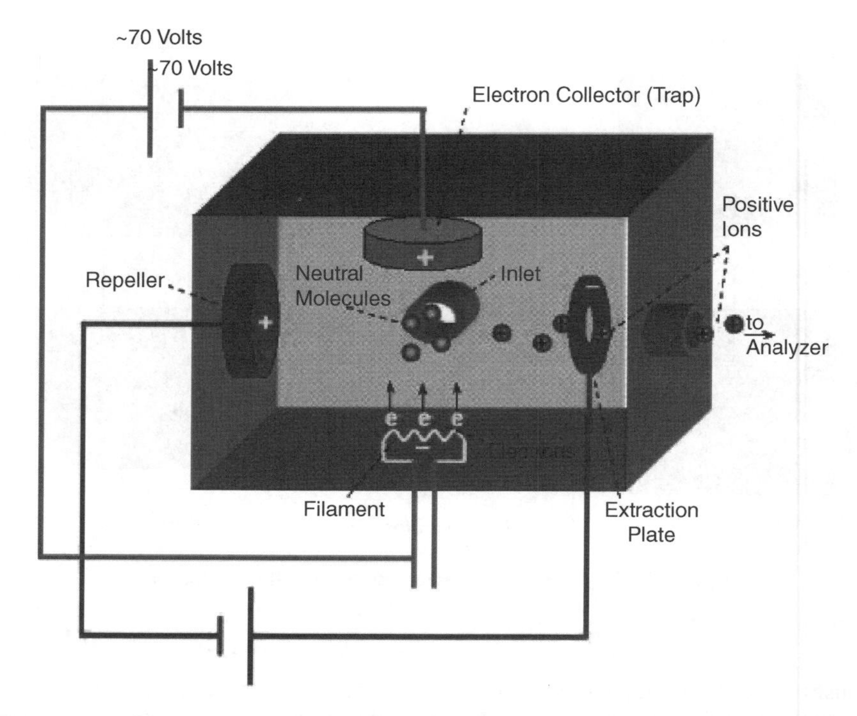

Figure 2.31 Schematic diagram of an electron impact ion source (Figure used by kind permission of Jim Roe, Department of Chemistry, University of Washington, Seattle, USA).

ionisation technique. This means that the molecules are broken up into smaller fragments. The resultant mass spectrum can therefore be detailed but gives excellent information on the structure of a molecule. In fact, for a given set of conditions, the spectrum can be a fingerprint of the particular compound under investigation and hence can be compared

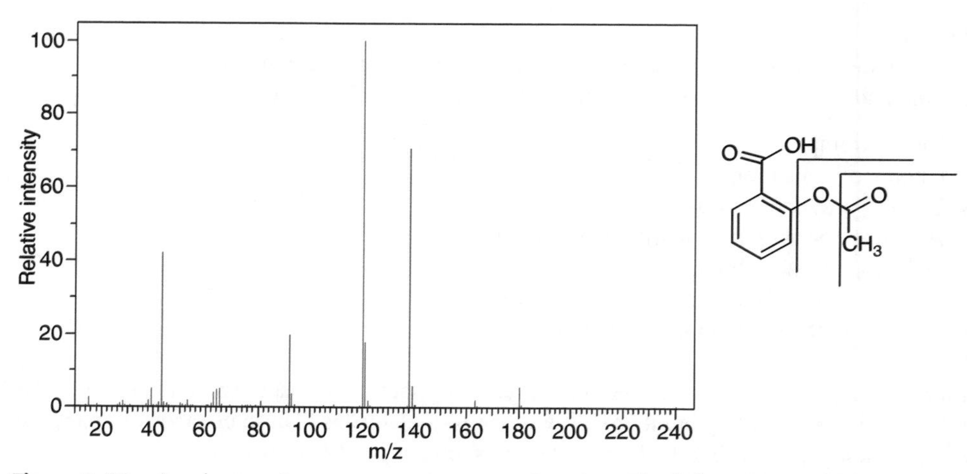

Figure 2.32 An electron impact mass spectrum of aspirin (The following m/z fragments can be seen: 43($C_2H_3O^+$), 92($C_6H_4O^+$), 120 ($C_7H_4O_2^+$), 138 ($C_7H_6O_3^+$), 180 (MH^+)).

to spectral libraries for definitive identification. The spectrum needs to be deconvoluted before the molecular weight of the compound (if unknown) can be deduced. The EI mass spectrum of aspirin is given in Figure 2.32.

Chemical ionisation (CI) is similar to EI in that it works well with gases but it is a less energetic process that begins with ionisation of methane (or other gas), creating radicals which in turn impact the sample molecules rendering them positively charged as MH^+ molecular ions. It is still a hard ionisation process and hence some of the molecular ions fragment but as less fragmentation occurs with CI than with EI, CI yields less information about the detailed structure of a molecule.

Electrospray ionisation (ESI) is a technique that takes place at atmospheric pressure and is considered to be a 'soft' ionisation process. It is very useful for liquids. Unlike 'hard' processes, the molecule is not normally fragmented and so the resulting mass spectrum is much simpler, the principal peak of which will be the pseudo-molecular ion, i.e. a protonated or sodiated peak. It is therefore much easier to decipher the molecular weight of a compound from an ESI source but there is less structural information given about the molecule, if any.

In ESI, the sample solution is mixed with a nebulising gas in the spray needle. This sample solution can be delivered by a syringe pump (direct infusion) or as eluent from a separation technique such as HPLC. The sample arrives in the spray chamber as a fine mist of droplets or spray. A drying gas, e.g. nitrogen, at a fairly high temperature causes the evaporation of any solvent from the droplets. A voltage gradient between the tip of the spray needle and the entrance to the transfer capillary, as well as a pressure difference from atmospheric pressure to vacuum, encourages appropriately charged ions to move into the capillary and on towards the skimmers. An ESI source is illustrated in Figure 2.33. Sometimes, the entrance to the mass analyser is placed at a 90° angle to the needle so that any ions entering the spray chamber that do not become ionised flow straight to waste. ESI accommodates a range of liquid flows.

One of the main attributes of ESI is its ability to multicharge large molecules, e.g. a 25 000 Da protein. If this protein is only singly protonated, it will only be detectable by a high m/z range instrument such as a time-of-flight MS. However, a protein this size is capable of accepting 10–30 protons under the right conditions, which results in a series of ions over a m/z range of approximately 600–2500. All the charge states will be seen in

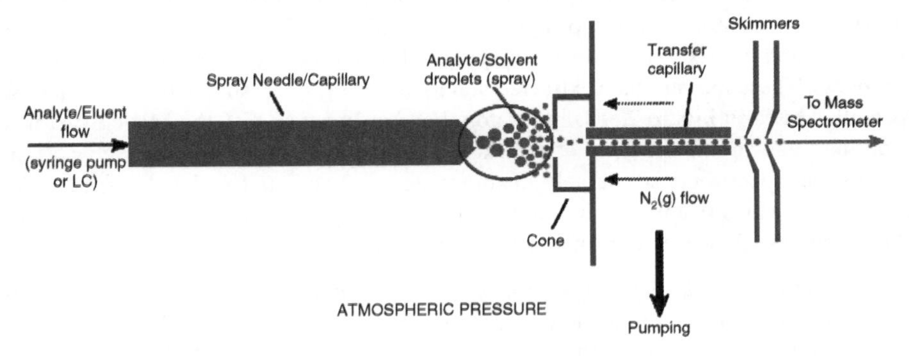

Figure 2.33 *Schematic diagram of an electrospray ionisation source (Figure used by kind permission of Dr Paul Gates, School of Chemistry, University of Bristol, UK).*

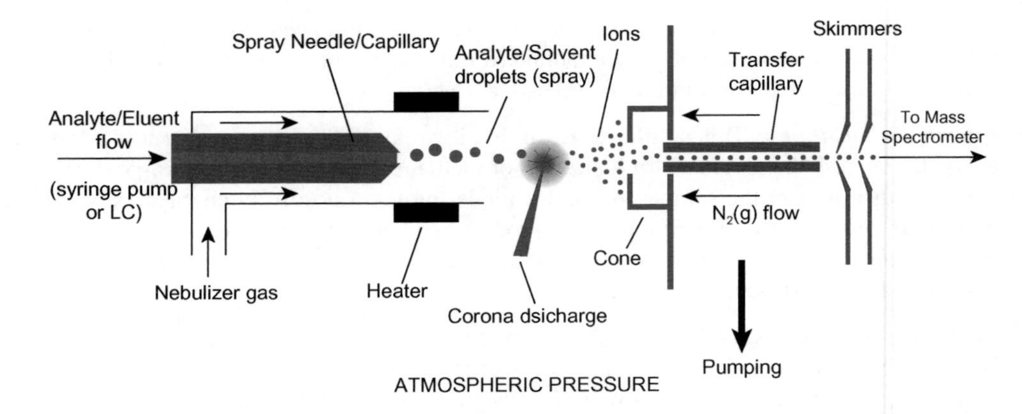

Figure 2.34 *Schematic diagram of an atmospheric pressure chemical ionisation source (Figure used by kind permission of Dr Paul Gates, School of Chemistry, University of Bristol, UK).*

an envelope of peaks and deconvolution yields the accurate molecular mass of the whole molecule.

Atmospheric pressure chemical ionisation (APCI) is a technique that also creates gas phase ions from the liquid sample. It too takes place at atmospheric pressure and uses a similar interface to that in ESI. As in ESI, the sample solution is mixed with a nebulising gas and the sample arrives in the spray chamber as a fine mist of droplets or spray. In APCI, an extra component – a corona discharge – is used to further ionise the analyte droplets in a manner similar to straightforward CI (Figure 2.34). While a small amount of fragmentation may occur, the technique is still considered to be a soft ionisation one. The gas-phase ionisation in APCI is more effective than ESI for analysing less polar species. ESI and APCI are complementary methods.

Fast atom bombardment is an ionisation technique that involves bombarding a solid spot of the analyte/matrix mixture on the end of a sample probe by a fast particle beam. The matrix (a small organic species like glycerol or 3-nitro benzylalcohol) is used to keep a homogenous sample surface. The particle beam is incident onto the surface of the analyte/matrix spot, where it transfers its energy bringing about localised collisions and disruptions. Both analyte ions and matrix ions are ejected (sputtered) from the surface as secondary ions by this process. These ions are then extracted and focused before passing to the mass analyser. This is illustrated in Figure 2.35. In fast atom bombardment (FAB), the particle beam is a neutral inert gas (argon or zenon) at 4–10 keV and the process is a comparatively soft one. The resulting spectra consist largely of protonated and sodiated molecular species (e.g. $[M+H]^+$ and $[M+Na]^+$) with some minor structural fragmentation. The low mass region of the spectra are, however, dominated by matrix and matrix/salt cluster ions.

Matrix-assisted laser desorption ionisation (MALDI) can analyse molecules with large mass >10 000 Da. The technique is similar to FAB and is also considered a soft ionisation process. The analyte is mixed with a matrix such as sinapinic acid or 2,5-dihydroxybenzoic acid that strongly absorbs UV radiation. A few microlitres of this analyte/matrix mixture is deposited on the sample plate and dried. The plate is inserted into the source region

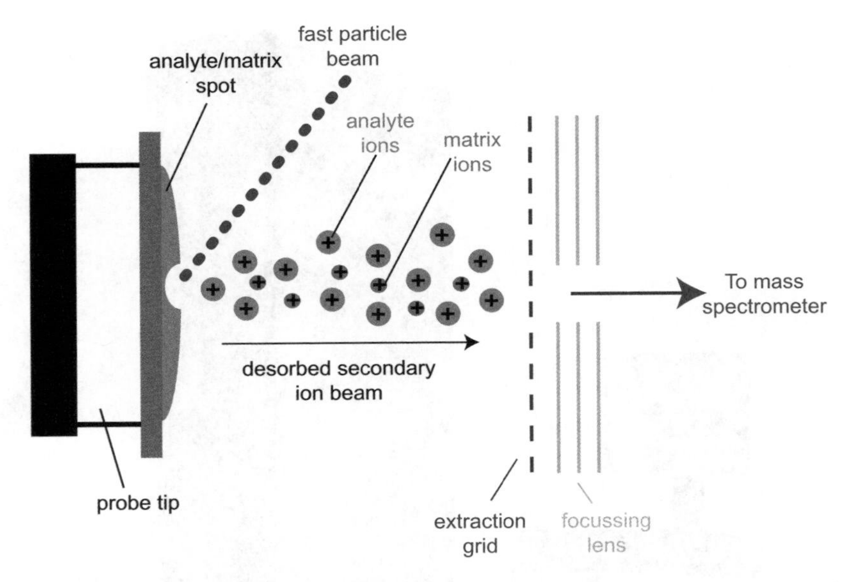

Figure 2.35 *Schematic diagram of a fast atom bombardment source (Figure used by kind permission of Dr Paul Gates, School of Chemistry, University of Bristol, UK).*

of the mass spectrometer and irradiated by a laser beam, e.g. nitrogen (337 nm). The matrix absorbs most of the energy from the laser pulse and transfers it to the analytes in the sample and both matrix and analyte ions are ejected into the gas phase as ions. A schematic of this ionisation process is shown in Figure 2.36.

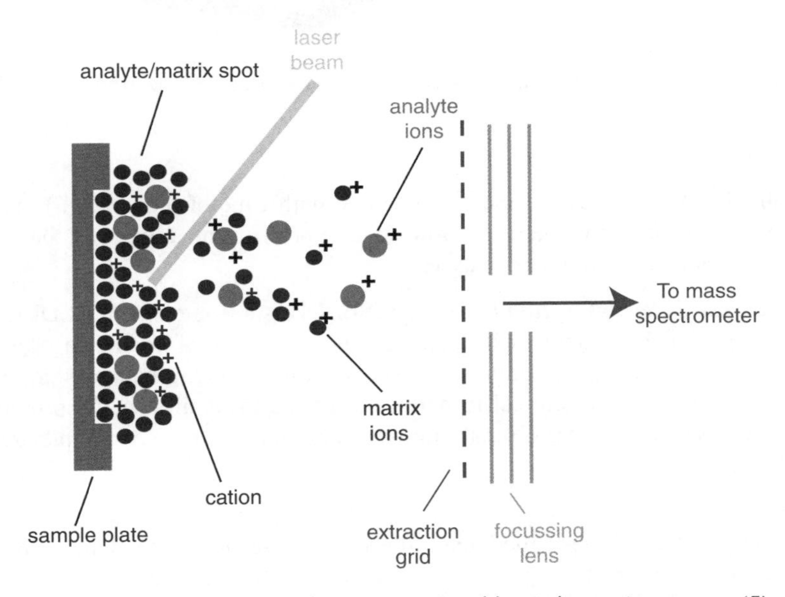

Figure 2.36 *Schematic diagram of a matrix-assisted laser desorption source (Figure used by kind permission of Dr. Paul Gates, School of Chemistry, University of Bristol, UK).*

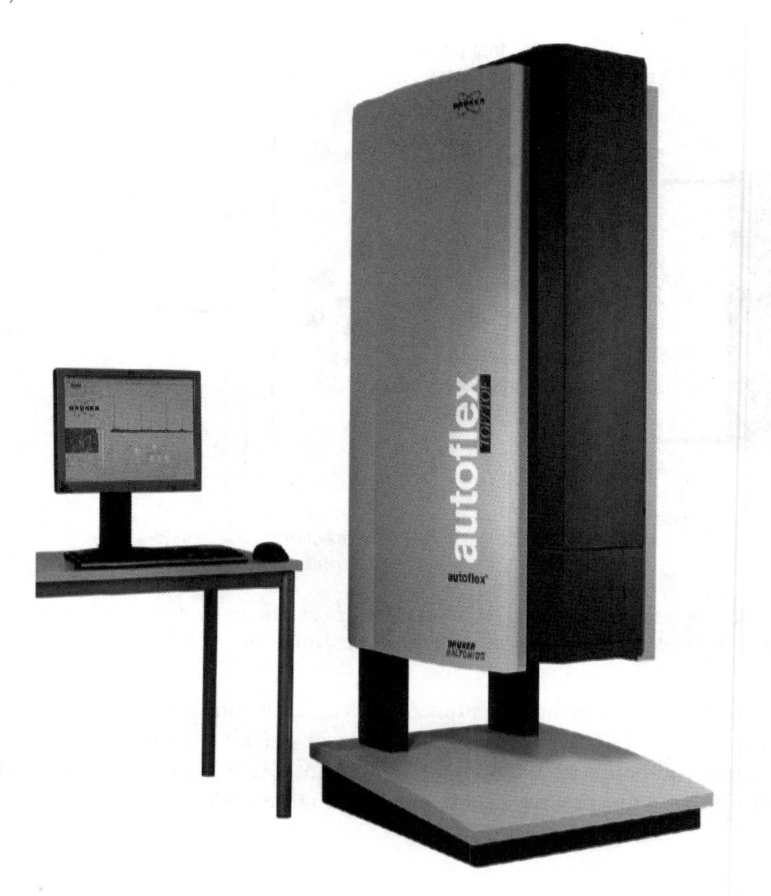

Figure 2.37 *A MALDI–TOF Molecular Imager (Reproduced by permission of Bruker Daltonics).*

Although MALDI sources are most compatible with time-of-flight (TOF) analysers (Figure 2.37), they can now operate at atmospheric pressure meaning that they can be interfaced to a wider array of mass analysers.

Surface enhanced laser desorption ionisation (SELDI) is a version of MALDI with the addition of a special chip from which sample components are ionised and desorbed. The chip can be a protein chip array composed of spots of different chromatographic surfaces (which are chemical or biochemical in nature) designed to retain specific proteins. Its main advantage is inclusion of this separation/selection step for crude or complex samples prior to MS analysis.

Discriminator

The mass analyser (sometimes called mass filter) is the discriminating element in a mass spectrometer and is where the ions are manipulated and sorted. There are a number of types of mass analyser in MS[38, 39] and among them are:

- Transmission quadrupole
- Quadrupole ion trap
- Time-of-flight
- Magnetic sector
- Fourier transform–ion cyclotron resonance.

The Transmission quadrupole is the most common type of mass spectrometer in use today. In a single quadrupole instrument, ions are guided through four metal rods or poles that have both DC voltage and AC voltages (RF) applied to them. For a given combination of voltages, only a single resonant ion will reach the detector. Other non-resonant ions never reach the detector. Thus the rods act as a mass filter. To scan all masses the amplitude of the RF and DC voltages are incremented stepwise while their ratio is held constant. A diagram of a single quadrupole is given in Figure 2.38.

The triple quadrupole instrument has three quadrupoles in series; these are often called Q1, Q2 and Q3. In full scan mode, a broad m/z range of ions passes through all three quadrupoles to the detector giving a mass spectrum. In 'selected ion monitoring', Q1 receives a mixture of all ions but only passes the ion of interest into Q2. Q2 is a collision

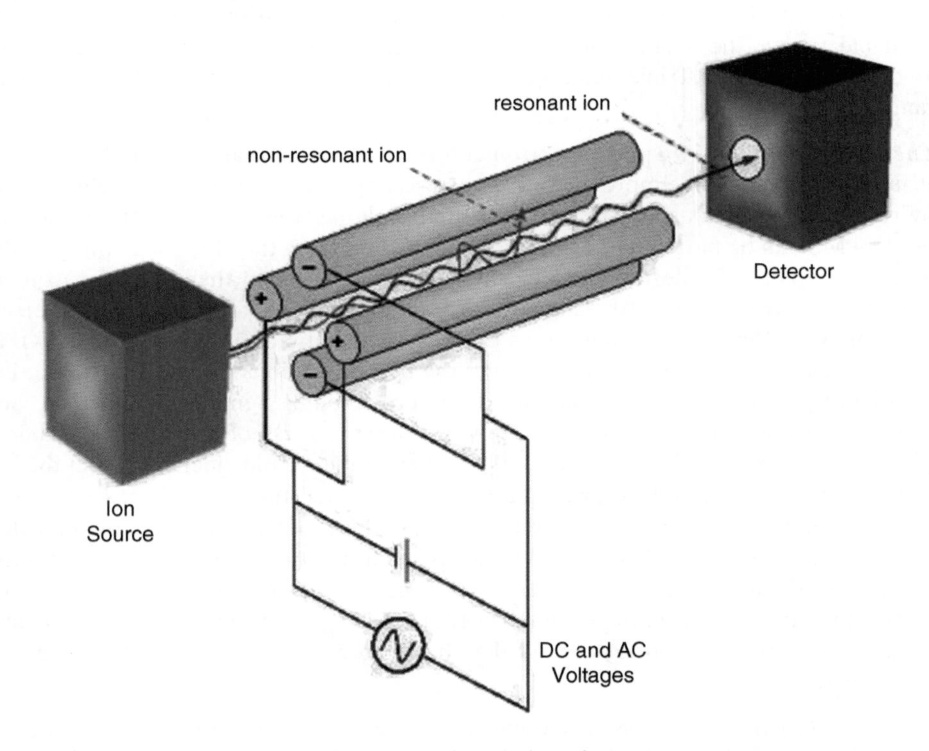

Figure 2.38 *Schematic diagram of a single quadrupole mass spectrometer (Figure used by kind permission of Jim Roe, Department of Chemistry, University of Washington, Seattle, USA).*

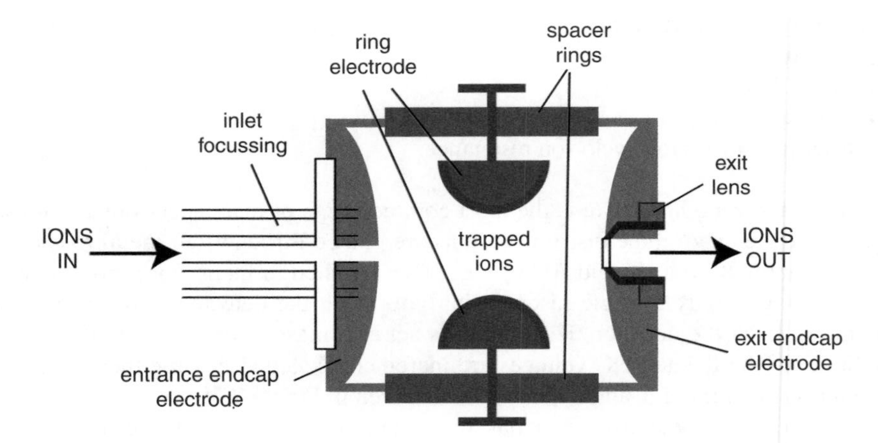

Figure 2.39 *Schematic diagram of an ion trap mass analyser (Figure used by kind permission of Dr Paul Gates, School of Chemistry, University of Bristol, UK).*

cell where the precursor ion is fragmented into product or daughter ions. Q3 allows the selected product ion through to the detector. There are other versions of this mass analyser available, such as the hexapole and octopole, which claim even greater control of the ions as they pass through. Triple quadrupole MS can provide sensitivity in the 10^{-12} molar range.

The quadrupole ion trap mass spectrometer is the most popular mass analyser today. It is often referred to simply as an ion trap. Using a 3-D quadrupole field created by ring electrodes, ions are trapped in a small volume where manipulations such as tandem MS can be performed on them. Then, by changing the electrode voltages, the ions are sequentially destabilised, ejected from the trap and then detected to form the mass spectrum. The entire sequence of trapping, storing, manipulating and ejecting ions is performed in a continuous cycle. A schematic of an ion trap is shown in Figure 2.39. The ion trap is gated so that the number of ions can be controlled, i.e. they can be allowed to accumulate until a strong enough signal is obtained or they can be prevented from entering the trap when there is a risk of overloading. Tandem MS is conducted by filling the trap with all available ions and ejecting all but the ion of interest. Helium is introduced as a collision gas and the trapped ions fragment. The fragments are then scanned out according to their m/z and detected or subjected to further MS. An example of a tandem mass spectral experiment is shown in Figure 2.40. The advantages of the ion trap mass spectrometer include compact size, the ability to trap and accumulate ions to increase the signal-to-noise ratio of a measurement, no transmission losses and the ability to carry out tandem mass spectral experiments.

A Time-of-Flight (TOF) mass spectrometer uses the differences in transit time through a flight/drift zone to separate ions of different masses. The principle is that smaller ions, being lighter, will reach the detector faster than heavier ions. It operates in a pulsed mode so ions must be produced or extracted in pulses. An electric field is used to accelerate all

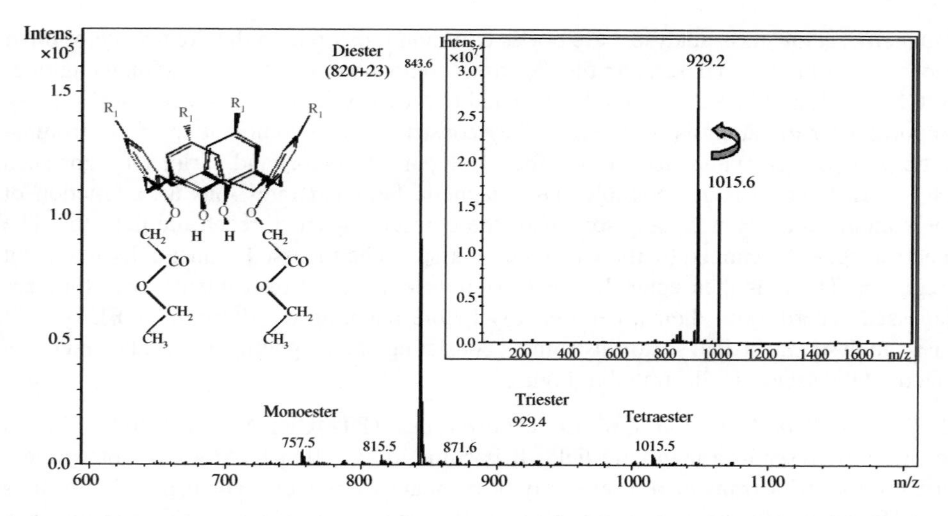

Figure 2.40 *A direct infusion ESI mass spectrum of a tetraester calix[4]arene molecule with sodiated molecular ion at 1016 m/z (The inset mass spectrum shows the fragmentation of the 1016 peak – yielding a daughter ion at 929 m/z representing loss of the most weakly held ester group (87 amu) from the tetraester). The predominant ion in the spectrum is that of the diester (structure shown) since this compound forms a strong sodiated adduct in the MS.*

ions into the field-free drift zone where lighter ions have a higher velocity than heavier ions and reach the detector sooner. A schematic of a TOF is given in Figure 2.41. It is most often seen in conjunction with MALDI as the ionisation source (Figure 2.37). There is no upper mass limit and if used in the reflectron mode and/or with orthogonal injection, its resolution is very high.

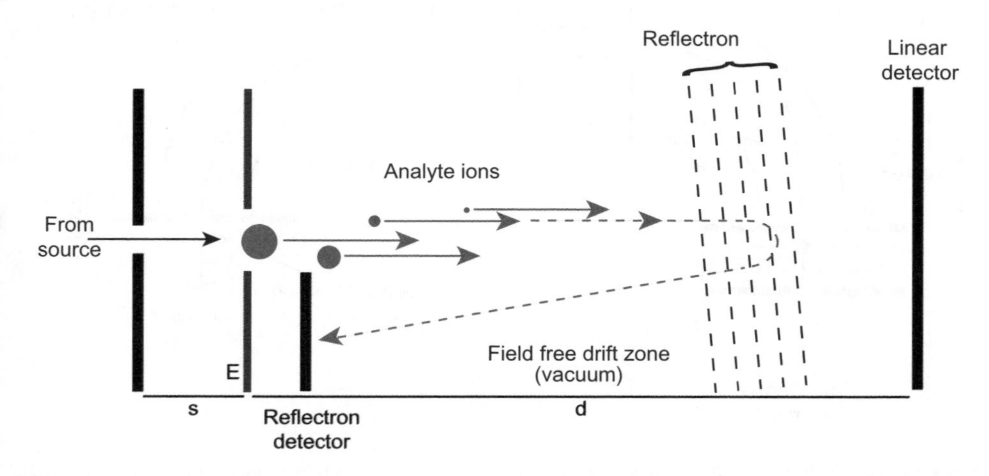

Figure 2.41 *Schematic diagram of a time-of-flight mass analyser (Figure used by kind permission of Dr Paul Gates, School of Chemistry, University of Bristol, UK).*

Magnetic sector mass analysers are not as common today due to the prevalence of other more modern mass analysers but they are very compatible with the ionisation techniques of CI and EI, and hence with the hyphenated (hybrid) system of GC–MS. They are also important for the analysis of dioxins. They consist of some combination of electromagnetic part (B sector) and electrostatic focussing part (E sector) and various designs such as BE, EBE and EB are possible. The magnetic field separates ions as a function of their momentum. Hence, only ions with the correct trajectory reach the detector. The electromagnet is scanned to allow different masses to be focussed sequentially into a slit (register). The ions then enter the electrostatic sector of the instrument where they are focussed according to their kinetic energy before reaching the detector. A BE type of mass analyser, i.e. a dual sector instrument consisting of a magnetic sector followed by an electrostatic sector, is illustrated in Figure 2.42.

A Fourier Transform–Ion Cyclotron Resonance (FT–ICR) MS functions like an ion trap analyser in a magnetic field. It is sometimes called FTMS. This instrument creates a Fourier transform of an array of resonant frequencies corresponding to ions of different m/z value stored in the magnetic ion trap. The ions in the trap move in cyclotron motion (circular orbits) in a plane perpendicular to the magnetic field. Excited ions make larger circular orbits and as they pass the receiver plates their frequencies are detected. The frequency of the motion of an ion is inversely proportional to its mass. Working at higher magnetic fields means better accuracy, sensitivity and resolution. It also has a very wide dynamic range, which enables low abundance species to be identified in the presence of high abundance compounds. A schematic of an FT–ICR is shown in Figure 2.43.

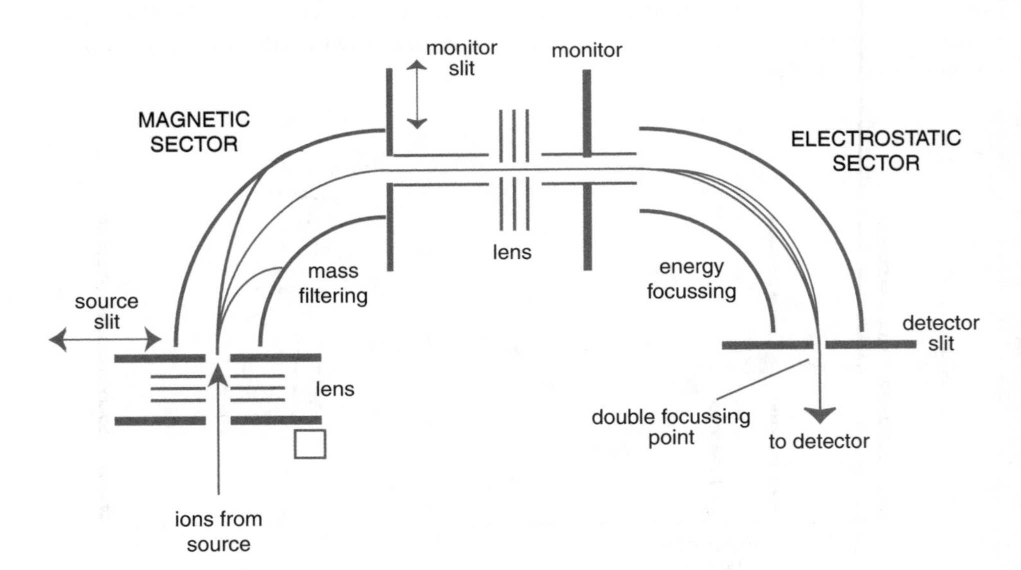

Figure 2.42 *Schematic diagram of a magnetic sector mass spectrometer (Figure used by kind permission of Dr Paul Gates, School of Chemistry, University of Bristol, UK).*

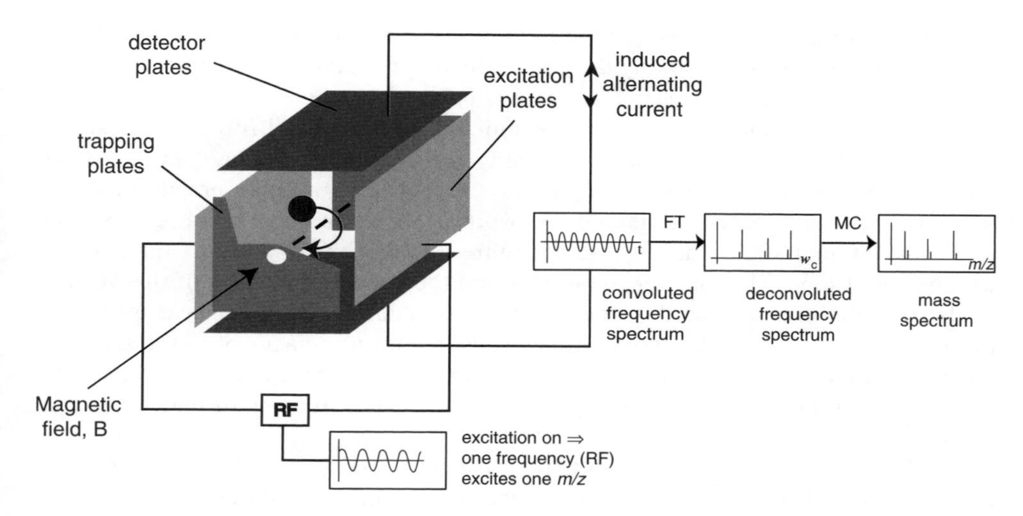

Figure 2.43 *Schematic diagram of a Fourier Transform–Ion cyclotron resonance mass analyser (Figure used by kind permission of Dr Paul Gates, School of Chemistry, University of Bristol, UK).*

Detector
At the mass detector, the signals are received and amplified. The abundances of all ions is calculated here.

Output
The data processing is carried out by a computer – qualitative analysis, quantitative analysis and structural elucidation of known (library searching) and unknown compounds.

Information Obtained

Mass spectrometry is an excellent technique for both qualitative and quantitative analysis. On the basis of molecular weight alone, many compounds can be identified (with the exception of isomers). Using a combination of molecular weight and fragmentation patterns, unambiguous identification is possible as well as structural elucidation of unknown compounds. MS can also be used for quantitative work using either internal or external standards for reference. Limits of detection are very low so sensitivity is normally excellent.

Developments/Specialist Techniques

There has been an explosion of new developments in mass spectrometry in recent years. One of the major ones is the reduction in size of the instruments so that many can now be used on the benchtop. This has resulted in prices coming down – initial and running costs used to be prohibitive in many cases. The development of atmospheric pressure ionisation sources is relatively new and has made hyphenation of MS with liquid separation techniques such as HPLC much more feasible. The selection of mass analysers has also grown and hybrids such as the LTQ–FT (linear trap quadrupole–FT–ICR) are now coming on stream, combining the best attributes of a number of analysers to create very powerful mass spectrometers.

Two new independently developed techniques called Dart[40] (direct analysis in real time) and Desi[41] (desorption electrospray ionisation) are making a huge impact on mass spectrometry. Together they remove the need for sample preparation and vacuum, speed up analysis time and can work in the open air. The sample is held in a gas or liquid stream at room temperature and the impact induces the surface desorption of ions. The ions then continue into the vacuum interface of the MS for analysis. Samples can be hard, soft or even liquid in nature. Ifa *et al.* have used Desi to image biological samples in two dimensions, recording images of tissue sections and the relative concentrations of molecules therein[42]. Jeol have launched a commercial Dart ion source for non-contact analysis of materials in open air under ambient conditions.

Novel methods for the direct mass spectral analysis of breath samples have been reported. Most of these detect volatile components but one has used extractive electrospray ionisation MS to directly detect non-volatile lipids, peptides and carbohydrates as well as volatile compounds in the breath of subjects[43].

Applications

Mass spectrometry is a suitable technique for both large and small molecules. It has important applications in proteomics for both the analysis of peptides and proteins[44, 45, 46, 47, 48]. Proteins can be analysed directly or their structure elucidated by peptide sequencing. There are extensive search libraries now available for characterising the protein or peptide of interest in the sample. MS is very useful for following the synthesis/quality control of oligonucleotides[49]. Rapid, unambiguous and comprehensive identification of metabolic changes in drug discovery can also be carried out using MS experiments[50, 51]. One interesting application of MS is for the fast detection of heroin on banknotes in drug trafficking cases[52]. Other applications include the determination of rosuvastatin in brain tissue[53], drug distribution in whole body sections using MALDI MS[54] and tandem mass spectrometric (MS^2) determination of dydrogesterone, an orally active synthetic progestogen, in human plasma[55].

One very important application of tandem mass spectrometry is its ability to distinguish between molecules of identical molecular weight on the basis of their fragmentation patterns; in fact, the daughter ions can then be used for subsequent quantitative measurement. The anthracyclines doxorubicin and epirubicin are epimers of each other and both have a molecular weight of 543. MS^2 was employed to quantitate them individually (Figure 2.44). While both samples, when injected separately, gave identical mass spectra with a molecular ion of 544 m/z, their fragmentation patterns resulting from isolation and fragmentation of 544 m/z, were quite different. These differences could then be exploited as a means of distinguishing the molecules from each other. This work has been used in the development of a quantitative LC/MS^2 method for epirubicin and its metabolites in human serum samples[56].

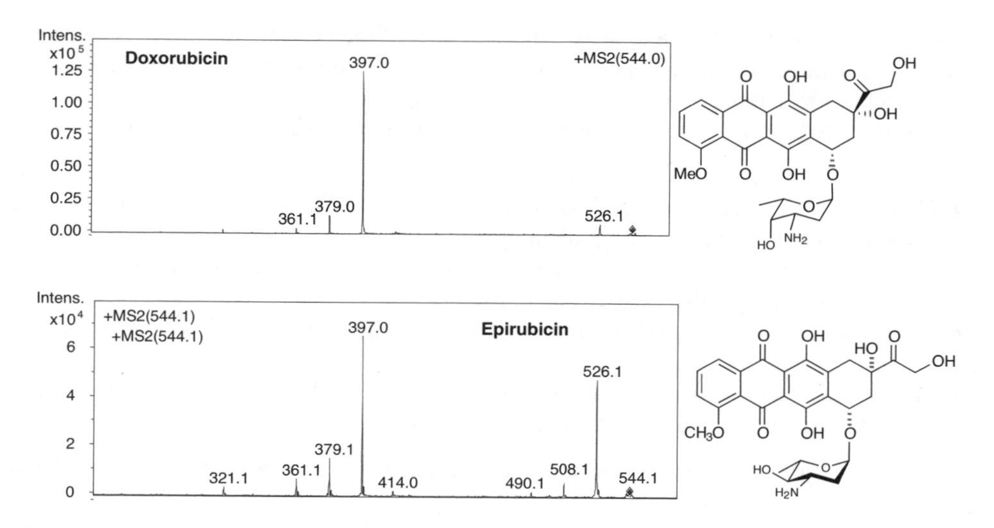

Figure 2.44 *Direct infusion MS^2 carried out on both doxorubicin (top) and epirubicin (bottom) (Fragmentation of doxorubicin [M+H$^+$] yielded an ion of 397 m/z, which corresponds to the loss of a sugar moiety from the parent ion. Fragmentation of epirubicin yielded ions of 526 and 397 m/z resulting from the loss of a water molecule and the sugar moiety respectively).*

2.2 Atomic Spectrometry

As atoms cannot vibrate or rotate as a molecule does, when energy is absorbed, electronic transitions take place and are seen as sharp lines as opposed to broad bands or peaks in the molecular spectra. These line spectra are specific for the element being studied and contain information on its atomic structure. Atomic spectrometry is applicable to most gas phase elements over a wide range of concentrations (sensitivity in the parts per trillion region) and involves detecting, measuring and analysing radiation that is either absorbed or emitted from the atoms or ions of the element of interest. In all these elemental techniques, the sample is decomposed by intense heat into hot gases consisting of free atoms and ions of the element of interest. If energy is provided to an atom, there are three possible transitions:

- Absorption – where optical radiation is absorbed by the electrons to excite them.
- Emission – where thermal or electrical radiation is absorbed by the electrons to excite them, and they then proceed to emit that radiation to return to the ground state.
- Fluorescence – where optical radiation is absorbed by the electrons to excite them and they then go on to emit some of this radiation.

In terms of this group of atomic spectrometric techniques, optical emission spectrometry (OES) is the oldest and most established. Atomic absorption spectrometry (AAS) and atomic fluorescence spectrometry (AFS) are two other important types of atomic spectrometry techniques. In this section, AAS and AFS will be discussed only briefly; elemental analysis based on ICP spectrometry will be discussed in more detail, specifically ICP–OES (inductively coupled plasma–optical emission spectrometry) and ICP–MS (inductively coupled plasma–mass spectrometry). ICP–OES is sometimes referred to as inductively coupled plasma–atomic emission spectrometry (ICP–AES).

The operating sequence of any instrument used for elemental analysis can be summarised as follows:

- Generation of a response by absorption of energy, followed by the production of a response in the form of emission of energy.
- Extraction of the useful signal information from the whole information using a device such as a wavelength dispersive system, filter etc.
- Conversion of the selected signal into electrical information using a detector for photons or ions.
- Processing of the electrical signal to obtain useful data with a good signal-to-noise ratio and software manipulation of the signal to yield analytical results.

Further information on the various techniques of atomic spectrometry is available elsewhere[57].

2.2.1 Atomic Absorption Spectrometry and Atomic Fluorescence Spectrometry

Atomic absorption spectrometry (AAS) is based on ground state free atoms absorbing visible or UV radiation at specific wavelengths. Under certain conditions, the amount of radiation absorbed can be correlated with the concentration of the atoms in the sample. As most atoms of an element will be in the ground state at temperatures below 5000 K, the temperature of the source must be high to encourage excitation and absorption. In AAS, the difference between the incident and exiting radiation over a very narrow wavelength range is detected and measured. Analytically AAS is a very easy and reliable technique yielding simple absorption spectra. It generally has fewer spectral interferences than OES and the absorption is almost independent of the temperature of the vapour produced. In AAS, there are a number of principal methods; these are based on how the sample is converted into an atomic vapour. The following are the most important and are discussed below:

- Flame atomic absorption spectrometry (FAAS)
- Graphite furnace atomic absorption spectrometry (GFAAS)
- Hydride generation atomic absorption spectrometry (HGAAS)

Atomic fluorescence spectrometry (AFS) is based on ground state free atoms absorbing light at specific wavelengths, as in AAS, and the subsequent emission of fluorescence at longer (characteristic) wavelengths. The amount of fluorescence emitted can be correlated with the concentration of the atoms in the sample. As in AAS, the spectra are simple and the technique easy to use.

Instrument

The components of a typical AAS instrument are shown in Figure 2.45.

Source

The radiation source in the case of this flame AAS is a line source hollow cathode lamp (HCL). These are the most commonly used sources in AAS. It can be designed for

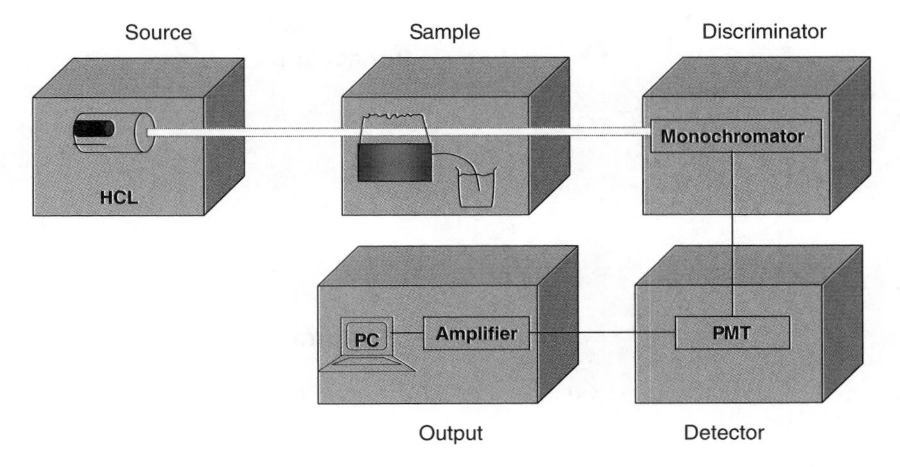

Figure 2.45 *Schematic diagram of a single-beam flame atomic absorption spectrometer (While the components of an atomic fluorescence spectrometer are similar, the emitted radiation is monitored perpendicular to the incident radiation).*

single-element or multi-element analysis. The single-element lamps have a higher energy output. It is also possible to use electrodeless discharge lamps and continuum sources. An example of a continuum source is a deuterium lamp or, for high resolution work, a high-pressure xenon arc lamp, which has much higher radiant power than a line source. For atomic fluorescence spectrometry, a HCL or a laser source (for increased sensitivity) can be used.

Sample

Subsequent to sample preparation and work-up, samples are supplied to the instrument in a form that depends on the atomisation technique used. In AAS, the function of the atomiser is to produce as many free atoms as possible for the purpose of sensitivity. There are three considerations when choosing the atomiser: the type of sample (solid, liquid or gas), the concentration of the sample and the size of the sample. The following three atomisation techniques are the most important.

Flame atomic absorption spectrometry (FAAS) is used primarily for liquids. The sample is converted into an aerosol by the nebuliser, atomised to free atoms, mixed with the combustible gases and ignited in the flame, which can rise to a temperature of 2600 °C. A long burner is used to increase the pathlength. The most commonly used flames are air–acetylene and nitrous oxide–acetylene mixtures.

Graphite furnace atomic absorption spectrometry (GFAAS) can be used for both liquids and solid or slurry samples. It is sometimes called electrothermal atomic absorption spectrometry (ETAAS). The sample is placed on a plate (known as a L'vov platform), which is made of solid pyrolytic graphite. The platform is placed in a small cylindrical graphite furnace tube (Figure 2.46), which can then be heated to 3000°C to vaporise and atomise the analyte. An electrothermal program with increasing temperatures at each stage is often used. The three steps are evaporation of solvent, ashing to decompose the matrix

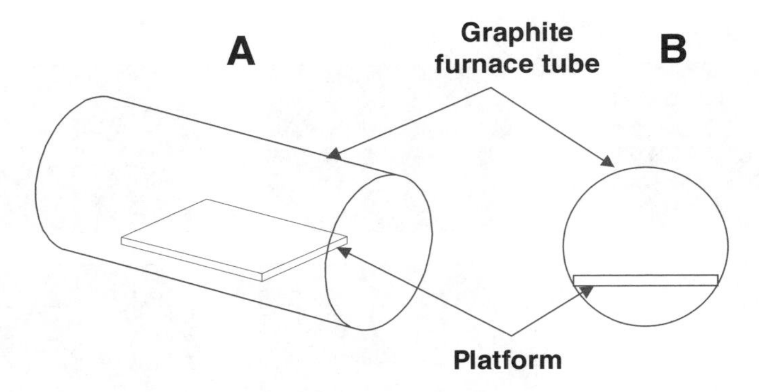

Figure 2.46 *Graphite furnace tube with sample L'vov platform (A is a side-on view and B is a longitudinal view).*

and atomisation where the molecules of analyte are converted into free gaseous atoms. The atoms absorb radiation and make transitions to higher electronic energy levels.

GFAAS has many advantages over FAAS, such as fast and complete atomisation of the sample (almost 100 % of the sample is atomised as opposed to 0.1 % in FAAS), low cost, the fact that the sample is easy to remove after analysis and, most importantly, that it gives 2–3 orders of magnitude better sensitivity. GFAAS is normally used in conjunction with an autosampler. Recent developments in GFAAS have been described elsewhere[58].

Hydride generation atomic absorption spectrometry (HGAAS) is used for samples that can be transformed into a hydride gas by chemical reaction, e.g. arsenic, germanium, lead and tin. Sodium tetrahydroborate is often used as the reducing agent. There are two types of atomiser tubes used in HGAAS: quartz tube atomisers and graphite tube atomisers. The first is used on-line and the second for preconcentrating the sample prior to analysis. Atomic fluorescence spectrometry (AFS) is mostly used in conjunction with hydride generation.

Discriminator

The selected line source radiation is separated at the monochromator from other emission lines and background emission. Diffraction gratings are most common. It is possible to have single-beam and double-beam spectrometers in AAS. The single-beam design sends all the radiation through the atomised sample so there is little loss of intensity. The double-beam design splits the radiation into a sample beam (which passes through the flame or furnace) and a reference beam (which does not). The beams are recombined before they enter the monochromator. Splitting of the beam takes into account any variations in the intensity of the source and sensitivity of the detector. The double-beam instrument is most commonly employed with a flame atomiser to avoid frequent recalibration, while the single-beam configuration is employed with the graphite furnace where the baseline is reset before each run. When a high resolution source is used, e.g. xenon arc lamp, the discriminator is composed of both a prism pre-monochromator and a grating monochromator. In AFS, the monochromator is placed at a right angle to the source and sample so that the incident radiation does not interfere with the emitted fluorescence.

Detector
The photomultiplier tube (PMT) is the detector of choice in many conventional AAS instruments. However, in a high resolution AAS instrument, a CCD detector is more useful.

Output
The ratio of the incident radiation over the exiting radiation gives a transmittance. The amplifier converts this transmittance signal into an absorbance one. The Beer–Lambert Law can then be used to calculate concentration values. There is often a need in AAS to carry out background correction to compensate for non-specific absorption due to the matrix.

Information Obtained

Qualitative information is obtained from the *wavelength* of light at which the radiation is absorbed (AAS) or fluorescence emitted (AFS). Quantitative information is derived from the *amount* of radiation absorbed or emitted, i.e. the intensity. Concentration measurements are usually determined from a working curve of intensity versus concentration after calibrating the instrument with standards of known concentration.

Developments/Specialist Techniques

The main thrust of development in AAS is towards simultaneous multi-element analysis without any compromise in sensitivity. To this end, both sample extraction techniques and chemometrics have contributed to the field. As mentioned in the section on GFAAS, solids can be analysed directly and this a great advantage in many application areas. Laser ablation is an evolving technique for solids whereby where a laser beam is focused onto the sample to vaporise it. The vaporised sample is then introduced into the plasma by argon carrier gas. A future area for atomic spectrometry is metallomics: the study of metals, metal species and their interactions, transformations and functions in biological systems[59]. Recent developments in AAS and AFS have been described elsewhere[60].

Applications

AAS and AFS are used to detect metals in environmental samples in both trace and major concentrations. Analysis of metals like lead, mercury and cadmium in various samples such as tuna fish[61], soils[62], mushrooms[63] and shellfish[64] is very common. AAS has been reported for the determination of drugs such as bromhexine, flunarizine and ranitidine hydrochlorides in pharmaceutical formulations down to concentrations of $<2\,\mu g/mL$[65]. AAS has been used to analyse cadmium and lead in blood samples[66]. The concentration ranges were 0.20–1.73 ppb and 12.0–65.7 ppb for cadmium and lead respectively. The techniques of AAS and AFS are also very popular in the analysis of herbal medicines[67, 68].

2.2.2 Inductively Coupled Plasma–Optical Emission Spectrometry

Principles

Optical emission spectrometry (OES) is based on the production and detection of line spectra emitted during the de-excitation of electrons in an atom when they fall from upper excited energy levels to lower energy levels. The affected electrons are 'outer' electrons, which are capable of absorbing and emitting radiation. Line spectra are specific for each element and therefore qualitatively indicative of that element. Line spectra intensities can also be used to determine the concentration of the element. Most elements in the periodic table can be analysed as long as they are dissolved prior to analysis. Another advantage is the extremely wide linear range (4–9 orders of magnitude), which means rarely having to dilute samples. Many elements can also be determined simultaneously.

While flames, arcs and sparks are still used as the sources for sample ionisation, plasma radiation sources have largely overtaken these due to a number of advantages, such as greater stability, higher temperature and the ability to excite many elements simultaneously and sensitively. Plasma is any gas that is appreciably ionised and therefore highly energetic, containing electrons and ions. The state-of-the-art plasma is based on argon. The inductively coupled plasma (ICP) is produced in a device known as a torch. Plasmas have flame-like shapes with brilliant white cores and are much hotter than traditional flames and furnaces. Hence plasmas can both break down samples and excite and/or ionise the free atoms. An optical emission spectrometer will consist of a radiation source, a sample introduction system, an optical dispersive system, a detector and electronics for data manipulation. Further information on ICP–OES is available elsewhere[69].

Instrument

The components of an ICP–OES instrument are shown in Figure 2.47.

Source
The source in the case of this instrument is the ICP itself, which is illustrated in Figure 2.48. The ICP is a high temperature plasma sustained with an external radiofrequency (RF) electric current, which acts to produce ions. The ICP reaches a much higher temperature

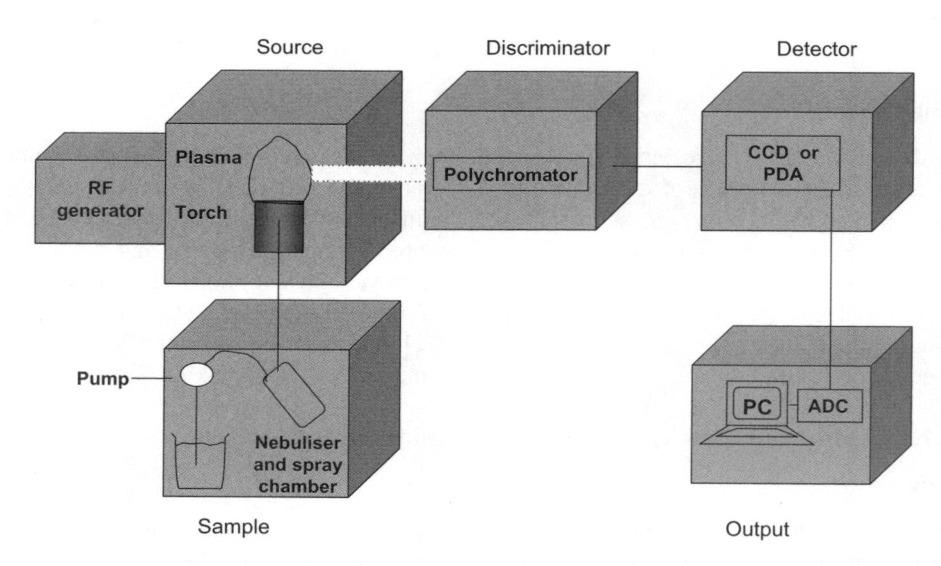

Figure 2.47 *Schematic diagram of an inductively coupled plasma–optical emission spectrometer.*

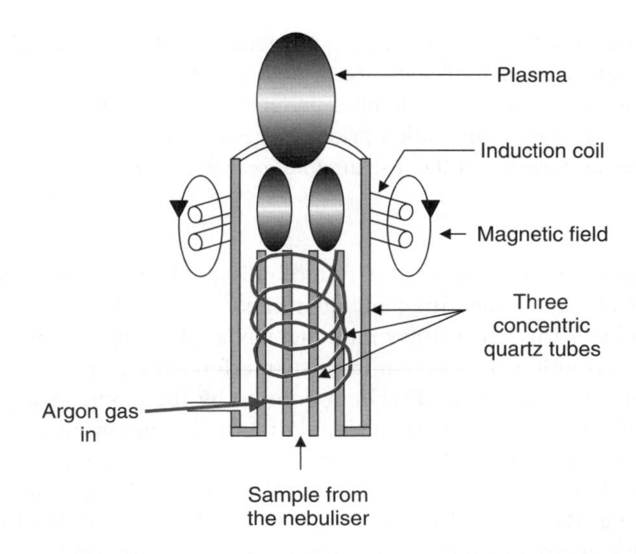

Figure 2.48 *Schematic of an inductively coupled plasma (ICP) torch.*

than flames and furnaces. The electric current is transferred to the plasma by an induction coil, wrapped around a configuration of three concentric quartz tubes which make up the torch. These tubes are termed the outer tube, which confines and insulates the plasma, the intermediate tube, which accelerates the plasma, and the inner injection tube for sample introduction. As flowing gases are introduced into the torch, the RF field is activated and the gas in the coil region is made electrically conductive. The ionised gas (cations and electrons) induces further ionisation of the gas atoms in a chain reaction. This sequence of events forms the plasma. The gas is normally argon, which is used because it is chemically inert, relatively cheap, capable of exciting and ionising most of the elements of the periodic table and the line spectra produced are simple and straightforward.

To prevent possible short-circuiting as well as meltdown, the plasma must be insulated from the rest of the instrument. Insulation is achieved by the flow of three gases through the system - the outer gas, intermediate gas and inner or carrier gas. The outer gas is typically argon or nitrogen. It helps maintain, stabilise and thermally isolate the plasma from the outer tube. Argon is also commonly used for both the intermediate gas and inner or carrier gas. The purpose of the carrier gas is to convey the sample to the plasma. The formation of the plasma is dependent upon an adequate magnetic field strength and the pattern of the gas streams follows a particular rotationally symmetrically pattern. The plasma is maintained by inductive heating of the flowing gases.

Sample

Following any necessary sample preparation and work-up, samples are supplied to the instrument as nebulised or vaporised sprays. An ICP generally requires that the elements for analysis are in solution and are delivered to the nebuliser by a peristaltic pump. The nebuliser transforms the aqueous solution into an aerosol (with the aid of the argon gas) in the spray chamber with an overall sample introduction efficiency of only a few percent. Once in the plasma, the analyte ions are heated to 4000–8000 K for 2–3 minutes and excited to emit light. An aqueous solution is preferred over an organic solution, as organic solutions require special manipulation prior to injection into the ICP. Solid samples are also discouraged, as clogging of the instrumentation can occur, but slurries or suspensions can be used.

Discriminator
The polychromatic radiation is separated into individual wavelengths so that emission from each element can be identified and quantified; a monochromator or polychromator is used to achieve this. Both allow multielement analysis but the polychromator can measure them all simultaneously. To minimise interferences, a number of characteristic lines are used for each element. The light emitted by the atoms of an element in the ICP must be converted to an electrical signal that can be measured quantitatively. The discriminator isolates spectral bands containing analyte lines and measures the light intensity with a detector such as a photomultiplier tube at the specific wavelength for each element line. Background measurement should be carried out at the location of the analyte wavelength and subtracted from the gross measurement to obtain the net line.

Detector
Even though the photomultiplier tube (PMT) has been around for a long time, it still has many advantages as a detector. It has a wide wavelength range (160–900 nm), a large dynamic range, a high amplification gain and low noise. However, as the PMT is a single detector, there has been a drive towards multichannel detectors such as a photodiode array (PDA), a charge-coupled device (CCD) or a charge injection device (CID).

Output
The intensity of the electron signal is compared to previously measured intensities of known concentrations of the element and a concentration is computed. Each element will have many specific wavelengths in the spectrum that could be used for analysis. Thus, the selection of the best line the analytical application in hand requires considerable experience of ICP wavelengths.

Information Obtained

Qualitative information is obtained from the *wavelength* of light at which the radiation is absorbed or emitted. Quantitative information is derived from the *amount* of radiation absorbed or emitted (intensity). A calibration curve is obtained by analysing standard solutions for the elements of interest and plotting their emission intensities versus their concentrations. Samples with unknown concentrations can be determined from the calibration curves. Precision in ICP–OES is generally very good with RSD values of $<1\,\%$. Accuracy depends on the availability of good standards and reference materials. Sensitivity of ICP–OES for most elements has hovered somewhere between flame atomic absorption spectrometry (FAAS) and graphite furnace atomic absorption spectrometry (GFAAS), which is the most sensitive of the three techniques. However, recent developments including ultrasonic nebulisation and axial measurement of the plasma have

Figure 2.49 *The Optima 5300 ICP-OES instrument (Reproduced by permission of Perkin Elmer Inc.).*

lowered the detection limits in ICP–OES to that of GFAAS. An example of an ICP–OES instrument is shown in Figure 2.49.

Developments/Specialist Techniques

Although liquid samples are best, if solid samples require analysis there have been a number of developments such as laser ablation that now allow this. Direct sample insertion is also possible where the solid sample is introduced into the plasma on a carbon electrode.

A new side-on-plasma (SOP) system has been developed by Spectro Analytical Instruments. It records the entire spectrum from 175–777 nm, enabling the simultaneous determination of more than 70 elements. Compared with axial plasma observation, it claims to be more precise over the ppm to percent concentration range.

Applications

Most of the elements of the periodic table can be determined by ICP–OES with limits of detection in the 0.1–50 ng/mL range. However, sensitivities for the alkali elements can be poor because of the very high plasma temperature. As ICP–OES can be used for any sample that can be put in solution, it has very wide applicability across a number of disciplines. It can be used for determination of nutrient levels in soil[70, 71], plants[72], food[73] and other sample types. Metals and other elements can be measured quantitatively in blood[74], urine and other biological fluids with sample preparation prior to analysis. Environmental samples probably make up the greatest volume of samples analysed by ICP–OES as water samples are especially well suited to the technique[75]. The levels of elements in ground waters, drinking waters, sea and waste waters is of huge regulatory importance.

2.2.3 Inductively Coupled Plasma–Mass Spectrometry

Principle

Inductively coupled plasma (ICP) can also be coupled with mass spectrometry (MS) to become ICP–MS. ICP–MS is a type of mass spectrometry that is capable of analysing of a range of metals and several non-metals at below one part in 10^{12}. Further information on ICP–MS is available elsewhere[76, 77, 78].

Instrument

With this inorganic analytical technique, the same types of mass spectrometers that are used for organic MS are used but as elemental analysis is the objective the mass range need only extend to 250 or 300 amu approximately. The quadrupole MS is probably the most commonly used system. A schematic of an ICP–MS is shown in Figure 2.50.

Sample
Although mainly used for liquid samples where the solution is nebulised into the plasma, solids can be introduced directly in the case of laser-produced plasmas. The microplasma is produced at atmospheric pressure and laser ablation of the sample occurs. Fine particles are taken up by a stream of argon for further volatilisation, atomisation and ionisation in the ICP.

Source
The ICP is the source as in the previous instrument, the ICP–OES. It replaces the usual ion sources that precede the mass analyser in MS.

Discriminator
The discriminator is the mass analyser of the mass spectrometer. The most commonly employed one for ICP–MS is the quadrupole although others are of course used such as the

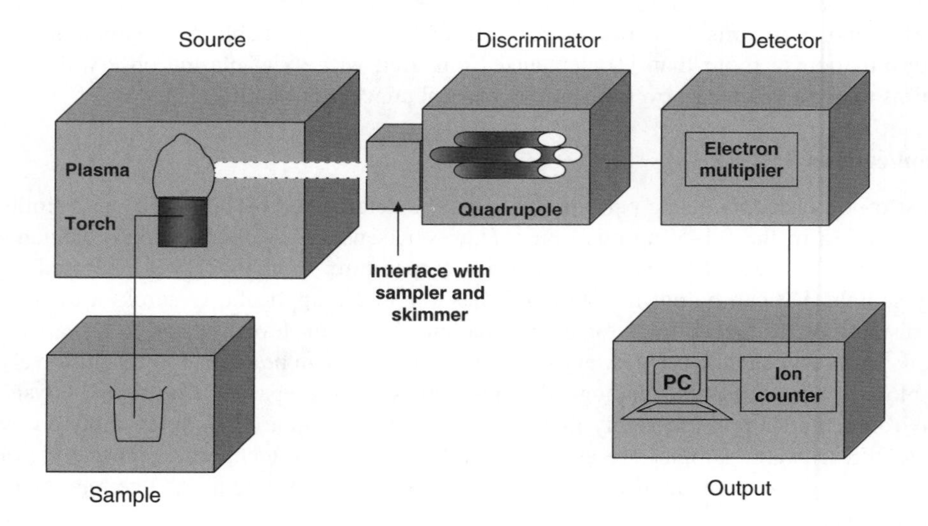

Figure 2.50 *Schematic diagram of an inductively coupled plasma–mass spectrometer.*

magnetic sector and time-of-flight. The interface between the ICP and the MS is a crucial part of the instrumentation as the ICP operates at atmospheric pressure and high temperature whereas the MS requires vacuum pressure and lower temperature. Thus the interface must reduce both the temperature and pressure prior to the ions reaching the MS. A common interface employs two cones made of copper or nickel, which are called the sampler and the skimmer. The orifices of these cones are <1 mm, the distance between them is <10 mm and they are aligned with the plasma. The tip of the sampler reaches into the central region of ions of the plasma and the pressure is reduced between the sampler and the skimmer by means of a rotary pump. Behind the sampler, the plasma becomes a supersonic beam which expands the plasma reducing the temperature. Ion optics keep the ions in alignment on their way to the high vacuum MS. The high vacuum is achieved by turbomolecular pumps.

Detector

The most commonly used detectors are based on the electron multiplier, especially the discrete dynode type. Here, the signal is received and amplified.

Output

The data processing is carried out by a computer: qualitative analysis, quantitative analysis and structural elucidation of known compounds (library searching) and unknown compounds.

Information Obtained

Qualitative information is related to the mass-to-charge ratio (m/z) at which the ions are detected. Quantitative information is related to the amount (intensity) of ions detected by the mass spectrometer. The concentration of a sample can be determined through calibration with elemental standards. Quantitative determinations can also be made using isotope dilution. An example of an ICP–MS Instrument is shown in Figure 2.51.

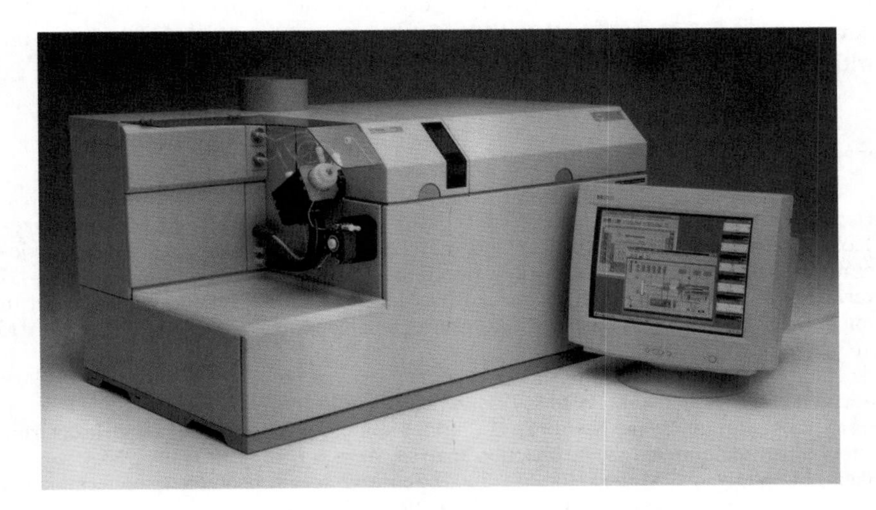

Figure 2.51 *Agilent 7500 Series ICPMS (© Copyright 2006 Agilent Technologies, Inc. Reproduced with permission).*

ICP–MS provides an efficient combination of precision (RSD $<4\%$), accuracy, number of elements that can be determined, sensitivity and sample throughput. The limits of detection can be as low as 1–10 ng/L, which is 1–3 orders of magnitude better than ICP–OES and equal to or better than GFAAS. For the metals lead, cadmium and mercury alone, the sensitivity is between 3 and 20 times better using ICP–MS compared to the method of AAS[79]. One interference issue is the presence of isobaric molecular species formed with argon which affects m/z values below 80. This problem can be partly solved by using a high resolution MS.

Developments/Specialist Techniques

There is now available laser ablation ICP–MS which allows solid samples to be analysed directly. It is an important technique for use in the forensic analysis of vehicle paint samples at the scene of hit-and-run crimes[80].

Applications

There are many applications for ICP–MS mainly involving the determination of metals. A geochemical survey of the northwestern part of the Thailand Gulf was carried out in order to define concentrations and distribution patterns of selected heavy metals (vanadium, chromium, cobalt, nickel, copper, zinc and wanium) by ICP–MS[81]. The results indicated the presence of two different sources of heavy metals in the studied environment. Vanadium and nickel enrichment also indicated that contamination by hydrocarbon discharge took place in the investigated area. Metals such as iron, magnesium, aluminium and calcium can be determined in more difficult matrices such as soil extracts, rock[82] and sediment samples[83]. Minor and trace elements in alloys, steel and aluminium can also be quantified, as can the levels of many elements in waste water and sludge samples. Even dust samples can be collected and analysed by ICP–MS. In one paper, ICP–MS was applied to the analysis of chromium, nickel, copper, arsenic, cadmium and tin in furnace flue dust[84]. The very low levels of detection make ICP–MS sensitive enough for the analysis of trace metal impurities in ultrapure materials such as quartz, tellurium, copper and water[85].

References

1. www.spectroscopynow.com/coi/cda/detail.cda?id=13397&type=Feature&chId=7&page=1
2. Gao, L.J., Chai, H.M., Sun, X.H. and Guo, Y.H. (2006) Simultaneous spectrophotometric determination of Fe(II) and Fe(III) by flow injection analysis. *Fenxi Shiyanshi*, **25 (9)**, 84–87.
3. Kawakubo, S., Naito, A., Fujihara, A. and Iwatsuki, M. (2004) Field determination of trace iron in fresh water samples by visual and spectrophotometric methods. *Anal Sci*, **20 (8)**, 1159–1163.
4. Melchert, W.R. and Rocha, F.R.P. (2005) A green analytical procedure for flow-injection determination of nitrate in natural waters. *Talanta*, **65 (2)**, 461–465.
5. Ensafi, A.A., Rezaei, B. and Nouroozi, S. (2004) Simultaneous spectrophotometric determination of nitrite and nitrate by flow injection analysis. *Anal Sci*, **20 (12)**, 1749–1753.
6. Nagai, M., Sugiyama, M. and Hori, T. (2004) Sensitive spectrophotometric determination of phosphate using silica-gel collectors. *Anal Sci*, **20 (2)**, 341–344.
7. Perrott, K.W. and Wise, R.G. (2000) Determination of residual reactive phosphate in soil. *Comm Soil Science Plant Anal*, **31 (11–14)**, 1809–1824.

8. Zhuang, B.Y. (2001) Spectrophotometric determination of adsorbed lead(II) on leaves by the reaction with xylenol orange (XO) in the presence of cetyltrimethylammonium bromide. *Lihua, Jianyan, Huaxue Fence*, **37 (9)**, 419–420.
9. Li, N., Min, S.G., Qin, F.L. *et al.* (2004) Nondestructive analysis of protein and fat in whole-kernel soybean by NIR. *Guangpuxue Yu Guangpu Fenxi*, **24 (1)**, 45–49.
10. Woo, Y.A., Terazawa, Y., Chen, J.Y. *et al.* (2002) Development of a new measurement unit (MilkSpec-1) for rapid determination of fat, lactose and protein in raw milk using near-infrared transmittance spectroscopy. **56 (5)**, 599–604.
11. Hendra, P. (1998) What is Raman Spectroscopy? *Int J Vib Spect*, [www.ijvs.com], **1 (5)**, 6–16.
12. Olcese, L.E., Palancar, G.G. and Toselli, B.M. (2001) An inexpensive method to estimate carbon monoxide and nitrogen oxide emissions from mobile sources. *Atmos Environ*, **35 (35)**, 6213–6218.
13. Kolb, C.E., Herndon, S.C., McManus, J.B. *et al.* (2004) Mobile laboratory with rapid response instruments for real-time measurements of urban and regional trace gas and particulate distributions and emission source characteristics. *Environ Sci Technol*, **38 (21)**, 5694–5703.
14. Sassenscheid, K. (2001) Infra-red laser measuring technology - new perspectives for process analysis and environmental analysis. *GIT Labor-Fachzeitschrift*, **45 (9)**, 962–963.
15. Nagarajan, R., Gupta, A., Mehrotra, R. and Bajaj, M.M. (2006) Quantitative analysis of alcohol, sugar, and tartaric acid in alcoholic beverages using attenuated total reflectance spectroscopy. *J Aut Meth Manag Chem*, **2006 (2)**, 1–5.
16. Ferrao, M.F. and Davanzo, C.U. (2005) Horizontal attenuated total reflection applied to simultaneous determination of ash and protein contents in commercial wheat flour *Anal Chim Acta*, **540 (2)**, 411–415.
17. Mazarevica, G., Diewok, J., Baena, J.R. *et al.* (2004) On-line fermentation monitoring by mid-infrared spectroscopy. *Appl Spec*, **58 (7)**, 804–810.
18. Sundarajan, N., Mao, D.Q., Chan, S. *et al.* (2006) Ultrasensitive detection and characterisation of post-translational modifications using surface-enhanced raman spectroscopy. *Anal Chem*, **78 (11)**, 3543–3550.
19. Hadjikakou, S.K., Xanthopoulou, M.N., Hadjiliadis, N. and Kubicki, M. (2003) Synthesis, structural characterisation and study of mercury (II) iodide complexes with the heterocyclic thioamides pyridine-2-thione (pytH) and thiazolidine-2-thione (tzdtH). Crystal structures of $[HgI_2(pytH)_2]$ and $[HgI_2(tzdtH)_2]$. *Can J Anal Sci Spec*, **48 (1)**, 38–45.
20. Raman spectroscopy: a complex technology moving from lab to the clinic – and before too long, the marketplace. *Optics Report*, Biotech section, 9 Jan 2006.
21. Moreira, A.B., Dias, I.L.T., Oliveira Neto, G. *et al.* (2004) Solid-phase fluorescence spectroscopy for the determination of acetylsalicylic acid in powdered pharmaceutical samples. *Anal Chim Acta*, **523 (1)**, 49–52.
22. Okumoto, S., Looger, L.L., Micheva, K.D. *et al.* (2005) Detection of glutamate release from neurons by genetically encoded surface-displayed FRET nanosensors. *Proc Nat Acad Sci USA*, **102 (24)**, 8740–8745.
23. Trevisan, M.G. and Poppi, R.J. (2003) Determination of doxorubicin in human plasma by excitation–emission matrix fluorescence and multi-way analysis. *Anal Chim Acta*, **493 (1)**, 69–81.
24. Loadman, P.M. and Calabrese, C.R. (2001) Separation methods for anthraquinone related anti-cancer drugs. *J Chrom B*, **764** (1–2), 193–206.
25. Abdellatef, H.E. (2007) Spectrophotometric and spectrofluorimetric methods for the determination of ramipril in its pure and dosage form. *Spectrochim Acta Part A*, **66 (3)**, 701–706.
26. Lee H-B, Peart, T.E. and Svoboda, M.L. (2007) Determination of ofloxacin, norfloxacin, and ciprofloxacin in sewage by selective solid-phase extraction, liquid chromatography with fluorescence detection, and liquid chromatography–tandem mass spectrometry. *J Chrom A*, **1139 (1)**, 45–52.
27. Bonnin, C., Matoga, M., Garnier, N. *et al.* (2007) 224 nm deep-UV laser for native fluorescence, a new opportunity for biomolecules detection. *J Chrom A*, In Press.
28. Svetlichny, V.Y., Merola, F., Dobretsov, G.E. *et al.* (2007) Dipolar relaxation in a lipid bilayer detected by a fluorescent probe, 4″-dimethylaminochalcone. *Chem Phys Lipids*, **145 (1)**, 13–26.

29. Kálai, T. and Hideg, K. (2006) Synthesis of new, BODIPY-based sensors and labels. *Tetrahedron*, **62 (44)**, 10352–10360.
30. Gronquist, M., Meinwald, J., Eisner, T. and Schroeder, F.C. (2005) Exploring uncharted terrain in nature's structure space using capillary NMR spectroscopy: 13 steroids from 50 fireflies. *JACS*, **127 (31)**, 10810–10811.
31. www.cis.rit.edu/htbooks/mri/index.html
32. Savukov, I.M., Lee, S.K. and Romalis, M.V. (2006) Optical detection of liquid-state NMR, *Nature*, **442 (7106)**, 1021–1024.
33. Bradley, D. (2006) NMR and the 3D world of proteins. *Chemistry World*, **3 (5)**, 40–44.
34. Kainosho, M., Torizawa, T., Iwashita, Y. *et al.* (2006) Optimal isotope labelling for NMR protein structure determinations. *Nature*, **440 (7080)**, 52–57.
35. Cartigny, B., Azaroual, N., Imbenotte, M. *et al.* (2004) Determination of glyphosate in biological fluids by ^{1}H and ^{31}P NMR spectroscopy. *Foren Sci Int*, **143 (2–3)**, 141–145.
36. Somashekar, B.S., Ijare, O.B., Nagana, G.A. *et al.* (2006) Simple pulse-acquire NMR methods for the quantitative analysis of calcium, magnesium and sodium in human serum. *Spectrochim Acta Part A*, **65 (2)**, 254–260.
37. Michaleas, S. and Antoniadou-Vyza, E. (2006) A new approach to quantitative NMR: Fluoroquinolones analysis by evaluating the chemical shift displacements. *J Pharm Biomed Anal*, **42 (4)**, 405–410.
38. Volmer, D.A. and Sleno, L. (2005) Tutorial – mass analysers: an overview of several designs and their applications, part I, *Spectroscopy*, **20 (11)**, 20–26.
39. Volmer, D.A. and Sleno, L. (2005) Tutorial – mass analysers: an overview of several designs and their applications, part II, *Spectroscopy*, **20 (12)**, 90–95.
40. Cody, R.B., Laramée, J.A. and Durst, H.D. (2005) Versatile new ion source for the analysis of materials in open air under ambient conditions. *Anal Chem.*, **77 (8)**, 2297–2302.
41. Takáts, Z., Wiseman, J.M., Gologan, B. and Cooks, R.G. (2004) Mass spectrometry sampling under ambient conditions with desorption electrospray ionization. *Science*, **306 (5695)**, 471–473.
42. Ifa, D.R., Wiseman, J.M., Song, O. and Cooks, R.G. (2007) Development of capabilities for imaging mass spectrometry under ambient conditions with desorption electrospray ionization (DESI). *Int J Mass Spectrom*, **259 (1–3)**, 8–15.
43. Chen, H., Wortmann, A., Zhang, W. and Zenobi, R. (2007) Rapid *in vivo* fingerprinting of nonvolatile compounds in breath by extractive electrospray ionization quadrupole time-of-flight mass spectrometry. *Angew Chemie*, **119 (4)**, 586–589.
44. McMahon, G. Mass spectrometry: peptides and proteins, *Encyclopedia of Analytical Science* (2nd ed), Eds P Worsfold, A Townshend and C Poole. ISBN: 012764100–9. Elsevier, 501–509 (2005).
45. Niwa, T. (2006) Mass spectrometry for the study of protein glycation in disease. *Mass Spectrom Rev*, **25 (5)**, 713–723.
46. Frewen, B.E., Merrihew, G.E., Wu, C.C. *et al.* (2006) Analysis of peptide MS/MS spectra from large-scale proteomics experiments using spectrum libraries. *Anal Chem*, **78 (16)**, 5678–5684.
47. Hernandez, P., Muller, M. and Appel, R.D. (2006) Automated protein identification by tandem mass spectrometry: issues and strategies. *Mass Spectrom Rev*, **25 (2)**, 235–254.
48. Bogdanov, B. and Smith, R.D. (2005) Proteomics by FTICR mass spectrometry: top down and bottom up. *Mass Spectrom Rev*, **24 (2)**, 168–200.
49. Hofstadler, S.A., Sannes-Lowery, K.A. and Hannis, J.C. (2005) Analysis of nucleic acids by FTICR MS. *Mass Spectrom Rev*, **24 (2)**, 265–285.
50. Deng, G.J. and Sanyal, G. (2006) Applications of mass spectrometry in early stages of target based drug discovery. *J Pharm Biomed Anal*, **40 (3)**, 528–538.
51. Balogh, M.P. (2005) Profiles in practice series: metabolism identification and structural characterization in drug discovery. *LC-GC Europe*, **18 (6)**, 330–336.
52. Sleeman, R., Carter, J. and Ebejer, K. (2005) Drugs on money and beyond: tandem mass spectrometry in the forensic sciences. *Spec Europe*, **17 (6)**, 10–13.
53. Cristoni, S., Brioschi, M., Rizzi, A. *et al.* (2006) Rapid communications in mass spectrometry. *Rap Comm Mass Spectrom*, **20 (22)**, 3483–3487.

54. Khatib-Shahidi, S., Andersson, M., Herman, J.L. *et al.* (2006) Direct molecular analysis of whole-body animal tissue sections by imaging MALDI mass spectrometry. *Anal Chem*, **78 (18)**, 6448–6456.
55. Abdel-Hamid, M.E., Sharaf, L.H., Kombian, S.B. and Diejomaoh, F.M.E. (2006) Determination of dydrogesterone in human plasma by tandem mass spectrometry: application to therapeutic drug monitoring of dydrogesterone in gynecological disorders. *Chromatographia*, **64 (5/6)**, 287–292.
56. Wall, R., McMahon, G., Crown, J. *et al.* (2007) Rapid and sensitive liquid chromatography-tandem mass spectrometry for the quantitation of epirubicin and identification of metabolites in biological samples. *Talanta*, **72 (1)**, 145–154.
57. Beaty, R. and Kerber, J. Concepts, Instrumentation and Techniques in Atomic Absorption Spectrophotometry. *Perkin Elmer*, (2nd ed) http://las.perkinelmer.com/content/RelatedMaterials/AA%20Concepts%20Book.PDF#search='single%20beam%20flame%20atomic%20atomic%20fluorescence (1993)
58. Butcher, D.J. (2006) Advances in electrothermal atomisation atomic absorption spectrometry: instrumentation, methods, and applications. *Appl Spect Rev*, **41 (1)**, 15–34.
59. Haraguchi, H. (2004) Metallomics as integrated biometal science. *J Anal Atom Spectrom*, **19**, 5–14.
60. Hywel Evans, E., Day, J.A., Palmer, C. *et al.* (2006) Atomic spectrometry update. Advances in atomic emission, absorption and fluorescence spectrometry, and related techniques *J Anal Atom Spectrom*, **21**, 592–625.
61. Emami Khansari, F., Ghazi-Khansari, M. and Abdollahi, M. (2005) Heavy metals content of canned tuna fish. *Food Chem*, **93 (2)**, 293–296.
62. Devai, I., Patrick WH Jr., Neue H-U *et al.* (2005) Methyl mercury and heavy metal content in soils of rivers Saale and Elbe (Germany). *Anal Letts*, **38 (6)**, 1037–1048.
63. Svoboda, L., Havlickova, B. and Kalac, P. (2006) Contents of cadmium, mercury and lead in edible mushrooms growing in a historical silver-mining area. *Food Chem*, **96 (4)**, 580–585.
64. Kwoczek, M., Szefer, P., Hac, E. and Grembecka, M. (2006) Essential and toxic elements in seafood available in Poland from different geographical regions. *J Ag Food Chem*, **54 (8)**, 3015–3024.
65. Khalil, S., Ibrahim, S.A., Zedan, F.I. and Abd-El-Monem, M.S. (2005) AAS determination of bromhexine, flunarizine and ranitidine hydrochlorides in pharmaceutical formulations *Chemia Analityczna (Warsaw, Poland)*, **50 (5)**, 897–904.
66. Castelli, M., Rossi, B., Corsetti, F. *et al.* (2005) Levels of cadmium and lead in blood: an application of validated methods in a group of patients with endocrine/metabolic disorders from the Rome area. *Microchem J*, **79 (1–2)**, 349–355.
67. Song, Q.G., Shi, J. and Zhao, K.L. (2005) Determination of trace bismuth in traditional Chinese medicines by hydride generation atomic fluorescence spectrometry. *Fenxi Kexue Xuebao*, **22 (1)**, 37–39.
68. Liu, Y.M., Wang, H., Han, J.T. *et al.* (2006) Determination of 15 trace elements in antineoplastic traditional Chinese medicine by atomic absorption spectrometry. *Guangpuxue Yu Guangpu Fenxi*, **26 (9)**, 1728–1731.
69. Boss, C. and Fredeen, K. Concepts, instrumentation and techniques in inductively coupled plasma-optical emission spectrometry. *Perkin Elmer*, (2nd ed) (1997).
70. Embrick, L.L., Porter, K.M., Pendergrass, A. and Butcher, D.J. (2005) Characterization of lead and arsenic contamination at Barber Orchard, Haywood County, NC. *Microchem J*, **81 (1)**, 117–121.
71. Pan, L., Qin, Y.C., Hu, B. and Jiang, Z.C. (2006) Determination of Co and Ni in environmental samples by low temperature electrothermal vaporization inductively coupled plasma optical emission spectrophotometry using diethyldithiophosphate (DDTP) as a chemical modifier. *Fenxi Shiyanshi*, **25 (8)**, 45–49.
72. Pendergrass, A. and Butcher, D.J. (2006) Uptake of lead and arsenic in food plants grown in contaminated soil from Barber Orchard, NC. *Microchem J*, **83 (1)**, 14–16.
73. Lee, H.S., Cho, Y.H., Park, S.O. *et al.* (2006) Dietary exposure of the Korean population to arsenic, cadmium, lead and mercury. *J Food Comp Anal*, **19**(Suppl.), S31–S37.

74. Chen, S.Z. (2004) Direct determination of trace elements in human serum by electrothermal vaporization ICP-AES. *Guangpuxue Yu Guangpu Fenxi*, **24 (11)**, 1441–1443.
75. Becker, J.S. (2005) Trace and ultratrace analysis in liquids by atomic spectrometry. *TRAC*, **24 (3)**, 243–254.
76. Beauchemin, D. (2004) Inductively coupled plasma mass spectrometry. *Anal Chem*, **76 (12)**, 3395–3416.
77. http://ewr.cee.vt.edu/environmental/teach/smprimer/icpms/icpms.htm
78. http://minerals.cr.usgs.gov/icpms/What_is_ICPMS.pdf#search='ICPMS'
79. Palmer, C.D., Lewis, M.E., Jr., Geraghty, C.M. *et al.* (2006) Determination of lead, cadmium and mercury in blood for assessment of environmental exposure: a comparison between inductively coupled plasma-mass spectrometry and atomic absorption spectrometry. *Spectrochim Acta Part B*, **61B (8)**, 980–990.
80. Deconinck, I., Latkoczy, C., Guenther, D. *et al.* (2006) Capabilities of laser ablation-inductively coupled plasma mass spectrometry for (trace) element analysis of car paints for forensic purposes. *J Anal Atom Spectrom*, **21 (3)**, 279–287.
81. Censi, P., Spoto, S.E., Saiano, F. *et al.* (2006) Heavy metals in coastal water systems. A case study from the northwestern Gulf of Thailand *Chemosphere*, **64 (7)**, 1167–1176.
82. Pitcairn, I.K., Warwick, P.E., Milton, J.A. and Teagle DAH. (2006) Method for ultra-low-level analysis of gold in rocks. *Anal Chem*, **78 (4)**, 1290–1295.
83. Grahn, E., Karlsson, S., Karlsson, U. and Duker, A. (2006) Historical pollution of seldom monitored trace elements in Sweden-Part B: Sediment analysis of silver, antimony, thallium and indium. *J Environ Mon*, **8 (7)**, 732–744.
84. Coedo, A.G., Padilla, I. and Dorado, M.T. (2005) Determination of minor elements in steelmaking flue dusts using laser ablation inductively coupled plasma mass spectrometry. *Talanta*, **67 (1)**, 136–143.
85. Balaram, V. (2005) Recent developments in analytical techniques for characterisation of ultrapure materials – an overview. *Bull Mater Sci*, **28 (4)**, 345–348.

3
Separation Instruments

Separation instrumentation is based mainly on chromatography, which is a procedure for separating the analyte(s) of interest from interferences (matrix) and other compounds in a sample mixture. All chromatographic techniques depend on the differing distributions of individual compounds between two immiscible phases – the mobile and the stationary phases. The stationary phase is fixed in a column or on a solid surface. In practice, the sample mixture is added to one end of the stationary phase and the mobile phase then passes through or over it carrying the sample. The mixture of compounds is eluted, the compound appearing first at the end of the stationary phase being that which has the smallest distribution into the stationary phase. As the separated compounds appear at the end of the stationary phase they are detected by means of a detector, which may be a general purpose detector or one that is specific for the analyte of interest. The actual identification and/or quantitation of the separated compounds is made by the detector. There are many types of chromatography, e.g. thin layer, gas and liquid chromatographies to name but a few.

Another separation technique is capillary electrophoresis. It is not based on chromatography but on electrophoretic separation, which depends on the mass and charge of the species in the mixture. Even though capillary electrophoresis is not a chromatographic process, it is often convenient to include it in the same category for easy comparison with the rest of the separation techniques.

3.1 Gas Chromatography

Principle

In gas chromatography (GC) a sample is vaporised and injected onto the start of the chromatographic column. The sample is transported through the column by the flow of an inert gas. The column contains the stationary phase, which is either a liquid or solid. The

Analytical Instrumentation: A Guide to Laboratory, Portable and Miniaturized Instruments G. McMahon

rate at which the molecules progress along the column depends on the strength of adsorption or distribution, which in turn depends on the type of molecule and on the column chosen. Substances are identified by the order in which they emerge from the column and by the residence time of the analyte in the column. Because molecular adsorption and the rate of progression along the column depend on the temperature, this is carefully controlled. A number of detectors are used in GC. The most universally applicable is the thermal conductivity detector (TCD) and the most widely used is the flame ionisation detector (FID). Further information is available elsewhere[1, 2].

Instrument

A schematic of a GC instrument is illustrated in Figure 3.1.

Source
Because the source (mobile phase) is a gas, the samples must be volatile and thermally stable. The flow must be controlled and constant, hence pressure and flow regulators may be employed. The analytes have no interaction with the mobile phase and are simply transported by it through the column. Typical carrier gases used are helium, argon, nitrogen and carbon dioxide, the choice of which is determined in part by the detector used.

Sample
Gas samples are usually injected directly into the mobile phase gas flow using a gas-tight syringe. Liquid samples are injected by syringe through a rubber septum and volatilised in the high temperature injection port, entering the column as a gas. Volumes range from nanolitres to microlitres depending on the column being used. There are two main types of injection – split and splitless.

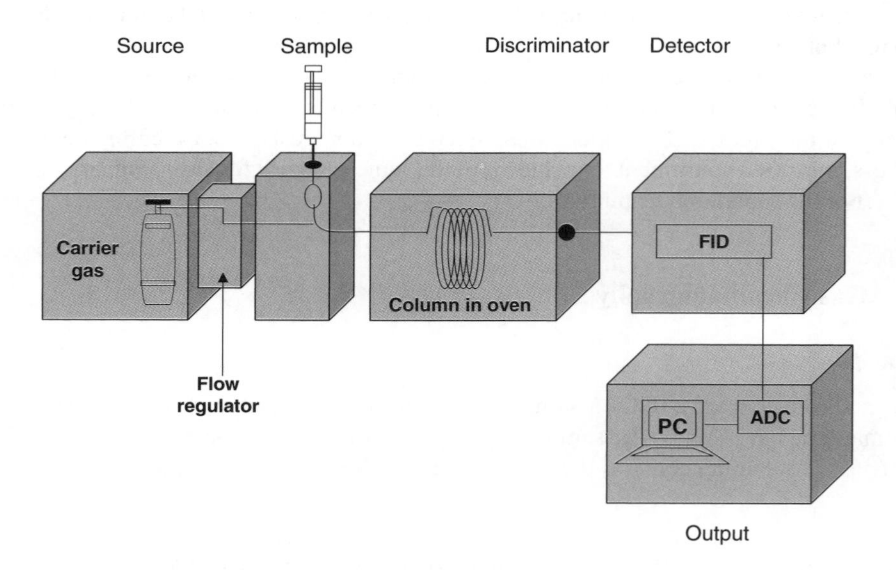

Figure 3.1 *Schematic diagram of a gas chromatograph.*

Split injection is used for concentrated samples where only a small portion of the sample volume (≤2 %) is introduced into the column. This is achieved by sending most of the sample to waste at the split vent. Injector temperature is high and the residence time of the sample in the injection port is short. It is used when resolution is most important and when sensitivity is not an issue.

Splitless injection is used for dilute samples and means introducing most of the sample into the column. Injector temperature is lower and the residence time of the sample in the injection port is longer (approximately one minute), so solvent trapping or cold trapping may be required. It is used when sensitivity is most important.

A third type of injection is on-column injection, where a sample is deposited directly onto the start of the column and the temperature is slowly raised just enough to volatilise the sample for chromatography but not so much that it decomposes. This mode of injection is used for quantitative work and for compounds that are thermally unstable above their boiling points.

Another means of sampling is headspace analysis, where a sample vial contains gas or vapour above the liquid, and that headspace is injected directly into the GC. This technique avoids liquid or solid probing. Traditional headspace GC relies on equilibrium being achieved in a closed vial between the liquid sample and its volatile components.

Temperature is a major factor in GC analysis and controlling it is important. Some methods require ramping/gradient control of temperature (temperature programming) during a separation to allow elution of compounds with a large range of boiling points in the same run. An oven controls the temperature of the column and the samples entering it. Compounds that are not volatile or thermally stable enough for direct analysis by GC can be derivatised with various functional groups to make them more amenable to this technique.

Discriminator

The column can be a packed capillary or, more usually now, an open tubular column. Open tubular columns are normally made of a narrow fused silica capillary where the walls are cased in polyimide for strength. They usually provide better resolution, more rapid separation and greater sensitivity than packed columns, but they cannot accommodate high sample amounts. Capillaries can be up to 100 m in length and have a narrow internal diameter, of 0.1–1 mm. There are three main types of open tubular columns – wall-coated, support-coated and porous layer, which are illustrated in Figure 3.2.

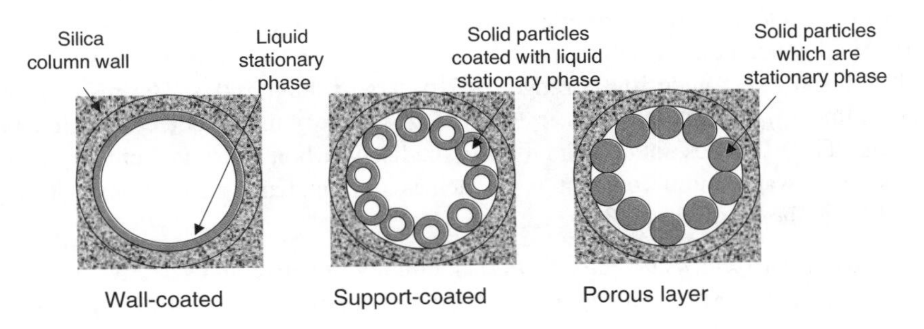

Figure 3.2 *The three types of open tubular column for GC.*

The wall-coated type has a layer of nonvolatile liquid stationary phase on the inside wall of the column. The support-coated type has solid particles attached to the inside wall of the capillary; the particles are coated with the liquid stationary phase. In either case, the liquid stationary phase must be chemically and thermally stable so that it does not 'bleed' off the capillary. The porous layer type has solid particles attached to the wall; the particles are themselves the stationary phase. In some capillaries, the stationary phase is covalently attached to the inner wall; this yields a more thermally stable 'bonded' column that is less likely to experience bleed.

Packed columns are usually made of stainless steel and are a few millimetres in diameter. The solid support particles can be the stationary phase or they can be coated with a liquid. To improve resolution and speed, the particles should be small and uniform in size. There are many kinds of stationary phase available in both packed and open tubular varieties of column covering the spectrum of polar to very nonpolar, which means a wide choice of capillaries when developing GC separations. Whichever type of column is chosen, it must be well equilibrated in the heated oven with mobile phase gas running through it before analysis begins.

Detector

A large number of GC detectors are available. In general, GC detectors are very sensitive and, thus, are ideal for trace analysis and environmental monitoring. The detectors that are more selective tend to be more sensitive. The most popular detector is the flame ionisation detector (FID). There are a number of possible detectors for GC, which include the following (the first four of these are based on ionisation processes):

- Flame ionisation detector
- Nitrogen phosphorus detector
- Electron capture detector
- Photoionisation detector
- Flame photometric detector
- Thermal conductivity detector
- Mass spectrometric detector.

*A flame ionisation detector (*FID) uses a hydrogen/air flame to decompose the carbon-containing sample molecules from the GC into ions by burning them and then measures the changes in current due to this. The FID detects most organic compounds when they are ionised and cause a voltage drop across the collector electrodes. The measured change is proportional to mass (number of carbon atoms) of the organic analyte. The background current and hence noise level are small. It is a destructive technique but is very sensitive (~100 pg) and has a wide linear range (6–7 orders of magnitude). The response of the FID is almost independent of flow rate and so it is a very useful detector with capillary columns. The FID does not respond to fully oxidised carbon atoms and ether groups. It is insensitive to water and permanent gases such as ammonia and sulfur dioxide A diagram of an FID is shown in Figure 3.3.

*The nitrogen phosphorus detector (*NPD) is a highly sensitive and selective detector that gives a very strong response to organic compounds containing nitrogen and/or phosphorus. It is capable of detecting as little as 10 pg of nitrogen or phosphorus, so it is commonly used to detect trace levels of pesticides, herbicides and drugs of abuse. The NPD is similar

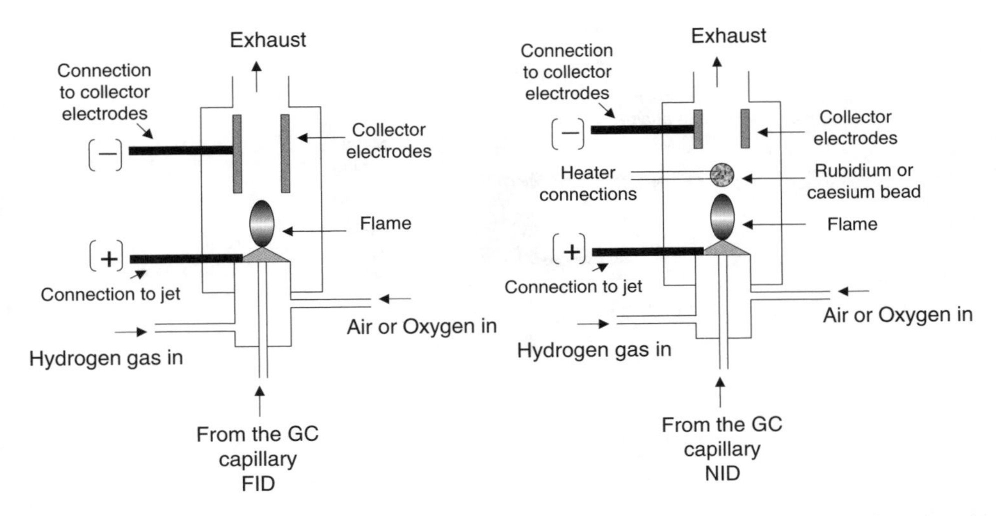

Figure 3.3 *Schematic diagrams of a flame ionisation detector (FID) and a nitrogen phosphorus detector (NPD) (The main difference between the two is the rubidium or caesium bead present in the NPD).*

in design to the FID, except that the hydrogen flow rate is reduced to about 3 mL/min and the actual sensor is a rubidium or caesium bead contained inside a small heater coil positioned near the jet orifice. The bead is heated and the gas mixture passes over it. The heated bead emits electrons by thermionic emission; these are collected at the anode and thus produce an ion current. When a solute containing nitrogen or phosphorus is eluted, the partially combusted nitrogen and phosphorus materials are adsorbed on the surface of the bead. This adsorbed material means that the emission of electrons is increased, which raises the anode current. Obviously nitrogen carrier gas cannot be used with this detector. A diagram of an NPD is shown in Figure 3.3.

The basic *electron capture detector* (ECD) consists of a small chamber (1–2 ml in volume) enclosing two metal electrodes. It can detect halogen-containing compounds, peroxides, nitro compounds and organometallic compounds among others but it is insensitive to hydrocarbons, amines and alcohols. The ECD contains a low energy radioactive β-ray source which is used to ionise the gas entering the chamber, producing electrons and a small background current. When analytes that have a high affinity for electrons enter the chamber, they capture some of the electrons and reduce the background current. This detector is probably the most sensitive of the GC detectors (about 50 fg) and is widely used in the analysis of halogenated compounds, in particular pesticides. A diagram of an electron capture detector is shown in Figure 3.4.

The photoionisation detector (PID) is used for the selective determination of aromatic hydrocarbons and unsaturated compounds such as aliphatics, aromatics, ketones, esters, aldehydes, amines, heterocyclics and some organometallics. It is relatively insensitive to saturated hydrocarbons and halocarbons. This device uses ultraviolet light as a means of ionising the analytes exiting the GC column and the ions produced by this process

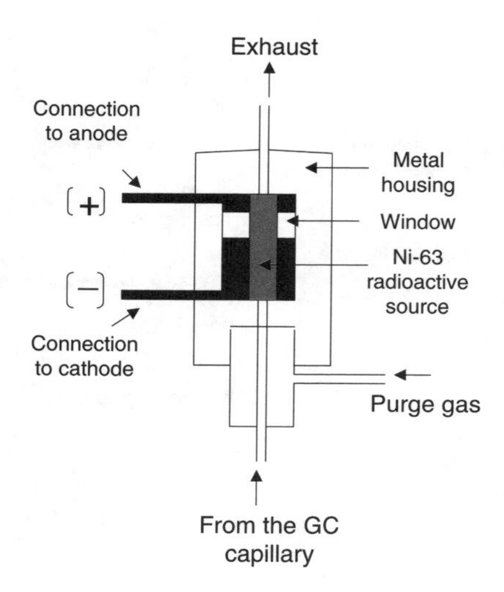

Figure 3.4 Schematic diagram of an electron capture detector.

are collected by electrodes. The current generated is therefore a measure of the analyte concentration. Since only a very small fraction of the analyte molecules are ionised, this detector is considered to be nondestructive and, as such, can be connected in series with another GC detector if required. It is a very sensitive detector (~2 pg). A schematic of a PID is shown in Figure 3.5.

*The flame photometric detector (*FPD) is a spectroscopic detector and is used for the selective detection of a number of elements by optical emission. The most commonly analysed elements are phosphorus and sulfur since these compounds, when partially

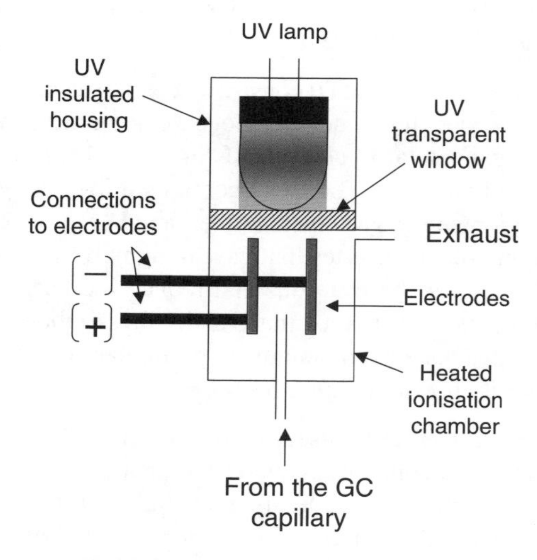

Figure 3.5 Schematic diagram of a photoionisation detector.

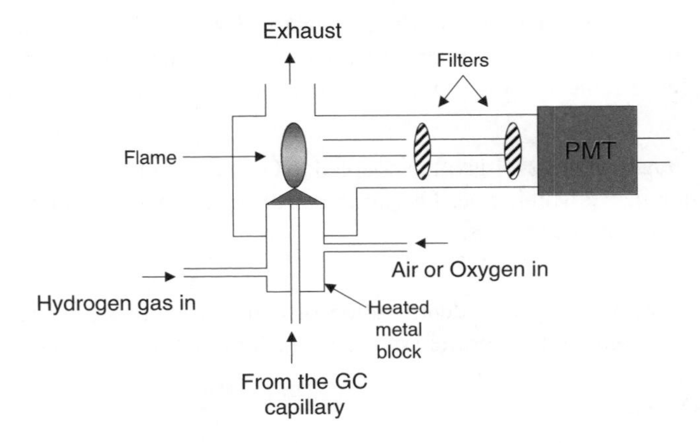

Figure 3.6 *Schematic diagram of a flame photometric detector.*

combusted in a hydrogen/air flame emit light at 536 and 394 nm respectively. It has similar sensitivity to the FID and can detect 100 pg of suitable compounds. A schematic of such a detector is shown in Figure 3.6.

The thermal conductivity detector (TCD) is the most universal and simple detector for GC (Figure 3.7). In the TCD, the thermal conductivity of helium or hydrogen carrier gas is relatively high but this is lowered in the presence of any analyte (organic or inorganic). When analyte exits the column, the heating filament gets hotter, the resistance, and hence the current, increases and changes the voltage. The measured voltage change is proportional to the concentration of the analyte. It is a nondestructive technique but is not as sensitive (only 1 ng) as other GC detectors. The TCD is

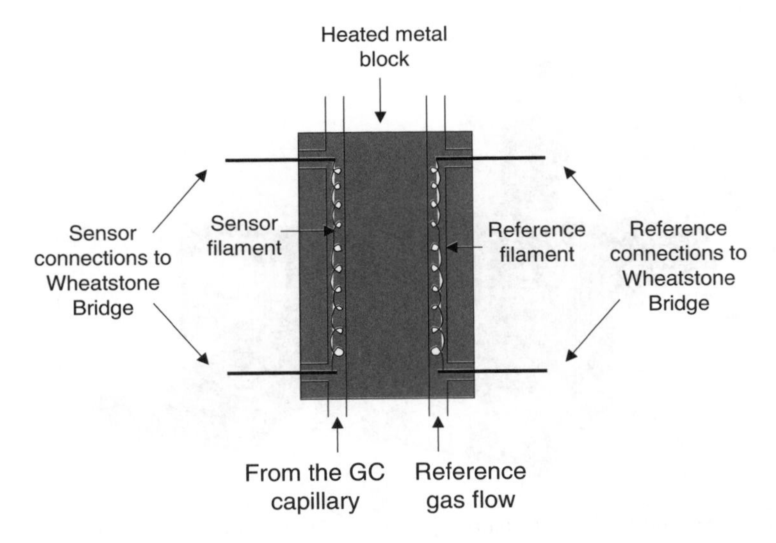

Figure 3.7 *Schematic diagram of a thermal conductivity detector.*

very flow and pressure sensitive and the sensor must be carefully thermostatted and fitted with reference cells to compensate for changes in pressure or flow rate. The device is simple, reliable and rugged. It is also often the detector of choice for process monitoring.

The mass spectrometric detector is often coupled to GC to provide a very sensitive separation and detection instrument in one. This system is discussed in more detail in the section on hyphenated (hybrid) techniques.

Output
A computer is used to carry out identification of compounds based on retention times from the chromatogram and integration of the peaks for quantitative work.

Information Obtained

Gas chromatography can be used as a qualitative or quantitative technique. Qualitative information can allow the identification of a substance on the basis of retention time comparisons between a standard and the sample. Alternatively, and more definitively, the use of a mass spectrometric detector enables the molecular weight of eluting components to be determined and as long as those molecular weights are unique, unambiguous identification is possible. When used quantitatively, the GC can employ external or internal standards to generate a standard curve. Then, the peak area of the sample peak on the chromatogram can be used to obtain a concentration. An example of a GC instrument is shown in Figure 3.8.

Developments/Specialist Techniques

Sample preparation is often an issue prior to injection for GC analysis. One useful method of sample preparation is thermal desorption, which is widely used for extracting and

Figure 3.8 *Agilent 6890 gas chromatograph (© Copyright 2006 Agilent Technologies, Inc. Reproduced with permission).*

isolating volatile and semivolatile compounds from various matrices. During the thermal desorption process, heat and a flow of inert gas are used to extract volatile and semivolatile organics retained in a sample matrix or on a sorbent bed. The analytes desorb into the gas stream and are ultimately transferred to the GC. Some thermal desorbers have two stages, i.e. they preconcentrate the analytes desorbed from the sample tube before releasing them into the analytical system in as small a volume of vapour as possible.

Another useful technique is solid phase microextraction. A fused silica fibre is attached to the base of a syringe with a fixed metal needle. The fibre is coated with a thin layer of stationary phase that is selective for the analytes of interest. The fibre is dipped into the liquid sample or into the headspace above the liquid for a period of time, allowing a fraction of the analyte to be extracted into the fibre. The fibre is then retracted into the syringe and the syringe injected into the injection port where the analyte is thermally desorbed from the fibre into the GC.

A third sample preparation method is 'purge and trap,' which aims to extract as close to all of the analyte as possible from the solid or liquid sample and is a deviation from headspace sampling. It works by bubbling a purge gas such as helium through the heated sample vial. The gas carries analyte up into an adsorption tube packed with selective stationary phase. After all the analyte has been trapped in the tube, the gas flow is reversed through the tube to remove any residual solvents. The tube is then directed to the injector port and, heated to desorb the analytes, which are then cold-trapped onto the head of the GC column. From there, the concentrated sample is heated for GC separation.

Other new developments in GC include the capability to analyse larger, more labile molecules using high GC column flow rates and the use of solvent-free syringeless injection, which allows inorganic materials to be introduced directly onto a heated injection port of a GC for thermal desorption[3]. There have been advances in headspace GC techniques also[4].

Multidimensional GC is a technique that is used when there are so many components in a separation that one column may not be able to separate them all adequately. In this case, a portion of the first column's effluent is transferred to a second column for analysis via column-switching apparatus. There are three main approaches for GC–GC depending on the application. The first is 'heart-cutting,' where an unresolved section of the separation is sent to a second column (perhaps with very different chemistry) for improved separation. The second is enrichment, where the first column preconcentrates the sample and the second column analyses the sample. The third is backflushing, where highly retained solutes that have not yet been eluted can be subjected to a reversal of flow on the same column, and hence be separated. GC–GC is advantageous if the sample is complex or the analytes are very different from each other and more than one assay would otherwise be required for good separation. Thermo Electron Corp. has released TRACE GCxGC, which is suitable for the analysis of target compounds in complex matrices. Data is presented as a structured 3-D chromatogram.

Hyphenated GC techniques are discussed at the end of this section.

Applications

Gas chromatography has found important applications in the fingerprinting of oil spills, where the pattern of peaks obtained from a sample can pinpoint the petroleum source, and can also be utilised to determine the long-term fate of the petroleum hydrocarbons[5].

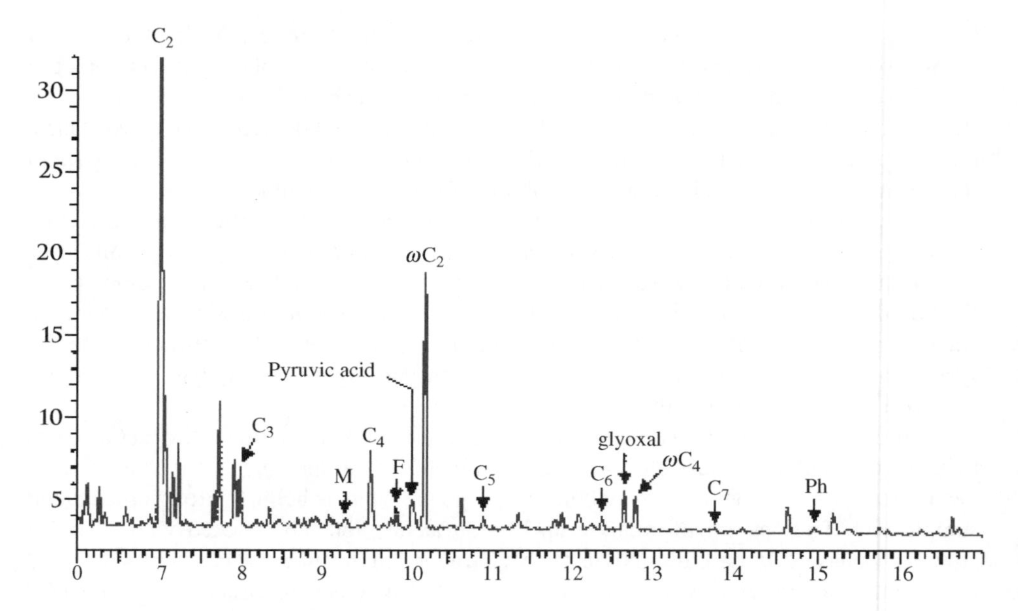

Figure 3.9 *Typical capillary gas chromatogram of the derivatives of dicarboxylic acids (C_2–C_9), ketocarboxylic acids (ωC_2, ωC_4, pyruvic acid), and dicarbonyl (glyoxal) isolated from sea water (Sample collected in the Bay of Marseille (northwestern Mediterranean Sea) in September 2005. Reprinted with permission from 'Determination of low molecular weight dicarboxylic and ketocarboxylic acids in seawater sample', Tedetti, M. et al., Analytical Chemistry (1 Sep 2006),* **78** *(17), 6012–6018[10]. Copyright (2006) American Chemical Society).*

Other common environmental examples of quantitative GC are in the determination of pesticides in water[6], dioxin levels in soil[7] and air pollutants. It is routinely used to examine levels of volatile organic compounds (VOCs), polycyclic aromatic hydrocarbons (PAHs) and polychlorinated biphenyls (PCBs). It is also a very important technique in the food industry, where it is used extensively for assay of fatty acids, flavours[8], sterols and residues such as insecticides, herbicides, preservatives, solvents[9] and veterinary drugs.

Through chemical derivatisation, more polar food compounds can be analysed, including sugars and carboxylic acids. Tedetti *et al.* reported the analysis of low molecular weight carboxylic acids by GC with prior derivatisation to dibutyl esters[10]. Low ppb concentrations were determined. A typical chromatogram of the separation of a number of these acids and related polar compounds from a sea water sample is shown in Figure 3.9.

3.2 High Performance Liquid Chromatography

Principle

High performance liquid chromatography (HPLC) is a form of column chromatography with lower resolution but wider applicability than GC. The acronym HPLC is often shortened to LC. The analyte is carried through a stationary phase in a packed column by a liquid mobile phase at high pressure, during which time the components separate from each other on the column. Traditionally, HPLC columns were polar, e.g. silica, and the mobile phases used were relatively nonpolar in nature. This mode of HPLC was called normal phase chromatography.

More common today is 'reversed' phase HPLC where the columns are hydrophobic, e.g. C-18 bonded phase. Solvents used in reversed phase HPLC include any miscible combination of water and various organic modifiers (the most common are methanol or acetonitrile). On a reversed phase column, hydrophobic analytes are better retained, eluting more readily as the proportion of the hydrophobic component of the stationary phase is increased. Often, a gradient of solvent composition passing through the column is used to separate mixtures, e.g. a water/methanol gradient. Following separation, the analytes are detected by the detector as they elute from the column. Trace organic and inorganic materials can be determined at concentrations of 10^{-6}– 10^{-12} g depending on the detector chosen. Further information on HPLC is available elsewhere[11, 12].

Instrument

A HPLC instrument consists of the components shown in Figure 3.10.

Source

The mobile phase and pump comprise the source. The mobile phase is chosen such that it has the optimal eluting power for the HPLC mode being used, low viscosity, high purity and stability, and such that it is compatible with the detection system. Solvents usually need to be degassed prior to use either by filtering under vacuum, use of an ultrasonic bath or degassing on-line. An optional solvent gradient mixer can precede the pump to enable mixing of solvents.

Liquid chromatography pumps can deliver a range of flowrates from nL/min to 10 mL/min depending on the type of pump purchased, e.g. standard or microflow. A standard pump used at a flowrate of 1 mL/min with a standard column of 250×4.6 mm dimensions and

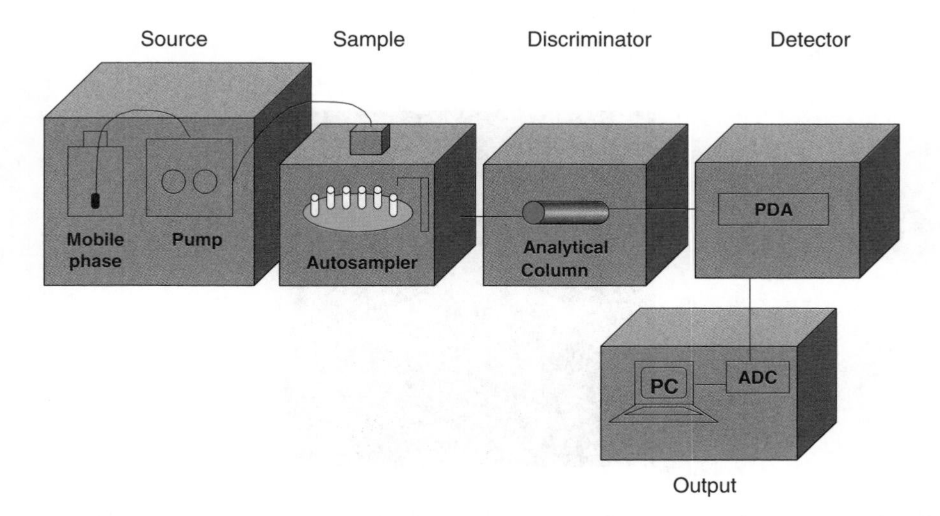

Figure 3.10 *Schematic diagram of a high performance liquid chromatograph.*

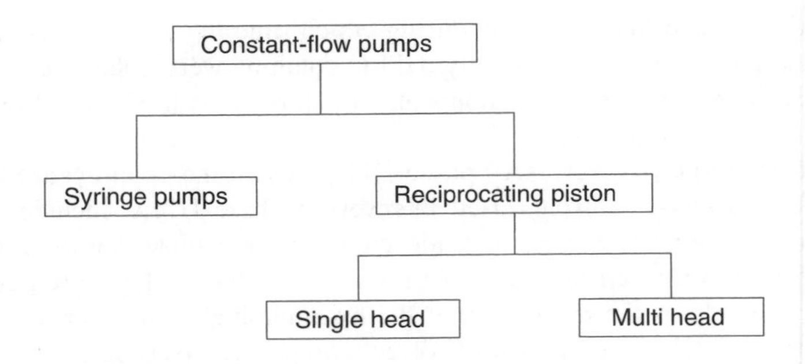

Figure 3.11 Types of constant flow pumps in HPLC.

3–10 μm particle size would tolerate a pressure of up to 3000 psi. The most commonly used pumps in HPLC are constant flow pumps, which can be subdivided into syringe pumps and reciprocating piston types (Figure 3.11). Syringe pumps deliver a smooth pulse-less flow but are limited by the volume they can hold; also during the refill of the syringe barrel, there will be no flow. Hence, reciprocating piston pumps are the most prevalent type in use today. However, because pistons having a fill stroke and a delivery stroke, pulsed flow results; this can be reduced with more than one piston, e.g. dual piston or triple piston pumps. In the dual head pump, each of the pistons is 180° out of phase with the other so that while one is filling, the other is delivering. This balances the pulsing somewhat and smoothes the flow. With more piston heads, the flow is smoothed even more. Nearly all new pumps also have pulse damping included.

Pumps can deliver solvent isocratically or via a gradient program (akin to temperature programming in GC). With a gradient program, the usual situation is to gradually increase the amount of organic component in the mobile phase over time so that the more nonpolar compounds are eluted in the same run as the more polar compounds.

Figure 3.12 *An injection valve for HPLC (Reproduced by permission of copyright owner: ©1998, Rheodyne, LLC).*

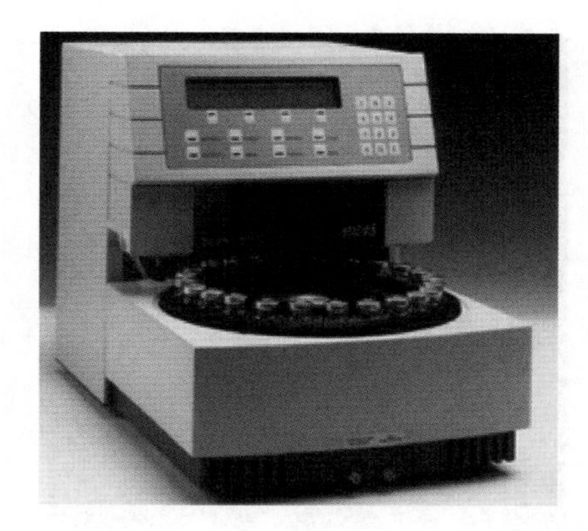

Figure 3.13 *The Smartline 3900 autosampler module (Reproduced by permission of Flowspek/Knauer).*

Sample

Samples can be injected via a sample loop using a valve (Figure 3.12). Once the sample has been introduced into the loop, the valve is switched (manually or by electric actuation) to deliver the sample plug onto the analytical column. Autosamplers allow more precise injection. In this case, a carousel or tray of sample vials can be left unattended for automatic injection (by compressed air or nitrogen) at predetermined intervals (Figure 3.13). The volumes injected are usually in the range 5–500 μL but lower and higher volumes can be injected under certain conditions if required.

Discriminator

Efficient analytical columns must have a homogeneous stationary phase of small particle size. In recent years, to reduce the volumes of solvents used in chromatography and to speed up analysis times, there has been a drive towards the use of narrower, shorter columns (Table 3.1). The use of such columns reduces the flow rate of the mobile phase and, hence, the overall amounts of solvent consumed. It also reduces the time spent in the column by the analytes. Another useful factor is that less sample is usually injected onto these columns, which can be very important where samples are precious, such as in the case of human blood samples and with proteomic separations. And if the particle size decreases with the length, there does not have to be an associated loss in efficiency and resolution. Connections between injector, column and detector should be of low volume and the inside diameters of components should

Table 3.1 *Dimensions of commercially available HPLC columns.*

Type of HPLC column	Internal diameter (mm)	Length (mm)	Particle size (μm)	Sample load (μg)	Flowrate (μL/min)
Standard bore	4.0–4.6	200–300	3–10	100–1500	1000–3000
Reduced bore	3.0–3.5	100–300	3–5	150–500	300–1500
Narrow bore	2.0–2.1	50–250	3–5	50–120	100–500
Microbore	0.5–1.0	50–250	2–5	10–50	10–200
Capillary	0.1–0.5	50–250	2–5	1–10	1–15

Figure 3.14 *Examples of the range of sizes of HPLC columns available (Courtesy of GL Sciences, Inc., USA).*

be of similar size. Temperature control may be required if the separation is sensitive to changes in temperature.

The types of stationary phase available are many and varied; the choice depends on the mode of HPLC being used (Figure 3.14). Reversed phase columns are the most common, accounting for over 75 % of separations today. The usual type of reversed phase column is packed with silica particles with nonpolar alkyl chains covalently attached to the silanol groups on the surface of the particles. Such silica columns are stable from pH 2–8 so for separations requiring alkaline conditions, polymer-based columns should be used instead. Not all silanol groups will be bonded due to factors such as steric hindrance, so those that remain are often coated in a process called end-capping. The bonded chains act like a liquid layer in which the analytes become distributed as they move through the column. Due to size and polarity differences between analytes in a mixture, they will partition in different ratios between the mobile and stationary phases, thus effecting a separation. During method development, a good rule of thumb is to choose a column with similar polarity to the compounds to be separated.

Normal phase columns contain a polar stationary phase, e.g. silica or alumina, and are used with non-polar mobile phases such as hexane. Guard columns sometimes precede the analytical columns to protect them from dust and impurities that may damage the column. HPLC separations are sensitive to temperature changes, so column ovens are sometimes employed to control the temperature of the separation. For example, at higher temperature, analysis speeds up and resolution improves.

There are a number of modes of HPLC available to the analyst. The main difference between them is the type of column used. Two have been mentioned already – reversed-phase and normal phase chromatographies. Five are be discussed briefly here, each having their own applications and suitabilities for certain types of compound. An example of an LC instrument is shown in Figure 3.15.

Reversed-phase chromatography is the most common of all the modes as it can usually be used to separate most types of compounds in HPLC. It is based on the principle that 'like dissolves like'. It uses a nonpolar column, e.g. C-18, and relatively polar mobile phases,

Figure 3.15 *The Series 200 HPLC Photo Diode Array System (Reproduced by permission of Perkin Elmer Inc.).*

e.g. a methanol–water mixture. Molecules partition themselves between the stationary and mobile phases to different extents depending on factors such as polarity, size and pKa.* Large non-polar molecules will elute last.

There are also versions of reversed-phase HPLC such as ion-suppression and ion-pair chromatography. Ion-suppression chromatography allows the separation of weak acids and weak bases under revered-phase conditions by using a buffer in the mobile phase. The pH of the buffer is chosen such that it renders the analyte uncharged or only partially charged, enabling the separation of such a compound in the normal way. Ion-pair chromatography allows the separation of strong acids and strong bases under revered-phase conditions by using a buffer and an added counterion (ion-pairing reagent) in the mobile phase. The pH of the buffer is chosen such that it renders the analyte charged, and the counterion is chosen so that it is oppositely charged to the analyte, thus enabling the two to 'pair' as a neutral complex during reversed phase separation.

Normal phase chromatography is used where very nonpolar molecules take too long to elute from a reversed-phase column. It, too, is based on the principle that 'like dissolves like' and uses polar columns, e.g. silica or alumina, and nonpolar mobile phases, e.g. toluene or hexane. Molecules distribute themselves between the stationary and mobile phases to different extents depending on factors such as polarity, size and pKa. Large nonpolar molecules will elute first.

Ion-exchange chromatography is useful for separating of compounds that are charged in solution or can be ionised in solution. It is based on the principle that 'opposites attract'. The columns used are based on cation-exchange or anion-exchange resins that have fixed negative charges on the resin which attract the ions of opposite charge. The strength of attraction to the resin is based on the size and charge on the molecules, thereby effecting separation of a mixture.

*pKa: the negative log of the of the acid ionisation constant. The lower the pKa, the stronger the acid. Conversely, the higher the pKa, the weaker the acid.

Size exclusion chromatography is used to separate compounds that differ in size from each other, e.g. a series of polymers. It is based on the principle of 'molecular sieving'. Special columns called gel filtration or gel permeation columns are employed. These columns contain particles with a defined pore size that allow only small molecules to penetrate them, thus the largest molecules in the mixture reach the detector first, followed by small molecules followed by solvent. It is often used for desalting applications as well as for determining the molecular weight of compounds.

Affinity chromatography is used mainly by biologists for purifying biomolecules such as proteins. The principle is based on the 'lock and key' model where biomolecules recognise and bind to other molecules in a very specific manner, e.g. antibody–antigen or enzyme–substrate interaction. The column is specially prepared by attaching an affinity ligand to the support. Alternatively, a molecular-imprinted polymer (MIP) can be prepared by synthesising it in the presence of a template molecule. Once the template is removed, the MIP now has imprints of the analyte built into it. If a mixture is applied to the column, only the species that is complementary to the ligand (or imprint) will bind to the column. This species can later be eluted, often in a concentrated volume.

Detector

A large number of LC detectors based on a variety of different sensing principles have been developed over the past thirty years. All LC detectors can be divided into two main types – bulk property detectors and solute property detectors. The former type is based on a characteristic of the mobile phase changing in some way as the analyte flows through and is based, therefore, on measuring the difference between the background without analyte and the background with analyte. The latter type of detector is based on some characteristic of the analyte being measured, generally against a zero or at least a low level background. There are many types of LC detectors; some of the most common ones are:

Bulk property detectors

- Refractive index detector
- Conductivity detector
- Light scattering detector.

Solute property detectors

- UV detector
- Fluorescence detector
- Amperometric detector
- Mass spectrometric detector.

Combination detectors.

The *refractive index* (*RI*) *detector* is the only detector in LC that responds to almost every compound. It is very useful for detecting species that are not ionic, do not absorb in the UV and do not fluoresce, e.g. polymers and carbohydrates. The detection principle involves measuring the difference in the refractive index of the background eluent and that of the eluent with analyte. The refractive index of an analyte is a function of its concentration so any change in concentration is reflected as a change in the RI. The

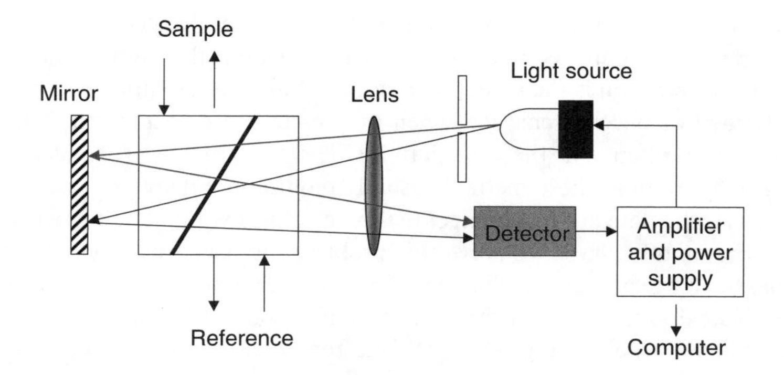

Figure 3.16 *Schematic diagram of a refractive index detector.*

greater the difference in RI between sample and mobile phase, the better the sensitivity. It is a nondestructive detector and relatively insensitive to dirt and air bubbles.

Refractive index detectors are stable, easy to operate and are often used in preparative chromatography where sensitivity is less important. However, since the RI detector is a pure differential instrument, any changes in the eluent composition require rebalancing of the detector so it is not suitable for gradient elution methods. Another limiting factor is that RI is very sensitive to temperature and pressure changes. Its main disadvantage is its poor sensitivity (100 ng) and narrow linear range. A diagram of a differential RI detector is shown in Figure 3.16.

The electrical *conductivity detector* measures the conductivity of the mobile phase. Conductivity detectors are universal and nondestructive and can be used in either direct or indirect modes. The conductivity sensor is the simplest of all the detectors, consisting of only two electrodes situated in a suitable flow cell. The basis of conductivity is the forcing of ions in solution to move toward the electrode of opposite charge on the application of a potential. To prevent polarisation of the sensing electrodes, AC voltages must be used and so it is the impedance (not the resistance) of the electrode system that is actually

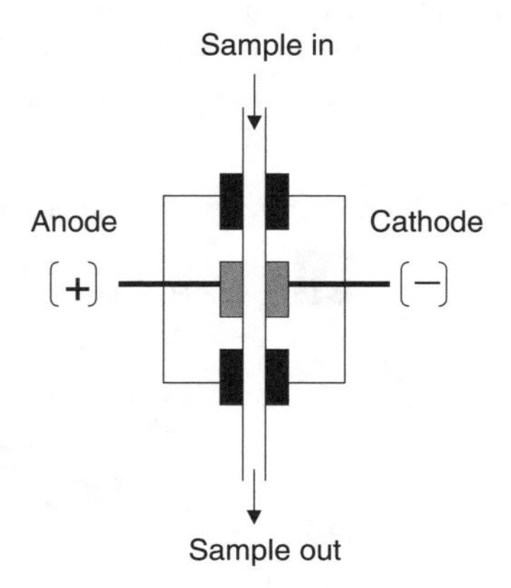

Figure 3.17 *Schematic diagram of a conductivity detector.*

measured. The frequency of the AC potential across the electrodes is usually around 10 kHz. An example of an electrical conductivity sensing cell is shown in Figure 3.17.

Conductivity detection is the detection mode of choice for organic or inorganic species that are charged or become charged when they enter the detector cell. The separation modes that most often use the conductivity detector are ion-exchange and ion-pair chromatography. Both of these methods usually require mobile phases containing strong electrolytes and the detector must detect the ionic solutes without being overwhelmed by the ions in the mobile phase. To solve this problem the background conductivity of the mobile phase can be suppressed. The pH of the mobile phase must be chosen carefully to maximise solute dissociation. Sensitivity is normally about 0.5–1 ng. Like the RI detector, this detector is not suitable for gradient HPLC. Temperature correction is often needed.

The evaporative *light scattering detector* is used to monitor compounds that are less volatile than the mobile phase of the HPLC. It has better sensitivity than the RI detector (0.1–1 ng). The system consists of three parts: the nebuliser, the drift tube and the light scattering cell. A diagram of the light scattering detector is shown in Figure 3.18. The analytical column outlet is connected directly to the nebuliser. The column effluent is nebulised (atomised) with air or nitrogen to form an aerosol mist. The mist then passes through the drift tube where the solvent evaporates leaving analyte droplets. These analyte droplets pass through a laser and the scattered light is detected at an angle by a photodetector. As the light dispersion is largely Raleigh scattering, the response should be proportional to the mass of solute present (it is sometimes referred to as a mass detector). The analytes must have higher boiling points than the mobile phase so that they remain while the solvent is vaporised. Reducing flow rate by using 2.1 mm ID columns should be considered when sensitivity is important. This detector is compatible with gradient elution.

UV absorbance detectors are nondestructive and respond only to substances that absorb radiation at the wavelength of the source light. They are classified as selective detectors because the mobile phase is chosen such that it exhibits little or no absorbance at the wavelength of interest. They are also relatively sensitive (0.1–1 ng) and can be used with gradient elution. While the UV detector is the most common and most widely applicable, these are also combinations of UV–Vis or UV–Vis–NIR spectrometric detectors.

A UV detector is generally found to be suitable for detecting of the majority of samples due to the large number of compounds that absorb in this wavelength range – aromatic rings, carbonyls, bromine, iodine and sulfur. For example, any compound that has a benzene ring will absorb at 205–225 and 245–265 nm. Because of this, and the fact that

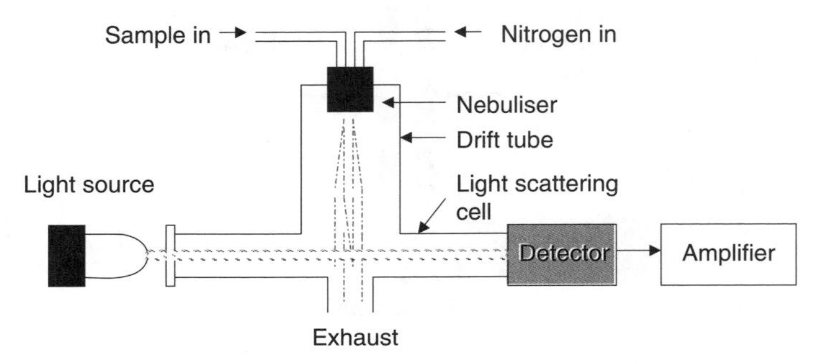

Figure 3.18 *Schematic diagram of a light scattering detector.*

the UV detector is relatively insensitive to temperature and gradient changes and is easy to use, almost 70 % of published HPLC analyses are performed with UV detectors.

In direct UV detection, the eluent exhibits little or no absorbance. When a UV-absorbing compound, dissolved in the eluent, passes through the detector, it absorbs some of the radiation, thereby preventing it from reaching the photodetector. The decrease in light is related to the concentration of the solute according to the Beer–Lambert Law. One absorbance unit corresponds to the depreciation of the light intensity by 90 % of the incident light. Molar absorptivity is dependent on the wavelength and chromatographic conditions (solvent, pH and temperature). It is a constant for that analyte at a specified wavelength. If a solute of interest does not contain a chromophore, indirect UV detection can be used instead. In this case, the eluent contains the chromophore (probe) giving a high background response. When a solute passes through the detector, a negative peak is displayed as no chromophore is seen. Typically, these negative peaks are inverted and displayed as positive peaks.

The detector itself consists of a small liquid flow cell through which the eluent from the column flows. UV light passes through the cell and hits the UV photodetector. The cell is usually made of quartz which is UV-transparent. To avoid band broadening, the flow cell volume is minimised: about 10–15 μL in the case of a standard flowcell or 6–8 μL in the case of a semimicro flowcell, the choice of which depends, among other factors on the column and the application.

Fixed wavelength detectors are detectors that do not allow the wavelength of the radiation to be changed. This detector is the least expensive, least flexible but most sensitive of the UV detectors. A low-pressure mercury vapour lamp, for example, emits very intense light at 253.7 nm and most UV-absorbing compounds have some absorbance at this wavelength. Other lamps that can be purchased are the low-pressure cadmium lamp, which generates the majority of its light at 225 nm, and the low-pressure zinc lamp that emits largely at 214 nm.

Multi-wavelength UV detectors utilise a narrow range of wavelengths to detect the solute. Most multi-wavelength UV detectors can also provide a UV spectrum of the eluted solute using a stop/flow setup. Their sensitivity is not as good as the fixed wavelength but they have the advantage of flexibility. There are two types of multi-wavelength detectors: the dispersion detector (variable wavelength detectors) that monitors the eluent at one wavelength only (selected by the analyst), and the diode array detector that monitors the eluted solute over a range of wavelengths simultaneously. The former passes the light from a broad emission light source through a monochromator, selects a specific wavelength

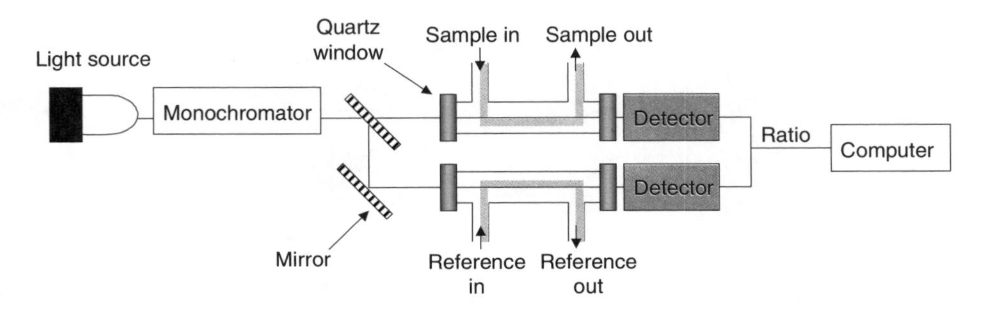

Figure 3.19 *Schematic diagram of a multi-wavelength UV detector.*

and allows it to pass through the detecting cell. Different wavelengths can be monitored during a run. A diagram of a multi-wavelength detector is shown in Figure 3.19.

The second type, the photodiode array (PDA) also uses a broad emission light source, but all the light is allowed to pass through the sensing cell and subsequently dispersed by means of a polychromator, with the dispersed light allowed to fall on an array of diodes. In this way the sample is exposed to all wavelengths. At the end of the run, the output from any diode can be selected and a chromatogram produced using the UV wavelength falling on that particular diode. With the PDA detector, either a single wavelength, a variety of wavelengths or the entire spectrum from the whole photodiode array can be monitored. PDAs can also be used to perform spectroscopic scanning to obtain a UV–Vis spectrum for each peak. This adds new analytical capability to LC because it permits identification as well as quantitative measurement. There are two major advantages of diode array detection. Firstly, it allows for the best wavelength(s) to be selected for analysis. Secondly, a PDA allows all the spectra under a peak to be obtained and its purity to be determined.

Fluorescence detectors are very selective and are about three orders of magnitude more sensitive (0.001–0.01 ng) than UV detectors as they measure the fluorescence of the analyte against an almost zero background. Being one of the most sensitive LC detectors, they are often used for trace analysis. Unfortunately, the response is only linear over a relatively limited concentration range. The multi-wavelength fluorescence detector contains two monochromators, one to select the excitation wavelength and the second to select the fluorescence wavelength or produce a fluorescence spectrum. The detector comprises a fluorescence spectrometer fitted with suitable absorption cell that is sufficiently small that the resolution of an LC column is not compromised. The grating selects the specific wavelength of the fluorescent light to be monitored. Light of the selected wavelength passes to a photodetector which monitors its intensity. A diagram of the fluorescence detector is shown in Figure 3.20.

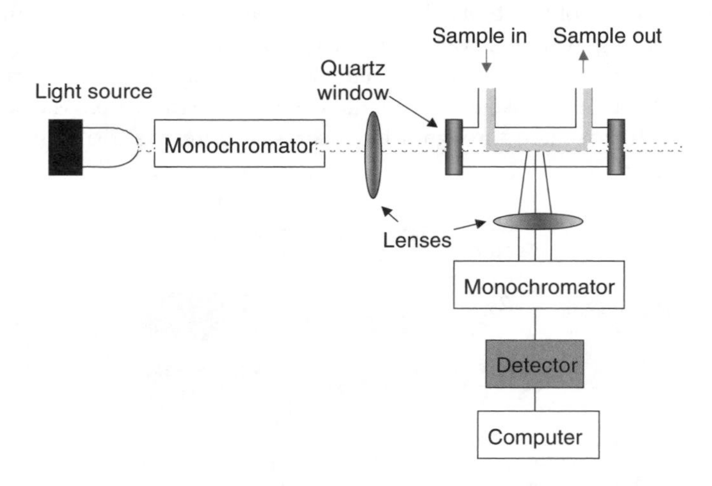

Figure 3.20 *Schematic diagram of a multi-wavelength fluorescence detector.*

The optical system of most fluorescent detectors is arranged such that the fluorescent light is viewed at an angle (usually 90°) to the exciting incident light beam. This minimises the amount of incident light that can interfere with the fluorescent signal. One big difference between absorbance and fluorescence detection is that fluorescence has a linear dependence on the excitation power, which means that sensitivity can be increased by using a more powerful source. The use of lasers as excitation sources offers a number of advantages, such as the increase in sensitivity seen with laser-induced fluorescence (LIF). In addition, because of the coherent and monochromatic nature of the laser beam, little light is lost to the surroundings. If a fluorescence detector is the detector of choice and the compound does not naturally fluoresce, it can be derivatised to do so. The mobile phase plays a very important part in the fluorescence of a molecule. Unless chosen with care, the mobile phase can quench the fluorescence of the molecule of interest. The pH of the mobile phase can also affect fluorescence, e.g. aniline fluoresces at pH 7 and pH 12 but at pH 2, where it is cationic, it does not fluoresce.

In an *amperometric detector,* current is measured between the working and auxiliary electrodes due to the reaction that takes place at the surface of the electrodes. The level of the current is directly proportional to the analyte concentration over six orders of magnitude, so this detector can be easily used for quantification. If the mobile phase is flowing past the electrodes, the reacting solute will be continuously replaced as the peak passes through the detector. It is a very selective and fairly sensitive detector (0.01–1 ng). However, it is sensitive to temperature and flowrate changes and is not suitable for gradient HPLC. A schematic of an amperometric flowcell is shown in Figure 3.21.

The purity of the eluent is very important, since the presence of oxygen, metals and halides may cause significant background current. The working electrode is typically glassy carbon or a precious metal such as platinum or gold. A major consideration when choosing an electrode material is its ability to maintain an active surface. Direct current amperometry is used for the analysis of catecholamines, phenols and anilines which are easy to oxidise. A single potential is applied and the current is measured. The current resulting from the oxidation or reduction of analyte molecules can be related to the concentration of the analyte. Pulsed amperometric detection is used for the analysis of carbohydrates and other nonchromophoric molecules such as alcohols, aldehydes and

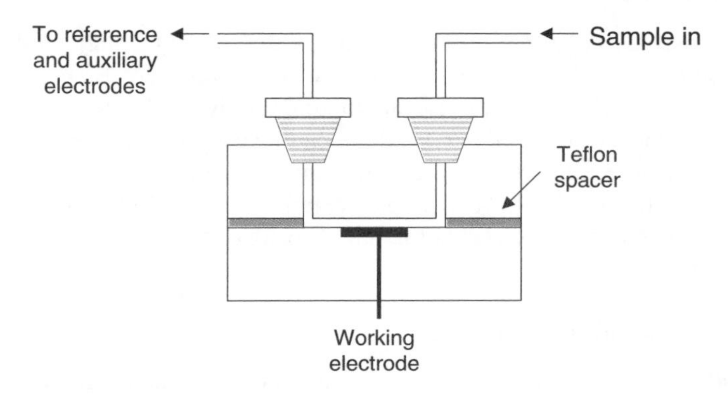

Figure 3.21 *Schematic diagram of an amperometric detector.*

amines. Alternative positive and negative potentials are repeatedly pulsed to maintain a stable and active electrode surface. The technique is robust, simple to use and inexpensive. It is also destructive.

The *mass spectrometric (MS) detector* is often coupled to HPLC yielding a very sensitive separation and detection instrument all in one, referred to as an LC–MS. A mass spectrometric detector is a universal, destructive detector that is used for the analysis of compounds with molecular weights up to and in excess of 100 000 Da. It is used to confirm the identity of a compound and can also be used to elucidate structures. Although mass spectrometry is not the most sensitive of LC detectors, it is only an order of magnitude less than LIF. The choice of interface is dependent on both the analysis and the instrumentation available. Soft ionisation methods such as electrospray are the most popular choice for interfacing with LC owing to their ability to generate gaseous ions from large, non-volatile and labile molecules. MS detection for LC may be performed using a single quadrupole (LC–MS) or a triple quadrupole (LC–MS–MS) mass spectrometer. Full scan spectra can be obtained where mass-to-charge ratios (m/z) are sequentially scanned to produce a spectrum. Further information on the use of the MS detector with LC is given in the section on hyphenated techniques.

Combination detector: The popularity of the UV detector, the conductivity detector and the fluorescence detector prompted Perkin-Elmer, a number of years ago, to develop a trifunctional detector called a TRIDET to detect solutes by all three method simultaneously in a single low volume cell. The UV absorption system consists of a low-pressure mercury lamp and a solid state photocell with quartz windows so it responds to light in the UV region. The mobile phase enters and leaves the detector cell through radial holes in the periphery of the stainless steel discs. The steel discs act as the electrodes for conductivity detection. Perpendicular to the cell is another photo cell that receives fluorescent light emitted at right angles to the UV excitation light. The column eluent can be continuously and simultaneously monitored by UV absorption, fluorescence and electrical conductivity. The TRIDET's main advantage is that it allows the analyst a choice of the three most useful detector functions in one device. This detector, although no longer commercially available, is still in use in many laboratories today.

Information Obtained

HPLC can be used as a qualitative or quantitative technique. Qualitative information can allow a substance to be identified on the basis of retention time comparison between a standard and the sample. Alternatively, and more definitively, the use of a mass spectrometric detector enables the molecular weight of eluting components to be determined and, as long as those molecular weights are unique, unambiguous identification is possible. When used quantitatively, HPLC can employ external or internal standards to generate a standard curve. Then, the peak area of the sample peak on the chromatogram can be used to obtain a concentration.

Developments/Specialist Techniques

Many manufacturers of HPLC instruments are now producing a high-speed, high throughput, smaller model for use with microbore or capillary columns. For example,

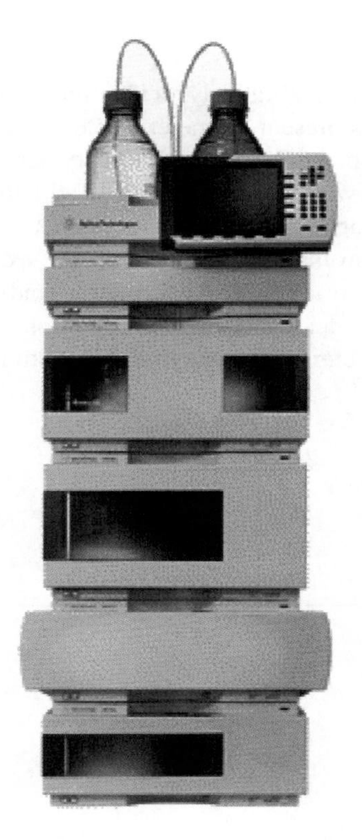

Figure 3.22 *Agilent 1200 Series Rapid Resolution LC system (© Copyright 2006, Agilent Technologies, Inc. Reproduced with permission).*

Shimadzu has produced the Prominence HPLC, Agilent has developed the 1200 Rapid Resolution HPLC (Figure 3.22), Thermo Scientific has the Accela LC system and Waters has brought out the Acuity UPLC. Jasco's X-LC system allows two systems to fit in the footprint previously occupied by one traditional system. It is also a modular system of detectors and autosamplers, which allows it to be customised. LC Packings has launched the UltiMate 3000, which is a nanoflow LC system for use with columns of 50 µm and larger. Modular systems are common now, e.g. Cecil Instruments produces both HPLC and ion chromatography systems and various detectors can be accommodated including UV–Vis, refractive index, conductivity and fluorescence.

Another development in HPLC analysis is the interfacing of on-line sample clean-up technologies at the front end of the HPLC analytical column. Using a switching valve, complex samples can be injected directly onto a sample preparation column, the stationary phase of which is designed to bind the analyte(s) of interest to the exclusion of the rest of the matrix, which runs to waste. Then the valve is switched and the mobile phase flows through the sample preparation column where it 'picks up' the analyte(s) and carries them onto the analytical column and through to the detector in the normal way. This automation of sample preparation saves time and effort and reduces errors. There are many applications of HPLC assays in conjunction with on-line sample clean-up methods in the literature[13, 14].

Applications

As with GC, HPLC can be used qualitatively for identification or quantitatively to determine how much of a compound is present. Some examples of the use of HPLC in qualitative work include identification of impurities and toxicity screening. Some examples of the use of HPLC in quantitative work are drug testing in athletes and in sports supplements[15], pharmacokinetic studies of drugs[16, 17], pharmaceutical assays and fatty acid analysis[18].

High pressure liquid chromatography is also an extremely useful technique in proteomics. It can be used to monitor the formation and/or breakdown of peptides in the presence of enzymes under experimental conditions. It can yield very important information about the character of the enzyme under study[19]. An example of a peptide

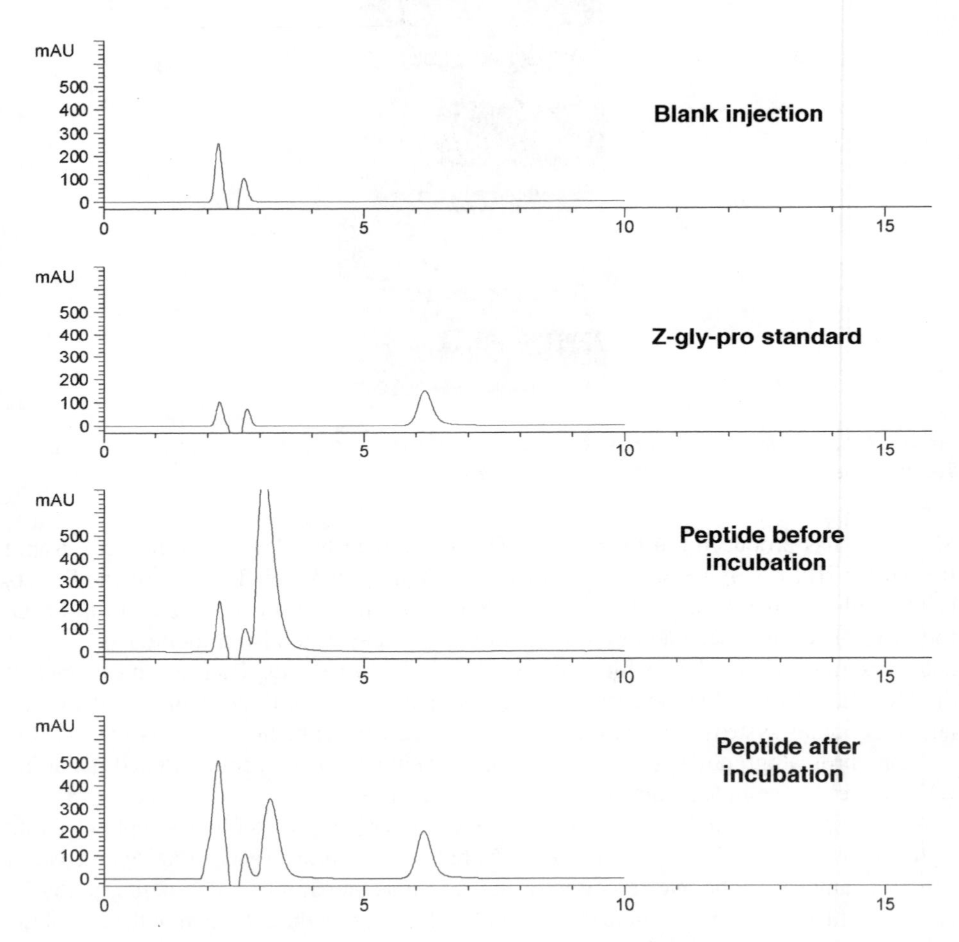

Figure 3.23 Chromatograms following LC separation with UV detection for incubated samples of peptide and enzyme (The predicted peptide cleavage fragment was injected alone in the second chromatogram. The peptide–enzyme mixture was assayed just before incubation and then after 20 hours of incubation. It can be seen clearly that the peptide peak has been reduced while the peak for the fragment has been formed[19]).

before and after incubation with an enzyme of interest is shown in Figure 3.23. A known peptide fragment (Z–gly–pro), which it was predicted would cleave off the peptide, was obtained as a standard and injected separately to check retention time. It can clearly be seen by comparing the before and after incubation chromatograms that the peptide has indeed been cleaved by the enzyme to produce Z–gly–pro.

3.3 Ion Chromatography

Principle

Ion Chromatography (IC) is a subset of one of the modes of HPLC – ion-exchange chromatography. It uses ion-exchange resins to separate ions based on their interaction with the fixed groups on the resin. Its greatest use is for the analysis of anions for which there are no other rapid analytical methods. It is also commonly used for cations and simple biochemical species such as amino acids. Aqueous solutions are normally required for analysis.

Ion chromatography can be subclassified into non-suppressed IC and suppressed IC. Suppressed IC uses a membrane suppressor or packed bed suppressor between the column and the detector to remove/suppress the mobile phase electrolyte and thereby enhance the signal from the analyte. In non-suppressed IC, special eluents are used, such as benzoate or salicylate to keep the background conductivity sufficiently low (suppressed). The equilibrium constant for the exchange reaction is known as the selectivity coefficient and provides a rough means for predicting the elution order of various ions. In general, the selectivity coefficients increase with increasing polarising power of the analyte ions, hence ions with a high charge and a small radius have the greatest affinity for the ion exchange column.

Instrument

An IC instrument consists of the modules shown in Figure 3.24.

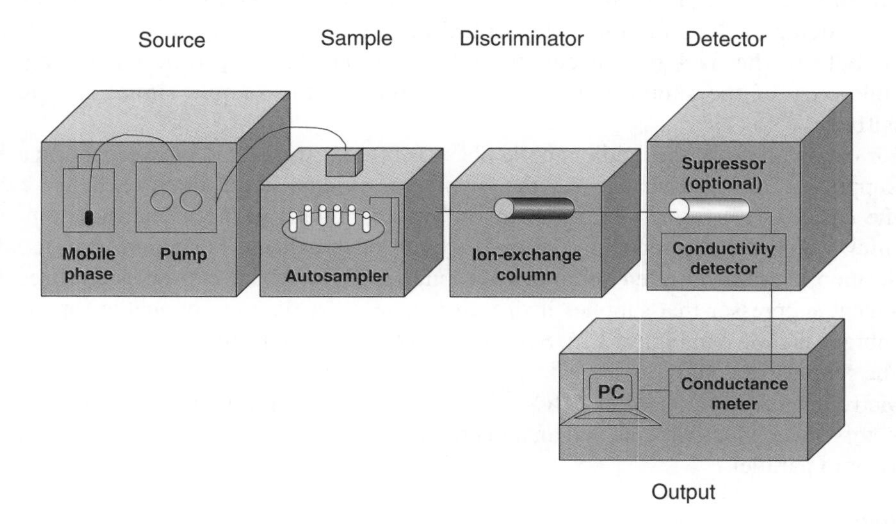

Figure 3.24 *Schematic diagram of an ion chromatograph.*

Source
As with HPLC, the mobile phase and pump comprise the source. Because the samples are ions, mobile phases are aqueous in nature and buffered. Solvents still need to be pure and degassed prior to use. The pump should be capable of delivering gradients. Once the charged analytes are attached to the exchange groups in the column, they must be eluted out using a buffer with a higher ionic strength or a different pH in order to weaken the electrostatic interactions between the analytes and the exchangers. The nature and concentration of the counterion also determine the ease with which the analyte ions will be displaced from the resin.

Sample
As with HPLC, samples can be injected manually by syringe, via a sample loop or utilising autosamplers.

Discriminator
The discriminator is the ion-exchange column where separation of the ions of interest takes place. For cation separation the cation-exchange resin is usually a sulfonic or carboxylic acid, and for anion separation the anion-exchange resin is usually a quaternary ammonium group. Different ions have different affinities for the resin and are therefore retained on the column for different lengths of time and hence separation is achieved.

Detector
Ions in solution can be detected by measuring the conductivity of the solution. The conductivity detector has been described in the previous section on HPLC detectors. In IC, the mobile phase contains ions that create some background conductivity but this background signal can be removed by converting the mobile phase ions to a neutral form or removing them with an eluent suppressor. This is called suppressed IC. In suppressed IC, a chemical suppressor is added between the analytical column and the detector. Its function is to neutralise the mobile phase so that it does not contribute to the background conductivity while at the same time enhancing the conductivity of the sample ions, thereby increasing the sample signal and hence sensitivity.

For cation analysis, the acidic mobile phase is often sulfuric acid or hydrochoric acid. In suppressed cation-exchange IC, the suppressor column is an anion exchange resin in the basic (OH^-) form, which exchanges with the anions in the acidic mobile phase forming water, thereby reducing the conductivity of the eluent. For anion analysis, the basic mobile phase is often sodium hydroxide (NaOH), which can be neutralised by an eluent suppressor that supplies hydrogen ions (H^+). A diagram of such a suppressor membrane is shown in Figure 3.25. Sodium ions (Na^+) are then either retained or removed by the suppressor column.

Metrohm has developed an MIC-9 Advanced IC system with two electrochemical detectors – a conductivity one and an amperometric one. They can be operated independently or in parallel.

Output
A computer is used to manipulate data, calculate results and display the chromatograms.

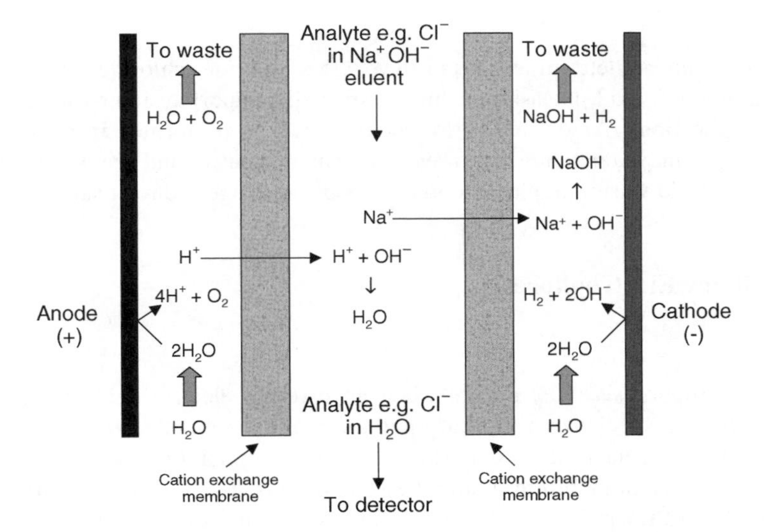

Figure 3.25 *Schematic diagram of a membrane suppressor for use in anion-exchange chromatography (H^+ ions replace the Na^+ ions yielding water as the background electrolyte which has low conductivity).*

Information Obtained

Ion chromatography can be used as a qualitative or quantitative technique. Qualitative information allows a substance to be identified on the basis of retention time comparisons between a standard and the sample ion. When used quantitatively, the IC can employ external or internal standards to generate a standard curve. Then, the peak area of the sample peak on the chromatogram can be used to obtain a concentration.

Developments/Specialist Techniques

With the use of new monolithic columns with narrow diameters for IC, the detection limits for ions have been decreased even further. The silica gel-based solid in the monolith is functionalised with the ion exchange material and it is highly porous with a high capacity. These columns have low back pressure so high flow rates, fast separations and well defined peaks mean better sensitivity (sub ppb).

Gradient IC is another new development which is only now possible due to the availability of powerful suppressors and the ability to generate changing eluents during the run.

Reagent-free IC (RFIC) is an up-and-coming technique and has been spearheaded by Dionex Corp.[20, 21]. Its ICS-3000 is an RFIC system that can perform dual analyses without increasing the system footprint or requiring a second autosampler. Older suppressors required regeneration periodically with acid (an extra reagent) which could take hours. When electric fields were introduced into suppressors, they enhanced the exchange of ions in the membrane as well as electrolysing the water to create hydrogen ions, rendering it acidic, and negating the need for the separate acid regenerant. The next step was to electrolyse the water to create hydroxyl ions for use as eluent, now negating the need for separate eluent. Hence the only solvent required for this type of IC is water.

Applications

Common ions can be determined at ppb levels, e.g. fluoride, chloride, nitrate, sulfate, sodium, ammonium, and potassium. This is especially important in environmental and industrial applications. Anion and cation analysis can be performed in many types of samples, e.g. groundwater samples, power plant waters, coastal and sea water samples[22], on air filters[23], solid waste samples, blood and food[24] and digested rock samples.

3.4 Capillary Electrophoresis

Principle

Capillary electrophoresis (CE) occurs when an external electric field is applied to a solution of charged species in a capillary and the ions are separated from each other as they move toward the electrode of opposite charge. Electrophoretic separation is based on solutes having different mobilities under these conditions. The velocity at which the ions move depends on the applied electric field, their shape and their environment. Strictly speaking, CE is not a chromatographic technique as it does not have a stationary phase but it has many similarities with the separation techniques in this category. Some of the differences in language used for the purposes of comparing HPLC and CE are shown in Table 3.2.

Capillary electrophoresis instrumentation includes a power supply, injector, capillary and detector. It is often set up as a modular system where the power supply, detector and other modules are bought separately off the shelf and configured by the scientist. The heart of the system is the capillary where separation occurs. Because capillaries in CE are mostly open tubular and not packed, resolution is excellent and peaks are very sharp. A high voltage is required to move the buffer through the capillary.

The electrophoretic process begins when the source buffer vial is removed and replaced with a sample vial. The sample is injected onto the top of the capillary. Separation of components occurs as the analytes and buffer migrate through the capillary under their own electrophoretic mobility and under the influence of electro-osmotic flow (EOF), which moves from anode to cathode. Eventually, each component migrates from the capillary as a narrow band (or peak).

Detection of the migrating components is important and can be either selective or universal depending upon the detector used. The response of the detector to each component is plotted as an electropherogram. The order of elution in the classic CE set-up is positive

Table 3.2 *Terminology for HPLC and CE.*

HPLC	CE
Chromatogram	Electropherogram
Column	Capillary
Pump	High-voltage power supply
Mobile phase or eluent	Electrolyte or buffer
Retention time	Migration time
Column retention factor	Electrophoretic mobility

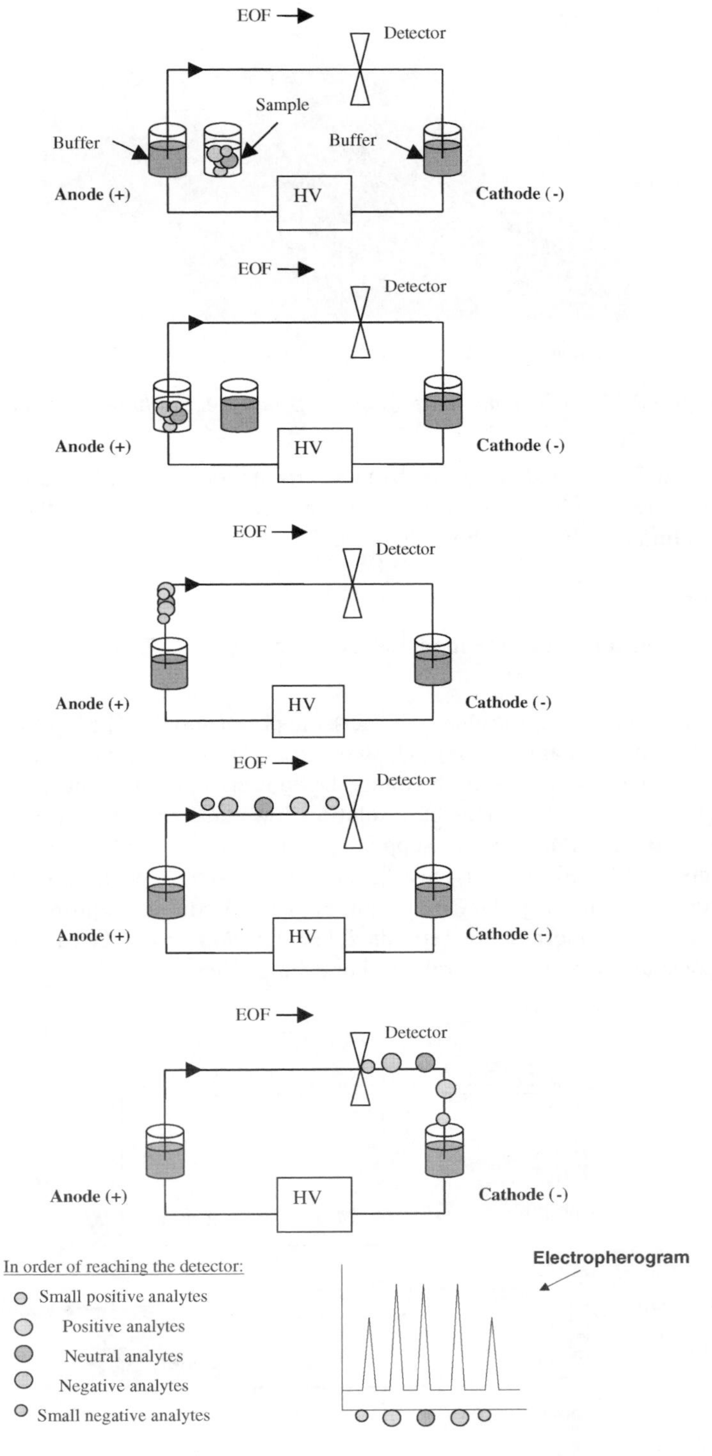

Figure 3.26 *The progression of capillary electrophoresis.*

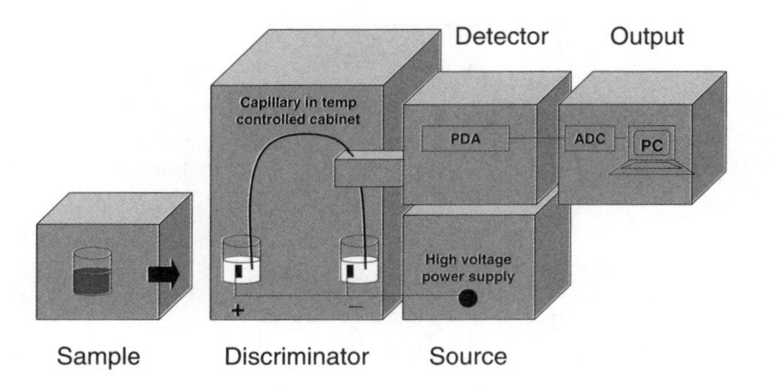

Figure 3.27 *Schematic diagram of a capillary electrophoresis instrument.*

ions, neutral molecules and lastly negative ions. Even though the anions are attracted back to the anode, the EOF (if strong enough) will still carry them to the cathode. A series of diagrams in Figure 3.26 shows how CE works.

Instrument

A CE instrument consists of the modules shown in Figure 3.27.

Source

The source is composed of the eluent (buffer) and power supply. The buffer choice is very important for good separation and pH, ionic strength and percentage organic modifier must be taken into account. A high voltage DC power supply is required to provide the electric field, because of which the bulk solution flow and electromigration of the charged analytes are possible. Most power supplies provide from −30 kV to +30 kV with very low currents of 200–300 mA. It must be possible to switch polarity in order to make the detector end of the capillary either anodic or cathodic as required. Stable voltage regulation is very important for reproducible migration times. Occasionally, constant power or constant current may be required or even gradients.

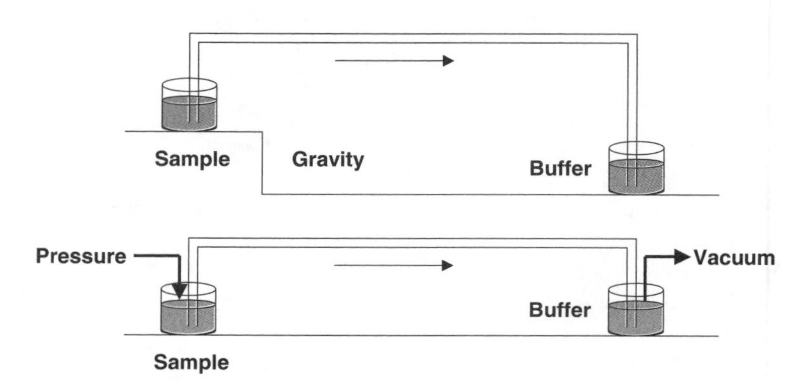

Figure 3.28 *The three types of hydrostatic injection used in CE.*

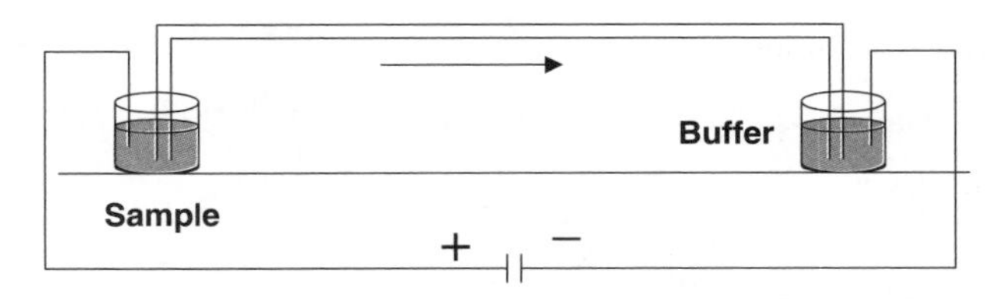

Figure 3.29 *Electrokinetic injection used in CE.*

Sample

Strictly speaking, the sample is introduced into the capillary rather than being injected into the capillary but the term injection is still used in CE. In CE, only small quantities (nL) of sample can be introduced onto the capillary if high efficiency is to be maintained. There are two main ways to achieve this – hydrostatic injection and electrokinetic injection. There are three types of hydrostatic injection – gravity, pressure and vacuum. These are illustrated in Figure 3.28. Gravity injection simply means that the sample vial is placed higher than the outlet buffer vial and so a small amount of sample will be pulled into the capillary under the influence of gravity. Pressure injection involves applying pressure to the sample vial to 'push' sample into the capillary, while vacuum applied to the outlet buffer vial will 'pull' sample into the capillary.

In the case of the electrokinetic injection (Figure 3.29), once the sample vial has replaced the inlet buffer vial, a small voltage is applied between the sample and outlet buffer vials which in effect starts a mini-electrophoretic separation. The sample ions move into the capillary in the usual order of positive ions, neutral molecules and negative ions. However, this particular type of injection means that positive ions are favoured (as they move first into the capillary being attracted to the cathode) and so the sample plug is not representative of the entire sample in the vial. However, if this is taken into account, this type of injection can be very useful.

Sample concentration can be an issue in CE. If greater sample loading is required, the sample can be 'stacked' during the injection process. This is a concentrating effect of the analyte zone and occurs when the sample is dissolved in a more dilute solution than the buffer. Sensitivity can also be enhanced by the use of a special technique called field-amplified sample injection (FASI). This works by hydrostatic injection of sample into the capillary until the capillary is almost full. Then the large amount of sample is focussed by applying a high voltage in the reverse direction, which removes water from the capillary. When the current drops to about 95 % of the original value, the voltage is switched back to the normal configuration and separation is performed.

Discriminator

The capillary and the electric field across it provide the discriminating element for electrophoresis to take place. The capillary has a buffer vial at each end and the inlet side of the capillary is the anode (positively charged) while the outlet side of the capillary is the cathode (negatively charged). The silica wall inside the capillary is covered in silanol groups (Si–OH), which above pH 2 are negatively charged ($Si–O^-$). Cations from the buffer cling closely to the negatively charged wall (compact layer). A mixture of ions (mostly cations) associates in a diffuse layer near the silica wall. The bulk solution in the capillary is composed of both buffer and analyte ions.

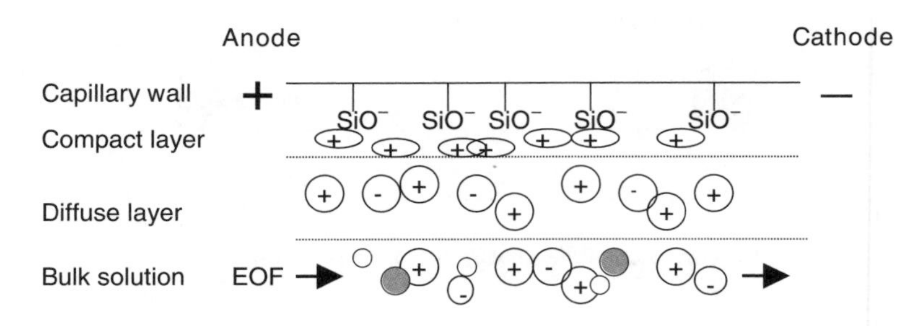

Figure 3.30 *Schematic diagram of how electroosmotic flow (EOF) is formed in a silica capillary.*

Under an applied electric field, the cations near the silica wall cause a net flow towards the cathode, pulling the bulk solution with them. This phenomenon is called electroosmotic flow (EOF) (Figure 3.30). Sample ions are separated from each other on the basis of their charge and mass (electrophoresis) and under the influence of the EOF. EOF can be eliminated if necessary but it is generally useful as, when it is flowing in the same direction as the analytes, it increases the speed with which they reach the detector and, when it is flowing in the opposite direction to the analytes, it improves their resolution.

EOF can be increased if desired by increasing pH, increasing applied voltage, reducing the ionic strength of the buffer, reducing the size of the buffer cation and anion or by reducing the amount of organic modifier in the electrolyte. More drastic measures can also be taken to manipulate the EOF. It can be reversed by switching the polarity of the voltage supply. Capillaries can be physically coated with polymer or chemically derivatised with functional groups to negate the effect of the silanols. Also, the capillary can be dynamically coated during a run by adding certain additives to the buffer. EOF can be measured by recording the migration time of an injected uncharged marker solute, which will be carried to the detector under the influence of only the EOF, e.g. methanol.

The most popular type of capillaries used are of fused silica and are commercially available with internal diameters ranging from 10 to 200 mm, although the most common dimensions are 25, 50 and 75 mm ID, and 350–400 mm OD. The longer and narrower the capillary, the better it is at dissipating heat generated during CE. Rinsing of capillaries with sodium hydroxide is required at the start of a run and sometimes in between injections to regenerate the surface. Equilibration with electrolyte is also good practice. There are a number of modes of CE available to the analyst besides the classic capillary zone electrophoresis that has been focused on up to now. The main differences between them are the type of capillary and/or buffer used. Each has its own applications and suitability for certain types of compound.

Capillary zone electrophoresis (CZE) is the most common of all the modes as it can separate a wide variety of positively and negatively charged species. It normally uses a bare fused silica capillary and relatively polar electrolyte, e.g. phosphate buffer. Molecules have different mobilities depending on their charge and size. Small anions will elute last.

Micellar electrokinetic capillary chromatography (MEKC or MECC) is the mode employed for separating of neutral analytes by CE. In CZE, neutral molecules migrate together as one unresolved peak. MEKC can also improve resolution for cation and anion separations too. It normally utilises a bare fused silica capillary. The electrolyte contains micelles which have a polar, negatively charged exterior and a nonpolar interior. The separation mechanism is based on partition of the analytes between the hydrophobic interior of the micelle and the aqueous buffer. Chiral separations can also be performed.

Capillary gel electrophoresis (CGE) aids in the separation of macromolecules such as proteins and nucleic acids whose mass-to-charge ratios do not vary much with size. In CGE, the capillary is filled with sieving material in the form of a gel or viscous polymer; larger molecules are held up by the gel while small molecules pass through easily (this is the opposite to the SEC mode in HPLC!). Under well-controlled conditions, the mobility of the analyte is inversely proportional to analyte size.

Capillary electrochromatography (CEC) is a hybrid of CE and HPLC, combining the separation power of reversed-phase HPLC with the high efficiency of CE. The capillary is packed with a stationary phase that is capable of retaining, and hence separating, solutes in a manner similar to column chromatography. However, there is no pressure pump and the analytes are carried through the capillary by EOF. The electrolyte is a combination of buffer for electrophoresis and eluent for the chromatographic mode.

Detector

Many of the detectors used in HPLC have been adapted for use in CE, e.g. UV–Vis, fluorescence and MS. In some cases, this has been a major challenge, since CE employs very narrow capillaries with low sample volumes and the detection takes place on this narrow capillary rather than in a separate flow cell. One advantage of UV–Vis detection with CE is the ability to exploit the shorter UV wavelength range (down to 185 nm) because the background electrolyte is normally just water. The added buffer can be selected so that it does not absorb at the low wavelength chosen.

Output

The data is collected, stored and analysed by computer.

Information Obtained

CE can be used as a qualitative or quantitative technique. An example of a CE instrument is shown in Figure 3.31. Qualitative information allows a substance to be identified on the basis of migration time comparisons between a standard and the sample. Alternatively, and more definitively, the use of a mass spectrometric detector enables the molecular weight of eluting components to be determined and, as long as those molecular weights

Figure 3.31 *Agilent Capillary Electrophoresis System (© Copyright 2006, Agilent Technologies, Inc. Reproduced with permission).*

are unique, unambiguous identification is possible. When used quantitatively, CE can employ external or internal standards to generate a standard curve. Then, the peak area of the sample peak on the electropherogram can be used to obtain a concentration.

Developments/Specialist Techniques

In CE, detection is performed on the capillary, so the detection cell is defined by the diameter of the capillary. Because sensitivity is proportional to path length (Beer–Lambert Law), the sensitivity of absorbance methods is limited. Extended path length capillaries such as the Z-cell and bubble cell have been developed to circumvent this issue. In the case of the bubble cell, a bubble is blown into the capillary in the region of the detector, giving an increase in sensitivity of 3–5 times. In the Z-cell, a longitudinal optical path length of 3–4 cm results and sensitivity is enhanced up to six times. Both of these designs are illustrated in Figure 3.32.

Another development in CE is the coupling or integration of sample preparation strategies such as liquid and solid phase extractions into commercial CE equipment[25].

Applications

Capillary electrophoresis is most useful as a technique for aqueous samples containing analytes that are charged or can be charged under the right conditions. It has the ability to separate large molecules that it may not be possible to analyse otherwise, such as

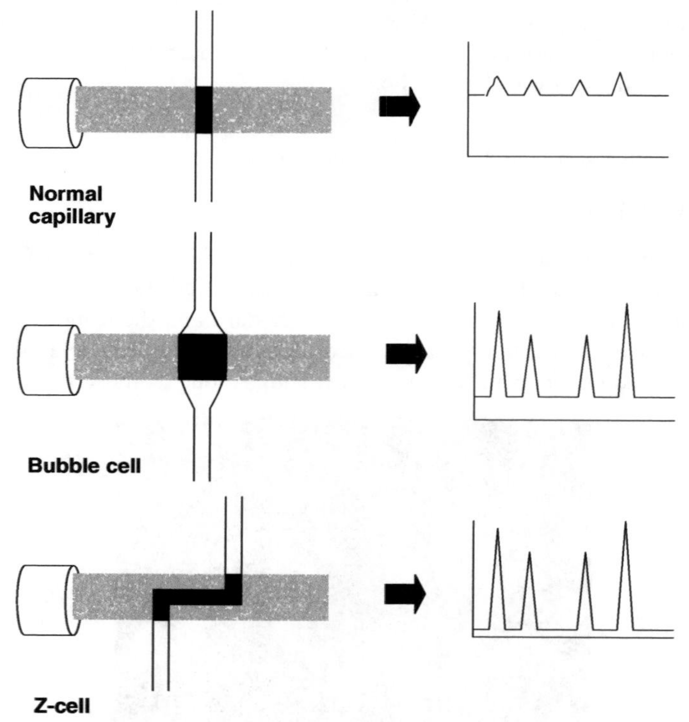

Figure 3.32 Schematic diagrams of a normal capillary, a bubble cell and a Z-cell (The increase in sample volume at the detector can be seen. This is translated into increased sensitivity on the electropherograms shown on the right).

proteins[26] and oligonucleotides. CE has many other applications, such as in drug impurity assays[27], metabolite screening[28], separation of enantiomers[29, 30], metal ions in samples[31], the analysis of peptides, polynucleotides and carbohydrates[32].

Capillary electrophoresis can also be used in a nonaqueous mode, where it is referred to as nonaqueous capillary electrophoresis (NACE). An example from the literature is shown below where the techniques of NACE and HPLC have been compared for the determination of herbicides and metabolites in water samples[33]. The target compounds were simazine (Sz), atrazine (Az), propazine (Pz), ametryn (Am), prometryn (Pm) and three metabolites deisopropylhydroxyatrazine (DIHA), deethylhydroxyatrazine (DEHA) and deethylatrazine (DEA). In both cases, an off-line solid phase extraction (SPE)

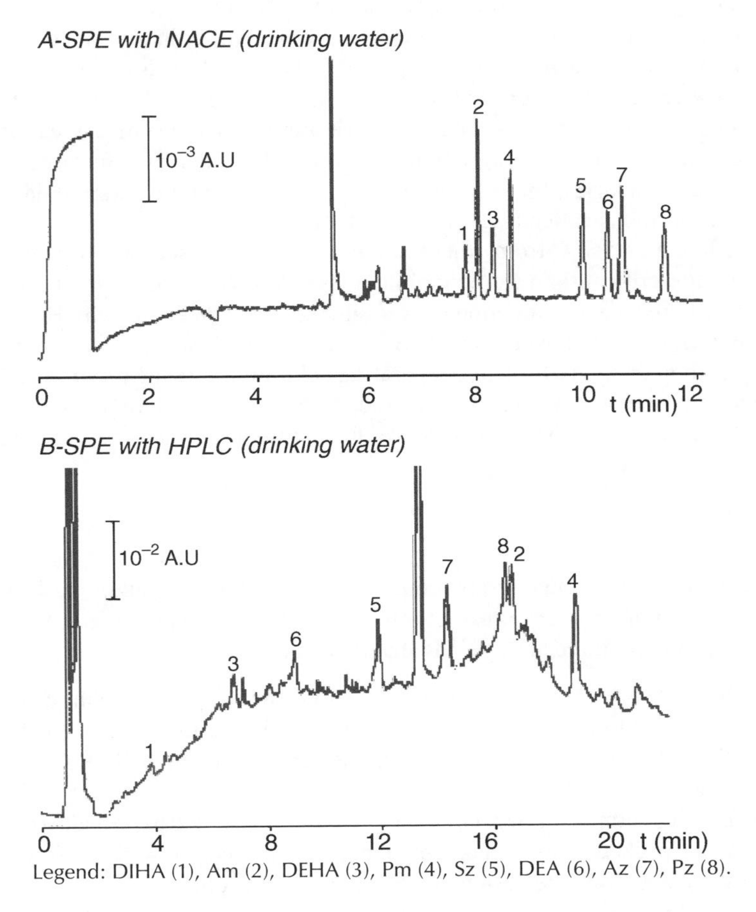

Figure 3.33 *Electropherogram (A) and chromatogram (B) corresponding to a fortified drinking water sample analysed with both SPE with NACE and SPE with HPLC methods (Five hundred milliliters aliquot of sample fortified at 0.50 µg/L for each triazine and polar metabolite. Reprinted from Journal of Chromatography A. Carabias-Martinez, R. et al., Comparison of a non-aqueous capillary electrophoresis method with high performance liquid chromatography for the determination of herbicides and metabolites in water samples,* **1122***(1–2), 194–201, Copyright 2006 with permission from Elsevier[33]).*

procedure and UV detection were used. The results obtained (Figure 3.33) showed that both methods gave the same results in the analysis of surface and drinking water samples. The detection limits were in the sub μg/L range for both methods.

3.5 Supercritical Fluid Chromatography

Principle

Supercritical fluid chromatography (SFC) is a hybrid technique of GC and HPLC that combines some of the best features of both methods. SFC is a relatively recent separation method, having only been commercially available since the 1980s.

For every substance there is a temperature above which it can no longer exist as a liquid, no matter how much pressure is applied. Likewise, there is a pressure above which the substance can no longer exist as a gas no matter how high the temperature is raised. These points are called the supercritical temperature (T_c) and supercritical pressure (P_c) respectively and are the defining boundaries on a phase diagram for a pure substance. Beyond these boundaries, the substance has properties that are intermediate between a liquid and a gas and is called a supercritical fluid.

What differentiates SFC from other separation techniques such as GC and HPLC is the use of this supercritical fluid as the mobile phase. Analytes that cannot be vaporised for analysis by GC and have no functional groups for sensitive detection with HPLC, can often be separated and detected using SFC. In SFC, the sample is carried by a supercritical fluid (typically carbon dioxide) through a separating column where the mixture is divided into unique bands based on the amount of interaction between the analytes and the stationary phase in the column. As these bands leave the column their identities and quantities are determined by a detector.

Instrument

A supercritical fluid chromatograph consists of a gas supply, usually carbon dioxide, a pump, the column in an oven, a restrictor to maintain the high pressure in the column and a detector. In general there are two possible hardware setups:

1. A GC-like setup with a syringe pump followed by a capillary column in a GC oven with a restrictor followed by a detector, where the pressure is controlled by the flow rate of the pump. A schematic of an SFC instrument with a GC-like setup is shown in Figure 3.34.
2. An HPLC-like setup with two reciprocating pumps designed to provide a mixed mobile phase with a packed analytical column placed in an oven followed by a detector, where the pressure and flow rates can be independently controlled.

Source

A gas supply and pump or flow regulator usually make up the source when a GC-like set-up is being used. The most common mobile phase for SFC is carbon dioxide; this is based on its low cost, low interference with chromatographic detectors and good physical properties. Other examples include nitrous oxide and ammonia. Supercritical fluids can

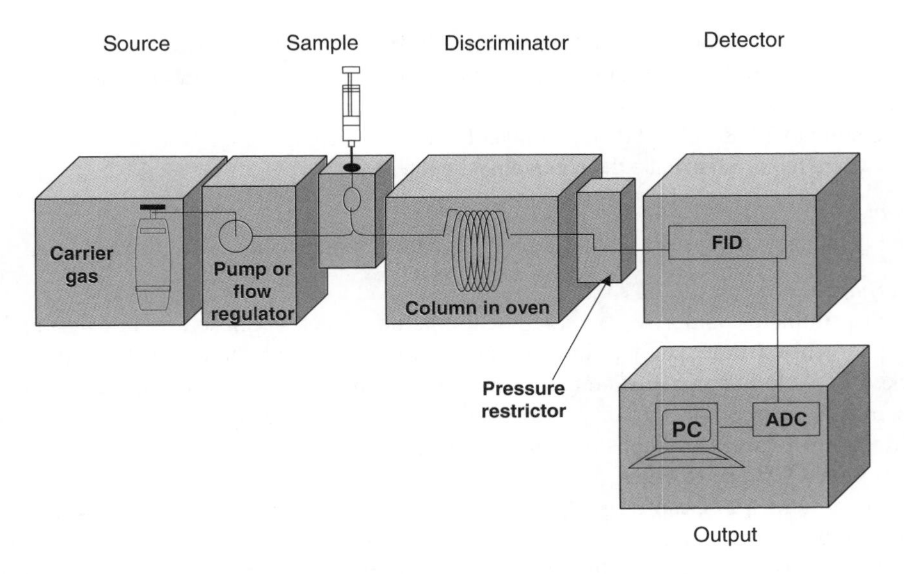

Figure 3.34 *Schematic of a supercritical fluid chromatograph.*

have solvating powers similar to organic solvents, but with higher diffusion coefficients, lower viscosity and lower surface tension. For nonpolar solutes on nonpolar stationary phases, separation can be achieved using pure carbon dioxide. Supercritical carbon dioxide has similar polarity to hexane so a disadvantage is its inability to elute very polar or ionic compounds. This can be overcome by adding a small portion of a second fluid called a modifier fluid. This is generally an organic solvent which is completely miscible with carbon dioxide (e.g. methanol or acetonitrile) but can be almost any liquid, even water. The addition of the modifier fluid improves the solvating ability of the supercritical fluid and sometimes enhances the selectivity of the separation. When polar modifiers are added to the supercritical fluid, then the separation can be considered to be similar to normal phase LC. Sometimes additives are also added, such as acids or bases.

Compared with LC, SFC can provide rapid separations without the use of organic solvents. SFC separations can be achieved faster than LC separations because the diffusion of solutes in supercritical fluids is about ten times greater than that in liquids (and about three times less than in gases). This results in a decrease in resistance to mass transfer in the column and allows for fast, efficient separations.

If the mobile phase is initially pumped as a liquid, it is made supercritical by heating it above its supercritical temperature before it enters the analytical column. In general, the type of high pressure pump used in SFC is determined by the column type. For packed columns, reciprocating pumps are used while for capillary columns, syringe pumps are used. The mobile phase passes through an injection valve where the sample is introduced into the supercritical stream and then into the analytical column. Part of the theory of separation in SFC is based on the fact that as the pressure is increased, the supercritical fluid density increases and correspondingly its solvating power is increased. Therefore, as the density of the supercritical fluid is increased, components retained in the column can be made to elute, which is similar to temperature programming in GC or using a solvent gradient in HPLC. The lower viscosity allows higher flow rates compared to HPLC. Gradients can be run in SFC and equilibration times are more rapid than HPLC. Common practice is to set the outlet pressure to 150–250 bar and column temperature to 30–80 °C and then to gradually increase the percentage of modifier and additives (where required) to elute the compounds of interest.

Sample
The injection system uses a valve like HPLC. Autosamplers can also be used.

Discriminator
The discriminator comprises the analytical column (which is housed in an oven) and a restrictor which immediately follows it. The column is usually a capillary GC column, but packed LC columns can also be used. Once the sample is injected into the supercritical stream it is carried into the analytical column. Different types of stationary phases are available with varying compositions and polarities. The ovens used in SFC are generally conventional GC or LC ovens.

Compared with GC, capillary SFC can provide high resolution chromatography at much lower temperatures. The ability to program the pressure provides a unique advantage over both GC and HPLC in that it allows the chromatographer to increase the solvating power of the solvent by increasing its density. The mobile phase is maintained supercritical as it passes through the column and into the detector by a pressure restrictor placed either after the detector or at the end of the column. The restrictor is a vital component as it keeps the mobile phase supercritical throughout the separation.

Detector
Supercritical fluid chromatography is compatible with both HPLC and GC detectors. As a result, optical detectors, flame detectors and spectroscopic detectors can be used. The FID is the most common detector used. However, the mobile phase composition, column type and flow rate must be taken into account when the detector is selected. Some care must also be taken such that the detector components are capable of withstanding the high pressures of SFC.

Output
A computer is used to process the data from the detector.

Information Obtained

Supercritical fluid chromatography can be used as a qualitative or quantitative technique. Qualitative information allows a substance to be identified on the basis of retention time comparisons between a standard and the sample. When used quantitatively, the SFC can employ external or internal standards to generate a standard curve. Then, the peak area of the sample peak on the chromatogram can be used to obtain a concentration. An example of an SFC instrument is shown in Figure 3.35.

Developments/Specialist Techniques

One important development of SFC is its use as an extraction technique – supercritical fluid extraction (SFE). SFE is widely applied in the food industry as the residual solvent

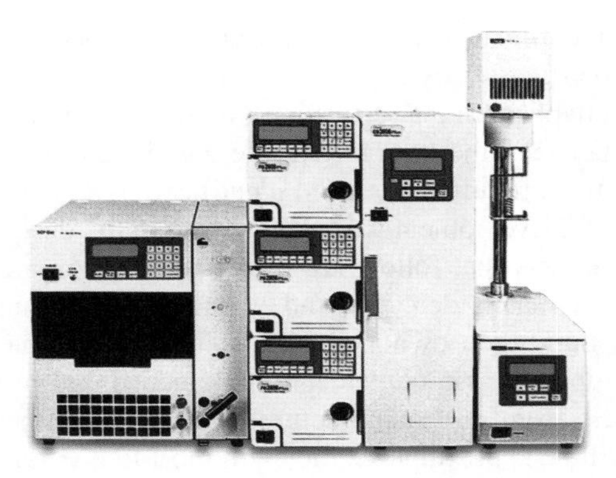

Figure 3.35 *A supercritical fluid chromatograph (Reproduced by kind permission of Jasco Corporation).*

can be easily removed after extraction. SFE is especially important in the decaffeination of tea and coffee. Other important areas are the extraction of essential oils and aroma materials from spices and for the extraction of hops in the brewing industry.

It is possible to buy a combined semi-preparative and preparative HPLC–SFC systems from Thar Technologies. Jasco Inc. has introduced a SFC–SFE–FTIR system.

Applications

Supercritical fluid chromatography is often used for compounds that are not ideally separated by either HPLC or GC. Some examples include the analysis of lipids and gasoline[34]. Also, where HPLC is proving too slow a technique, SFC can be and has been used for chiral separations where the narrow peaks and efficiency and speed offer advantages[35, 36]. Another big application for SFC is in preparative chromatography where the more environmentally friendly solvents make it cost efficient on a large scale. The tea and coffee industries rely heavily on SFE for commercial decaffeination processes. The extraction process for coffee involves forcing supercritical carbon dioxide through the green coffee beans. The gas-like behaviour allows it to penetrate deep into the beans where it dissolves 97–99 % of the caffeine present. The caffeine is recovered following removal from the carbon dioxide. The technique is also employed to produce nicotine-free tobacco and purify essential oils and flavours from herbs and other natural products.

3.6 Hyphenated (Hybrid) Instruments

Hyphenation can create more powerful instruments with better capabilities which yield more identification information on more complex samples in a faster time. The most important of these instruments are probably the GC–MS and the LC–MS. The first hyphenated technique was really LC with PDA, closely followed by GC–MS, which is almost routine now. GC and LC require the coupling of further analytical techniques for many reasons: due to coelution of compounds, for increased sensitivity or selectivity or where more definitive identification is required. Some detectors that follow LC or GC do not require perfect resolution as they can resolve the components of interest themselves.

The main issue with hyphenating a separating instrument with an identification instrument is the interface between the two. Often, the conditions for a good separation are at odds with the conditions required for good identification and detection by the second instrument. Hence, much development work has gone into the design of the interfaces so that some of the instruments that follow are now available commercially. In fact some companies offer 'mix and match' detectors and other modules of an instrument so that an analyst can effectively build their own analytical suite. This can be a very cost effective way of purchasing equipment as modules can be detached and re-attached as required with little downtime. If from one manufacturer, communication between the various pieces of equipment will be kept simple as they will most likely 'talk' to each other using the one software platform.

Applications for hyphenated techniques are wide ranging. GC–MS and LC–MS for example are extremely important in toxicology, doping control and metabolite identification owing to their high sensitivity and specificity[37].

3.6.1 Hyphenated Gas Chromatography Techniques

Gas chromatography (GC) has already been described and is an excellent separation technique for compounds that are volatile, thermally stable, generally of small molecular weight and nonpolar in nature. The common detectors for GC are the FID, NPD, ECD, TCD and FPD, all of which have been discussed in Section 3.1 (Figure 3.36). However, sometimes it is desirable to have a more powerful detector attached to a GC instrument and, as such, the following combinations are possible: GC–infrared spectrometry, GC–mass spectrometry and even GC–infrared–mass spectrometry. Further information on GC tandem techniques can be found elsewhere[38].

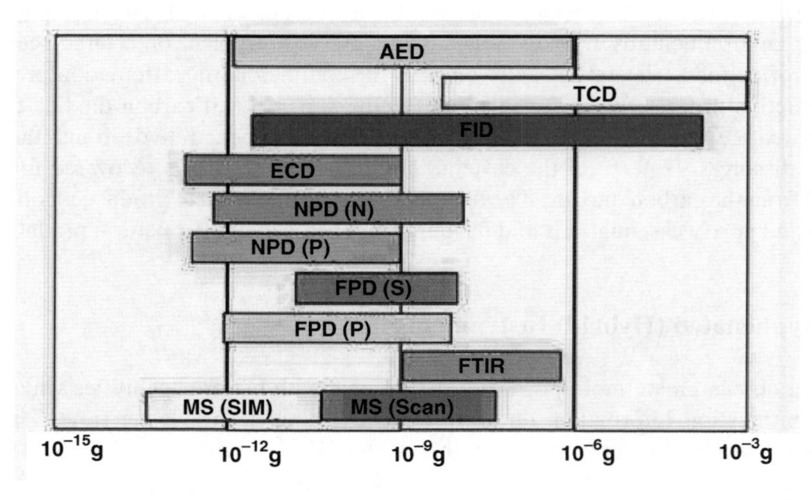

Figure 3.36 *The sensitivities and ranges of GC detectors (Figure used by kind permission of Jim Roe, Department of Chemistry, University of Washington, Seattle, USA).*

Gas Chromatography–Infrared Spectrometry

Gas chromatography–infrared spectrometry (GC–IR) combines the separating power of GC with the nondestructive identification capabilities of IR spectometry.

Interface issues:

- GC effluent is composed of gas phase compounds in carrier gases such as helium.
- GC compounds are usually volatile, thermally stable, of small molecular weight and fairly non-polar.
- IR spectrometry usually works best with solid samples.

Interface solutions:

1. Adapt the IR spectrometer to examine gases as they elute.
   One approach has been the use of a low-volume, internally reflecting, heated lightpipe (flow cell) interface between the gas chromatography and the infrared spectrometer (see Figure 3.37). The lightpipe has IR-transparent alkali halide windows at each end so the IR gas phase spectra of the eluent are recorded as they emerge from the column.
2. Convert the GC effluent to solids.
   The second approach uses a cryo-trapping interface where the column eluent is continuously deposited on a moving cooled plate. The condensed sample spots on the plate are brought under the IR beam for analysis in rotation. This method can be more sensitive than the lightpipe as it allows repeated samples to be collected, hence improving the signal-to-noise ratio.

How GC–IR works

The most common GC–IR system is that using the lightpipe interface. It begins with introduction of the sample into the GC inlet, where it is vaporised immediately and swept onto the column by the carrier gas (usually helium). The sample flows through the column

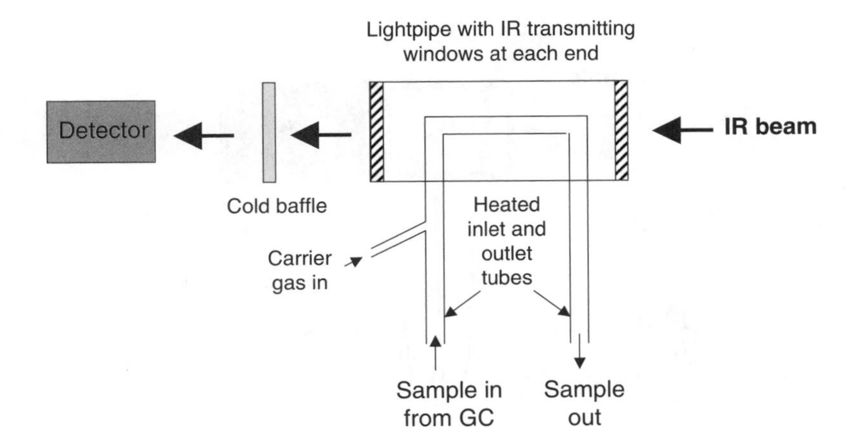

Figure 3.37 *Schematic diagram of a lightpipe interface between a gas chromatograph and an infrared spectrometer.*

experiencing the normal separation processes. As the various sample components emerge from the column opening, they are mixed with a carrier gas and flow into the lightpipe interface, where they are kept heated by an oven as they flow into and out of the lightpipe. The IR beam is focused onto the entrance to the lightpipe and then emerges at the exit where it is focused onto a cooled IR detector. A cold baffle between the exit and the detector ensures that there is no interference from any IR radiation from the hot oven. Fully integrated GC–IR instruments used to be available commercially but nowadays, the interface is generally purchased separately and the instrument is assembled in a modular fashion.

Gas Chromatography–Mass Spectrometry

Gas chromatography–mass spectrometry (GC–MS) combines the separating power of GC with the uniquely powerful detection capabilities of MS (Figure 3.38). With this coupled technique, it is possible to separate, identify and quantify components in a mixture. As this is quite an established hyphenated technique, there is a choice of mass spectrometric ionisation techniques and analysers available to couple with GC.

Interface issues:

- GC effluent is composed of gas phase compounds in carrier gases such as helium.
- The GC effluent is at atmospheric pressure.
- The MS is at high vacuum.
- MS may need to remove the carrier gas without removing analytes.

Interface solutions:

1. Use of a membrane.
   A semipermeable membrane which is selective for the analyte that travels on to the MS can be used at the interface. This is used in particular for packed column GC where the carrier gas flow rates are higher than those used with capillary columns.

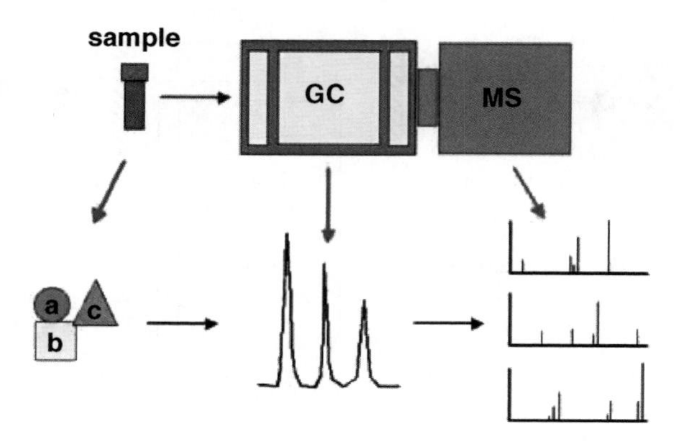

Figure 3.38 *The GC–MS process (Figure used by kind permission of Jim Roe, Department of Chemistry, University of Washington, Seattle, USA).*

2. Open split interface.
   Another option is an open split interface which sucks the helium out before the sample goes on to the ionisation chamber. These can concentrate the analytes which gives a sensitivity advantage.

3. Direct coupling interface.
   This is the most common method, where the GC capillary is inserted *directly* into the ion source via a length of fused capillary tubing and a vacuum tight flange. This works because the MS is already fairly compatible with the GC gaseous analytes at slow GC flow rates, i.e. up to 2 mL/min. In this case, the ionisation source is the interface and the only issue is the creation of a vacuum, which can be achieved by vacuum pump. Suitable ion sources are described below.

Suitable ionisation sources

- Electron impact
- Chemical ionisation.

These gas ionisation sources have been described in Section 2.1.5 on mass spectrometry. They are particularly suitable for the analysis of mixtures of volatile and low molecular weight compounds (<800) such as hydrocarbons, essential oils and relatively nonpolar drugs. Chemical derivatisation, e.g. trimethylsilylation, can often be employed to increase the volatility of compounds containing polar functional groups (–OH, –COOH, $-NH_2$ etc) so that GC–MS can be used.

Suitable mass analysers

- Quadrupole/triple quadrupole
- Ion trap
- Time-of-flight
- Magnetic sector.

These analysers have been described in Section 2.1.5 on mass spectrometry. They are suitable for the analysis of the types of mixtures normally separated by GC, e.g. volatile, nonpolar, low molecular weight compounds.

How GC–MS works

The sample solution is introduced into the GC inlet, is vaporised immediately because of the high temperature (~250 °C) and is swept onto the column by the carrier gas (usually helium). The sample flows through the column experiencing the normal separation processes. As the various sample components emerge from the column opening, they flow into the capillary column interface, which creates the vacuum, and then into the ion source. Once ionised, a small positive potential is used to repel the positive ions out of the ion source. The next component is the mass analyser (filter), which separates the positively charged particles according to their mass. After the ions are separated, they enter the mass detector and then go on to an amplifier to boost the signal. Fully integrated GC–MS systems are commercially available from a number of manufacturers (such as the system shown in Figure 3.39). Many of these have interchangeable ionisation sources and mass analysers.

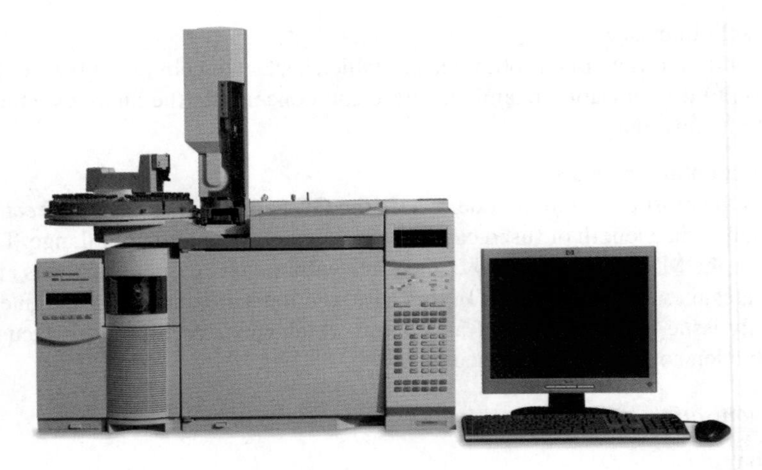

Figure 3.39 Agilent 5975 Inert MSD and Agilent 6890 GC with PC (© Copyright 2006, Agilent Technologies, Inc. Reproduced with permission).

Gas Chromatography–Infrared–Mass Spectrometry

Because GC–MS can have some difficulties in distinguishing between isomers or because structural information is required, some instruments have IR too (before the MS). This allows even compounds of the same mass to be separated, identified and quantified. In the technique of GC–IR–MS, the first detector is nondestructive so the sample passes into the MS unchanged. Connecting detectors in series in this way allows the analyst two ways of 'seeing' the molecules that are eluting. In a parallel arrangement, the flow from the GC is split such that most of it (up to 99 %) is directed into the FTIR detector and the balance into the MS. In a serial arrangement, make-up gas is added first for FTIR detection and afterwards the flow is split before entering the MS.

3.6.2 Hyphenated Liquid Chromatography Techniques

Liquid chromatography (LC) has already been described and is an excellent separation technique for compounds that are nonvolatile, thermally unstable and relatively polar in nature. The usual detectors for LC are based on refractive index, conductivity, amperometry, light scattering, UV and fluorescence, all of which have been discussed in Section 3.2. However, sometimes it is desirable to have a more powerful detector attached to an LC instrument and, as such, the following combinations are possible: LC–infrared spectrometry, LC–atomic spectrometry, LC–inductively coupled plasma–mass spectrometry, LC–mass spectrometry, LC–UV–mass spectrometry, LC–nuclear magnetic resonance and even LC–nuclear magnetic resonance–mass spectrometry.

Liquid Chromatography–Infrared Spectrometry

Liquid chromatography–infrared spectrometry (LC–IR) combines the separating power of LC with the identification and quantitative capabilities of FTIR.

Interface issues:

- LC effluent is often strongly absorbing in the IR region.
- LC compounds are usually non-volatile and thermally unstable (opposite to GC).
- IR usually works best with solids.

Interface solutions:

1. Adapt the IR to examine liquids as they elute.
   In a similar manner to the lightpipe in GC–IR, a flow cell interface can be used where the eluent flows into a transmission flow cell with IR windows at each end and the IR beam passes through the cell perpendicular to the flow and continuously records IR spectra. This only works if the solvents do not absorb so strongly in the IR region of interest and mask the analyte transmission/absorption, or if the mobile phase absorption spectrum is subtracted (which only works well when the LC method is isocratic). In fact, the solvents used in normal phase HPLC are more compatible with IR than the more polar ones used in reversed phase HPLC.
2. Convert the LC effluent to solids.
   The solvent elimination approach is better than the one above as full spectral information from the analytes can be obtained and the chromatographic conditions do not have to be modified as much. In practice, the eluting compounds are deposited onto potassium bromide pellets, the mobile phase is evaporated and the pellets are then transferred to the FTIR for spectral analysis by diffuse reflectance. Hence the solvents used in the mobile phase must be more volatile than the analytes being studied. A microdispenser for interfacing LC to IR or Raman which uses the solvent elimination approach has been reported[39].
3. Convert the LC effluent to solids on a position-encoded rotating disc.
   Another type of interface is the system commercially available from Laboratory Connections, which is similar to that above. The sample is deposited as it elutes (sprayed from a nebuliser) onto a position-encoded, slowly rotating germanium disc. The disc is transferred to a reflectance sampling stage where it is rotated at the same rate as before in order to reconstruct the chromatogram, all the while collecting spectra. Potassium bromide discs cannot be used in reversed phase HPLC so zinc–selenium or calcium fluoride can be employed instead.

How LC–IR works
Based on the Laboratory Connections interface, the sample is injected into the LC from where it flows through the column and experiences the normal separation processes. As the various sample components emerge from the column opening, they are automatically sprayed onto the position-encoded, rotating germanium disc. Each sample is collected sequentially and its IR spectrum obtained in that same sequence. LC–IR instruments are generally configured according to the analyst's requirements using separate LC, interface and IR modules.

Liquid Chromatography–Atomic Spectrometry

Liquid chromatography–atomic absorption spectrometry (LC–AAS) and liquid chromatography–optical emission spectrometry (LC–OES) combine the separating power of LC with the identification and quantitative capabilities of atomic spectrometry techniques. They are of particular importance in speciation studies. To improve the sensitivity of LC–AAS, hydride generation is often employed, i.e. LC–HGAAS.

Interface issues:

- LC effluent is a liquid phase.
- LC compounds are usually nonvolatile and thermally unstable.
- Samples for AAS and OES must be thermally stable at the high temperatures employed and must be in the gas phase for analysis.
- HGAAS requires samples that can be transformed into a hydride gas by chemical reaction in an atomiser. Sodium tetrahydroborate is often used as the reducing agent.

Interface solutions:

1. Samples are nebulised/atomised as they elute from the LC.
   The LC eluent flows into a nebuliser which converts the liquid into an aerosol. For HGAAS, each element of the sample must also be atomised and converted into a hydride.

How LC–HGAAS works
One method employed for analysis of metals is where the sample solution is introduced into the LC and carried by mobile phase through the column. The sample experiences ion-exchange separation processes. As the various metals emerge from the column opening, they are mixed on-line with the required hydride generating reagents, and from there they are transferred into the atomiser. Various tin and arsenic species have been analysed in this way using a HPLC connected to a continuous hydride generation system connected to an electrothermal atomiser[40], a schematic of which is shown in Figure 3.40. These tandem analytical systems are not available commercially and are generally purpose built for the particular application.

Liquid Chromatography–Inductively Coupled Plasma–Mass Spectrometry

Because the toxicity of some elements (especially organometallics) can depend on the species in which they are present rather than on their total concentration, the individual species may require chromatographic separation followed by ICP–MS detection. Hence, LC–ICP–MS has evolved as an important analytical technique for speciation analysis. Sometimes a UV detector is included in the system so that the spectra from both detectors can be overlaid for extra analytical information.

Liquid Chromatography–Mass Spectrometry

Liquid chromatography–mass spectrometry (LC–MS) combines the separating power of LC with the uniquely powerful detection capabilities of MS. With this technique, it is now possible to separate, identify and quantify components in a mixture. As with GC–MS, there is a choice of mass spectrometric ionisation techniques and analysers available. LC–MS can also be taken a step further as LC–MS^n (otherwise known as LC–tandem MS) where ions can be subjected to fragmentation so as to gain information on their structures.

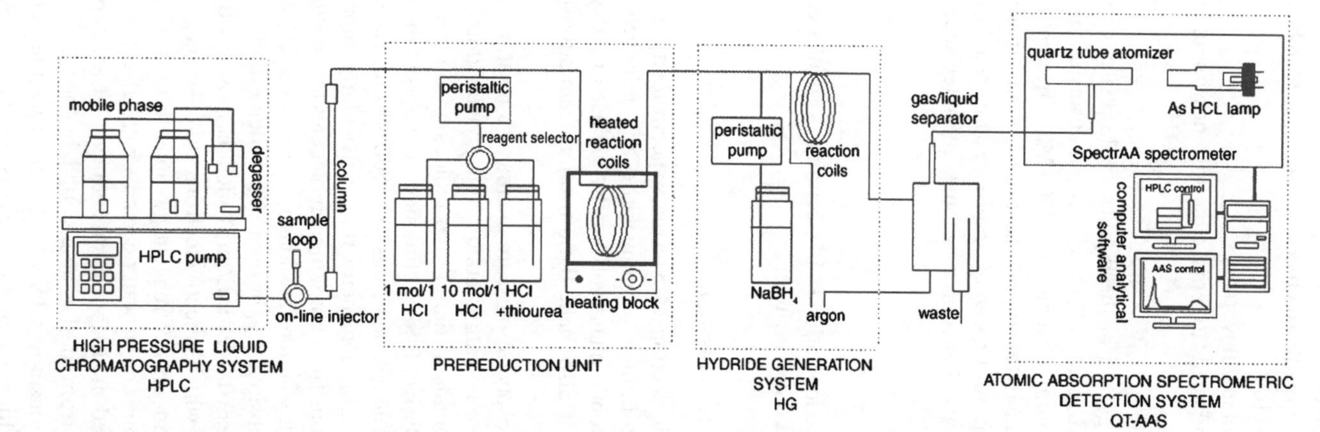

Figure 3.40 *Schematic of an LC–HGAAS system (Reprinted from Analytica Chimica Acta, Niedzielski, P. 'The new concept of hyphenated analytical system: Simultaneous determination of inorganic arsenic(III), arsenic(V), selenium(IV) and selenium(VI) by high performance liquid chromatography–hydride generation–(fast sequential) atomic absorption spectrometry during single analysis',* **551***(1–2), 199–206*[125]*. Copyright 2005 with permission from Elsevier).*

Interface issues:

- LC effluent is liquid phase and at atmospheric pressure.
- LC compounds are usually nonvolatile and thermally unstable (can be difficult to ionise).
- LC flow rates are typically high.
- MS works best with gases and low flow rates.
- The MS is at high vacuum and needs to remove the solvent without removing analytes.

Interface solutions:
The interface must be able to get rid of the liquid mobile phase, convert the relatively involatile and/or thermally labile analytes into a 'gas' and transfer the 'gas' from atmospheric conditions to a high vacuum. Compared to GC–MS, different ionisation methods must be used for these kinds of analytes (liquid phase, nonvolatile, thermolabile) as EI and CI are not suitable. However, in most cases the capillary is inserted *directly* into the ion source and, in this case, the ionisation source becomes the interface; suitable ionisation sources are described below.

Suitable ionisation sources

The use of atmospheric pressure ionisation (API) techniques enables conversion of liquid phase analytes to the gas phase. They include:

- Electrospray ionisation
- Atmospheric pressure chemical ionisation

These sources are described in Section 2.1.5 on mass spectrometry. They are particularly suitable for the analysis of mixtures of nonvolatile and relatively polar molecules. Additionally, they get rid of mobile phase solvents by the use of a drying gas and some heat. These ionisation chambers also make the transition from atmospheric to very low pressure an easy one.

Electrospray (ESI) and atmospheric pressure chemical ionisation (APCI) offer 'soft' ionisation that results in little or no fragmentation. They yield simple spectra where the principal peak usually corresponds to the pseudo-molecular ion (+H or +Na), making the molecular weight easy to derive. Both techniques can be used in negative or positive mode. ESI can only accommodate low flow rates (0.5–500 mL/min) so narrow or capillary HPLC columns should be used. Solution chemistry is particularly important in ESI as this technique does not create gas ions from the sample molecules to the same extent as APCI. ESI simply converts the ions in solution into ions in the aerosol. Highly aqueous phases or those with very nonpolar solvents, e.g. hexane, are difficult to electrospray. APCI can accommodate higher flow rates (0.1–2.0 mL/min) such as those used in regular HPLC. The corona discharge needle which injects electrons into the 'gas' can be positioned to optimise proportions and amounts of reagent gas ions formed. Certain solvents, e.g. acetonitrile, and a high concentration of water can adversely affect APCI ionisation. It is now possible to purchase mixed-mode ionisation sources that are capable of generating both an ESI and an APCI spectrum for the same sample in one run.

A third API source is the ionspray interface. It combines the best features of both ESI and APCI. It can accept higher flow rates and is therefore sometimes called high flow electrospray. This source can tolerate highly aqueous mobile phases and gradient elution.

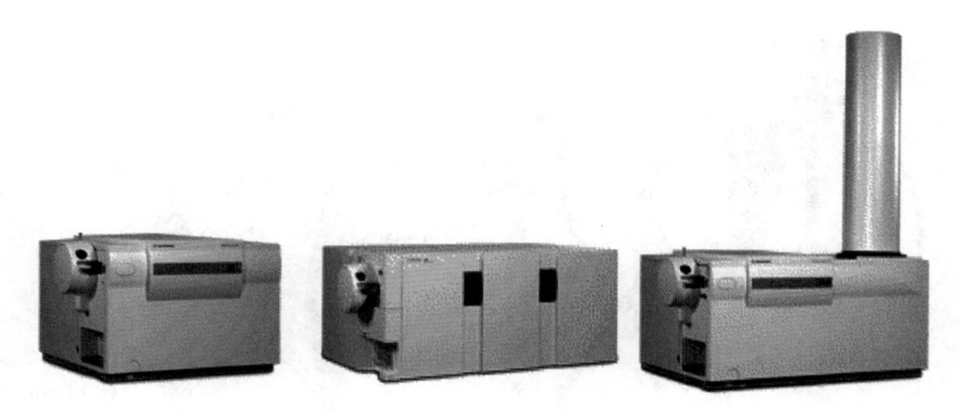

Figure 3.41 *Agilent 1200 Series LCMS selection of detectors – ion trap, triple quadrupole and TOF with multimode source (© Copyright 2006, Agilent Technologies, Inc. Reproduced with permission).*

Suitable mass analysers

- Quadrupole/triple quadrupole
- Ion trap
- Time-of-flight.

These analysers are the same as those that suit GC–MS and have been discussed in Section 2.1.5 on mass spectrometry. All are capable of quantitative analysis and all can be interfaced easily with HPLC. Many manufacturers have a large selection of analysers that can be chosen when buying the instrument and can be easily changed for another if required (Figure 3.41). They can be used in the total ion current (TIC) mode where all ions are detected and recorded graphically with respect to time. The MS can also be used in selected ion monitoring (SIM) mode where only the m/z values of interest are selected by the MS or multiple reaction monitoring (MRM) mode where a number of peaks and their fragments are detected. These modes increase the S/N ratio and hence improves sensitivity.

LC–tandem MS enables the study of fragmentation patterns and hence elucidation of structures known or unknown. A triple quadrupole can carry out MS^2. Ion trap can carry out MS^n, though with each successive stage of isolation and fragmentation, some sensitivity is lost. Using an ion trap analyser (Figure 3.42), ions of one m/z or a range of m/z values can be collected and stored in the trap. The packets of ions are contracted, allowing a greater number to be stored and enabling faster, more efficient ejection. This improves resolution and sensitivity. For tandem MS, all but the ions of interest are ejected from the trap; this is the isolation step. The energy of the trapped ions is increased and helium is introduced. Collisions with the helium cause fragmentation. The fragments are stored, scanned out of the trap according to their m/z and detected. Fragmented ions can be stored and subjected to further rounds of MS. MS^n is very useful for characterising small molecules and for sequencing peptides and proteins. In proteomics for example, the tryptic digest peptides from a complex protein are drawn into the MS as they elute from an LC column and can be analysed by MS and MS–MS.

Hybrid MS systems were mentioned in Section 2.1.5 but there are now also hybrid LC–MS systems appearing on the market. For example, Shimadzu Scientific Instruments has

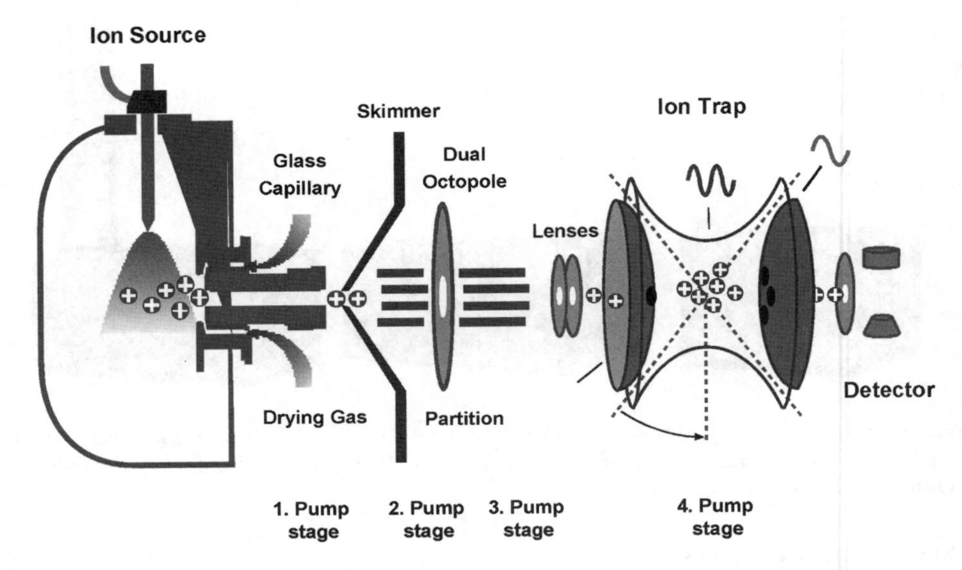

Figure 3.42 Schematic of ion trap MS system with orthogonal ionisation source (ESI or APCI) attached (Courtesy of Bruker Daltonics).

introduced the LCMS–IT–TOF instrument that couples atmospheric pressure ionisation with both ion trap (IT) and time-of-flight (TOF) mass analyser technologies. The ion trap is not only used to focus ions before ejection into the TOF, but it also supports fragmentation. Applied Biosystems/MDS Sciex has launched the 3200 Q Trap LC–MS–MS which combines a triple quadrupole with a linear ion trap for greater sensitivities in either mode.

How LC–MS works

The sample solution is introduced into the LC and carried by the mobile phase through the column. The sample experiences the normal separation processes. As the various sample components emerge from the column opening, they flow through the ionisation interface where the analytes are converted into a gaseous aerosol and ionised, and from there they move into the mass analyser. The most frequently used ionisation methods are electrospray ionisation (ESI) or atmospheric pressure chemical ionisation (APCI), as these convert liquids to gases for MS detection. Once ionised, the ions move to the mass analyser (filter), which separates the charged particles according to their mass. After the ions are separated according to their masses, they enter the mass detector and then an amplifier to boost the signal. Fully integrated LC–MS systems are available commercially from a number of manufacturers. As with GC–MS instruments, many of these have interchangeable ionisation sources and mass analysers. Much progress in interfaces and mass analysers for LC–MS instruments is being made[41].

In proteomics, the technique can be used to investigate the products of enzyme–peptide reactions. These can be cleavage products[42] or newly formed products[43]. An example of an application for LC–MS is shown in Figure 3.43, where the LC effects the separation of a mixture and then the mass spectrum for each peak can be obtained and displayed to confirm identity of the breakdown products. A peptide (Z–gly–pro–tyr–OH) was mixed with an enzyme and, following incubation the mixture was subjected to LC–MS

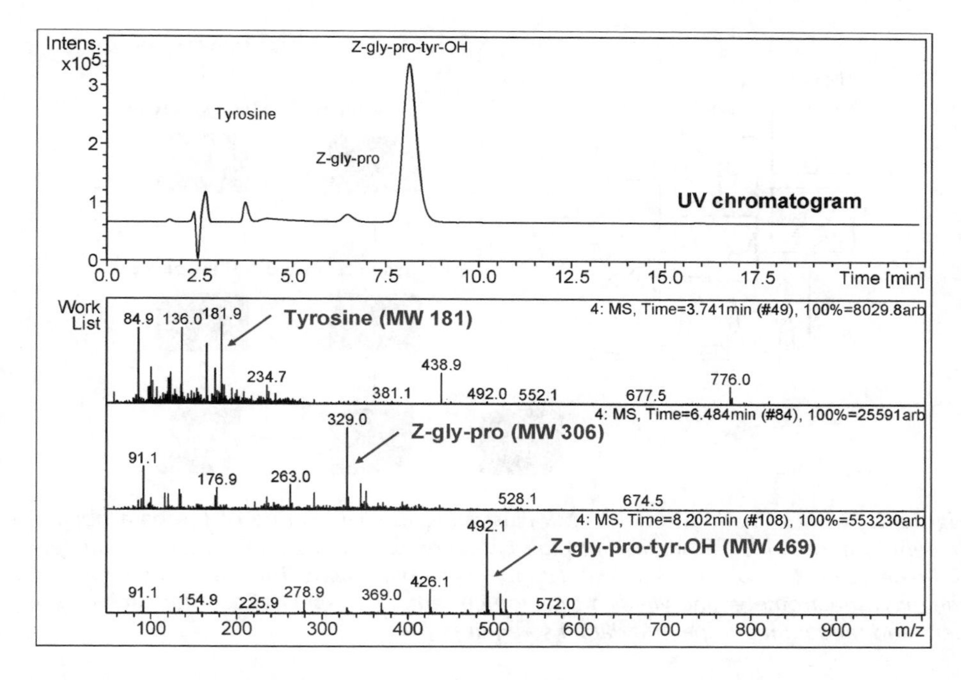

Figure 3.43 *Results obtained for an enzyme-peptide reaction using the LC–MS technique.*

analysis. It can be easily seen how the mass spectra give the identity of each of the peaks, confirming that cleavage of the peptide into fragments of Z–gly–pro and tyrosine took place during incubation.

Liquid Chromatography–UV–Mass Spectrometry

Because LC–MS can have some difficulty in distinguishing between isomers, some instruments have UV detection too (before the MS). This allows compounds of the same mass to be separated, identified and quantified. In LC–UV–MS, the UV detector is nondestructive so the sample passes into the MS unchanged. Connecting detectors in series in this way allows the analyst two ways of 'seeing' the molecules that are eluting. The UV detector 'sees' any chromophores and delivers a UV chromatogram. The MS detects any ions and delivers a total ion current trace. These two traces can be superimposed and it can be immediately seen which compounds respond best to which detector. Sometimes, a molecule is only 'seen' by one detector and not the other. As well as the traces obtained above, further useful information can be obtained from the two detectors in series. The MS can yield a mass spectrum for any compound of interest and the UV detector (if it is a PDA) a UV spectrum for any compound of interest. This is illustrated in Figure 3.44.

Having the two detectors in series can prove invaluable in determining the presence and quantities of compounds in a mixture, especially when they are closely related in structure[44]. For example, calix[4]arenes are useful sensor molecules and novel derivatives of these are made to improve selectivity for ions[45]. Calix[4]arene analysis by HPLC is typically challenging since the molecules are very non-polar; even more difficult is the separation of isomers of the calix[4]arenes. In one experiment to synthesise a diester

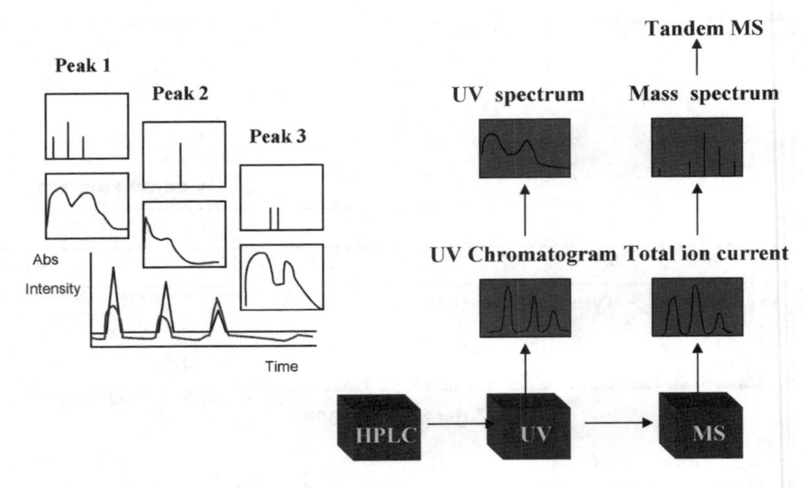

Figure 3.44 Schematic of LC–UV–MS illustrating the 3-D nature of the data obtained (Eluent from the HPLC flows first into the UV detector yielding a UV chromatogram and, if the detector is a PDA, UV spectral data for each component. The eluent then flows into the mass spectrometer and yields a total ion current, mass spectral data for each component and tandem mass spectral data if so equipped.).

substituted calix[4]arene, it was found that under certain conditions a mixture of the diester product, monoester product and the starting material was formed. In fact, there were even two isomers of the diester product created: a 1,2 and a 1,3 substituted compound: The UV chromatogram and the single ion current (SIC) for each ion are shown in Figure 3.45.

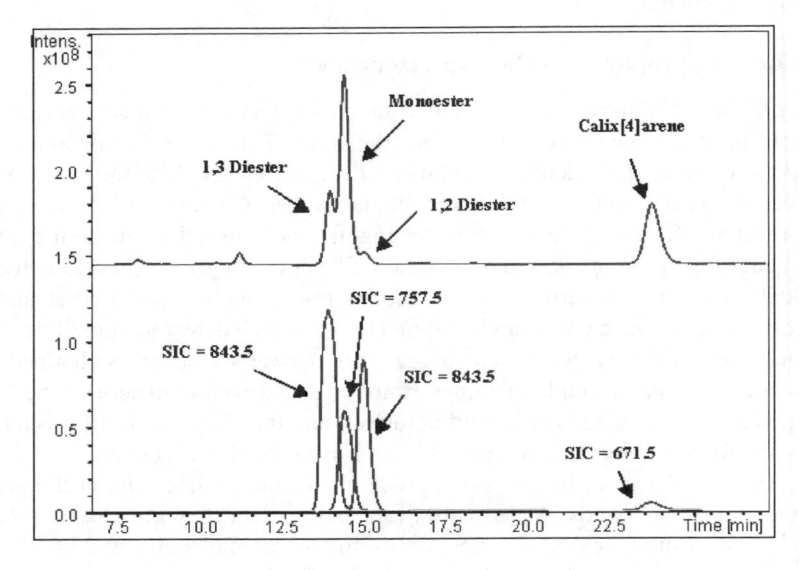

Figure 3.45 A UV chromatogram and the single ion currents (SIC) for a mixture of closely related calix[4]arene compounds.

There are a few very interesting features to note. Firstly, if the mass spectrometer was the only detector coupled to the LC, the fact that there were two isomers of m/z 843.5 present may not have been picked up. Secondly, if it was necessary to rely on the ion currents for quantitation, the concentration of the diester would have been overestimated since it ionises much more easily than the monoester. If the UV was the only detector coupled to the LC, it would not have been known that the peaks either side of the monoester were isomers of the same diester compound. With information from both detectors, it was possible to show definitively that there were four separate compounds in the mixture and, since the UV spectra of the related compounds are so similar, quantitative measurement was carried out based on the UV chromatogram and identification carried out based on the MS.

Liquid Chromatography–Nuclear Magnetic Resonance

Liquid chromatography–nuclear magnetic resonance (LC–NMR) combines the separating power of LC with the structural elucidation capabilities of NMR. It is now possible to separate and elucidate the structures of components in a mixture on-line using this technique. It is most often used to identify natural products, novel compounds and metabolites.

Interface issues:

- LC solvents can give strong signals in the NMR.
- LC is a flowing system.
- NMR is a relatively insensitive technique.
- NMR works best with deuterated solvents but these are too expensive to use for LC separations.
- NMR experiments normally take place in a static mode.

Interface solutions:

1. Use flowing mode with background subtraction.
   The NMR needs to remove the solvent without removing analytes or needs to suppress the solvent signals to improve sensitivity of the weak NMR signals. Solvent suppression is even more challenging in gradient elution, where the resonance frequencies change with the solvent composition. A recent system that works well for solvent suppression uses an NMR difference probe. The technique is based on dual beam background subtraction where the reference and sample signals are collected simultaneously and subtracted from each other automatically.
2. Use stopped flow analysis.
   This is where the LC pump is stopped, and the sample in the NMR cell collected and subjected to static NMR measurements. This enrichment step improves sensitivity (S/N ratio) since the eluted component can be repeatedly scanned or a number of fractions can be combined or preconcentrated to increase sample size/concentration.

How LC–NMR works

HPLC is connected on-line to a 'flow' NMR instrument. The sample mixture is separated on the HPLC column into its components and each flows into the NMR probe/cell.

These cells are usually millimetres in diameter and have a volume of 50–500 μL. An NMR spectrum is carried out on each peak (component). There are a number of modes of operation including continuous flow, stopped flow, peak parking and peak trapping. In continuous mode, the flow is sampled in real time as it moves through the detection coil and all components are sampled. In stopped flow, the pump is stopped and NMR experiments carried out on a static sample. Previous data such as UV spectra might be used to signal when to stop the flow. In peak parking, peaks of interest are held in off-line sample loops. In peak trapping, solid phase extraction cartridges are used to extract/concentrate samples.

Liquid Chromatography–Nuclear Magnetic Resonance–Mass Spectrometry

While MS gives some structural information on the basis of molecular weight and principal fragments, NMR yields much more detailed information on structure in terms of the environment of the atoms and on stereochemistry. MS is a much more sensitive

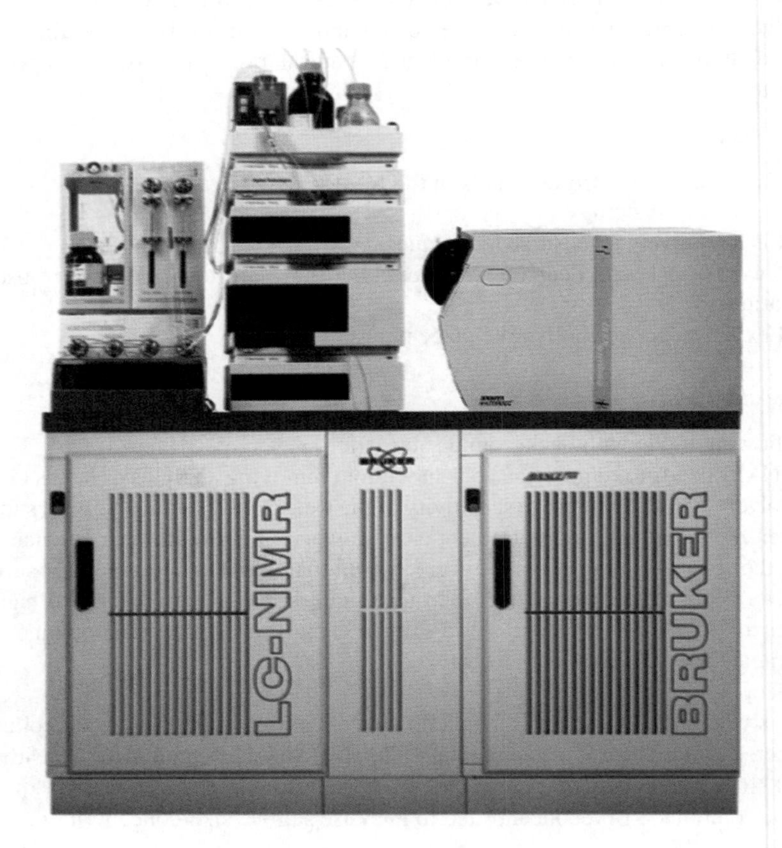

Figure 3.46 *An LC–NMR–MS system showing the LC, the MS, the NMR–MS interface and the NMR itself (The magnet is not shown. Image courtesy of Bruker BioSpin Limited).*

method (low ng or less) while NMR is relatively insensitive (μg in flowing mode). In terms of the range of possible experiments, NMR has advantages over MS with the number of nuclides that can be investigated and 2-D studies that can be carried out in stopped flow mode, while in MS there is a more a limited range of fragmentation and collision studies that can be done. However, by combining the two techniques, their complementary data give the 'best of both worlds' and allow more definitive assignment of structures, especially to novel molecules and for compounds where complex rearrangements can occur on standing. However, this hyphenated technique is still extremely expensive to install and is mainly a research tool. An example of an LC–NMR–MS instrument is shown in Figure 3.46.

Interface issues
These are the same as for LC–MS and LC–NMR instruments, but combined. There are also some extra issues such as the fact that the magnet can also have a negative affect on the MS data if it is not adequately shielded and solvent incompatibilities between the techniques. Solvents need to be carefully chosen so that they do not appreciable increase the background in the NMR spectra. For example, deuterated solvents will reduce the background NMR signals but the knock-on effect of this is the formation of $[M+D]^+$ ions instead of $[M+H]^+$ ions in the mass spectrum. Buffers should be volatile for MS detection and contribute little or no signal in the NMR spectrum.

Interface solutions
There are a few ways of linking the techniques of LC–MS and LC–NMR. The most common method is in a parallel mode by splitting the flow, e.g. 50:1, so as to direct the majority of it to the NMR due to its relative insensitivity. This means that the analytes are detected simultaneously by both detectors and possibly also by UV, which may actually be used as the trigger to begin detection by the NMR and MS modules. Alternatively, the rapidly acquired MS data can be used to direct the NMR experiments or vice versa. A second method of interfacing the two techniques is to use the serial mode or stopped flow mode, which enables more sensitive NMR experiments to be carried out. A recent development in stopped flow NMR is the inclusion of in-line solid phase extraction (SPE) after the LC. The SPE acts as a fraction collector for individual compounds. This trapping/washing step can improve sensitivity several fold. A third method is fraction collection, where samples from the LC are collected in a loop for analysis later, perhaps after certain data have been reviewed.

The LC and MS are kept close to the NMR to minimise diffusion and band broadening but, at the same time, the MS needs to be far enough away from the NMR so as to minimise unusual magnetic effects. The actual distance depends on the magnet – if it is shielded, they can be closer. Solvents used in LC must be compatible with both MS and NMR. This can prove challenging. For example, acetonitrile in D_2O that is pH-adjusted with sodium phosphate is often used for NMR as does not mask analyte signals. However, phosphate is not compatible with MS. Formic acid is suitable for MS but does give a signal in NMR, fortunately it is at 9 ppm and away from most analytes. Deuterated solvents can be used to eliminate NMR signals altogether but these solvents can cause H/D exchange in MS thereby yielding false molecular masses.

3.6.3 Hyphenated Capillary Electrophoresis Techniques

Capillary electrophoresis (CE) has already been described and is an excellent separation technique for compounds that are nonvolatile, thermally unstable, large in molecular weight and polar. The usual detectors for CE are the same as those for LC, all of which have been discussed. However, sometimes it is desirable to have a more powerful detector attached to a CE instrument and one that is increasing in importance is CE–MS.

Capillary Electrophoresis–Mass Spectrometry

Capillary electrophoresis–mass spectrometry (CE–MS) combines the separating power of CE with the uniquely powerful detection capabilities of MS. This hyphenated system enables the analyst to separate, identify and quantify components in a mixture. As with GC–MS and LC–MS instruments, there is a variety of mass spectrometric ionisation techniques and analysers available.

Interface issues:

- CE effluent is a liquid buffered phase and at atmospheric pressure.
- CE compounds are usually polar, nonvolatile and thermally unstable.
- MS works best with gases.
- The MS is at high vacuum and needs to remove the buffer without removing analytes.
- Every compound has mass and therefore, as long as it can be ionised, will respond to a MS detector.

Interface solutions:
The interface must be able to get rid of the liquid aqueous buffer, convert the relatively involatile and/or thermally labile analytes into a 'gas' and transfer the 'gas' from atmospheric conditions to a high vacuum. In most cases, the capillary is inserted *directly* into the ion source and in this case the ionisation source is the interface; suitable ones are described below. Different ionisation methods must be used for these kinds of analytes (polar, non-volatile, thermolabile) as EI and CI, which are used in the GC–MS instrument interface, are not suitable here. Those used for LC–MS are often compatible with CE–MS.

Suitable Ionisation Sources

As with LC–MS, the use of atmospheric pressure ionisation (API) techniques enable conversion of liquid phase analytes to the gas phase. They include:

- Electrospray ionisation
- Atmospheric pressure chemical ionisation.

These sources are particularly suitable for the analysis of mixtures of nonvolatile and relatively polar molecules. They get rid of aqueous buffered solvents by using a drying gas and some heat. These ionisation chambers also make the transition from atmospheric to very low pressure an easy one. ESI and APCI offer 'soft' ionisation, resulting in little or no fragmentation. They yield simple spectra where the principal peak usually corresponds to the pseudo-molecular ion (+H or +Na), making the molecular weight easy to derive. Both can be used in negative or positive mode.

Suitable mass analysers

- Quadrupole/triple quadrupole
- Ion trap
- Time-of-flight.

These analysers are the same as those that suit GC–MS and LC–MS instruments. Tandem MS can also be carried out using suitable mass analysers such as the triple quadrupole (MS^2) and ion trap (MS^n).

How CE–MS works
The sample solution is introduced into the CE and carried by buffer through the capillary, which is under the influence of an electric field. The sample experiences the normal electrophoretic separation processes. As the various sample components emerge from the capillary opening, they flow through the interface, where the liquid eluent is converted into a gaseous aerosol, and then into the ion source. The most frequently used ionisation methods are electrospray ionisation (ESI) or atmospheric pressure chemical ionisation (APCI) as these convert liquids to gases for MS detection. This part of the process is much easier than with LC eluent as the flow rates are slower and the analytes are already ions. The ions then move to the mass analyser (filter), which separates the charged particles according to their mass. After the ions are separated according to their masses, they enter firstly the mass detector and then an amplifier to boost the signal before being collated at the computer. CE–MS is particularly suitable for proteins[46, 47], peptides[48] and glycoproteins[49, 50]. An example of a CE–MS instrument is shown in Figure 3.47.

Figure 3.47 *Agilent CE–MS System (The CE can be interfaced with any of the Agilent MS or ICP–MS detectors. © Copyright 2006, Agilent Technologies, Inc. Reproduced with permission).*

References

1. Gas Chromatography from on-line Chrom-Ed Series by Scott RPW www.chromatography-online.org/3/contents.html
2. Gas Chromatography Detectors from on-line Chrom-Ed Series by Scott RPW www.chromatography-online.org/2/contents.html
3. Novel applications and new developments in gas chromatography. Patel-Predd, P. LC-GC Meeting Report from Pittcon 2006, March 2006.
4. Zhu, J.Y., Chai, X.S. (2005) Some recent developments in headspace gas chromatography. *Current Anal Chem*, **1**, 79–83.
5. Reddy, C.M., Eglinton, T.I., Hounshell, A. *et al.* (2002) The West Falmouth oil spill after thirty years: The persistence of petroleum hydrocarbons in marsh sediments. *Environ Sci Technol*, **36 (22)**, 4754–4760.
6. Zhang, T.Z., Chen, X.M., Liang, P. and Liu, C.X. (2006) Determination of phenolic compounds in wastewater by liquid phase microextraction coupled with gas chromatography. *J Chromatogr Sci*, **44 (10)**, 619–624.
7. Martinez, K., Abad, E. and Rivera, J. (2006) Surveillance programme on dioxin levels in soils in the Campo de Gibraltar (southwest Spain). *Chemosphere*, **65 (3)**, 382–389.
8. Aubert, C. and Pitrat, M. (2006) Volatile compounds in the skin and pulp of Queen Anne's pocket melon. *J Ag Food Chem*, **54 (21)**, 8177–8182.
9. Uematsu, Y., Hirata, K., Suzuki, K. *et al.* (2002) Survey of residual solvents in natural food additives by standard addition head-space GC. *Food Add Contam*, **19 (4)**, 335–342.
10. Tedetti, M., Kawamura, K., Charriere, B. *et al.* (2006) Determination of low molecular weight dicarboxylic and ketocarboxylic acids in seawater samples. *Anal Chem*, **78 (17)**, 6012–6018.
11. Liquid Chromatography from on-line Chrom-Ed Series by Scott RPW www.chromatography-online.org/4/contents.html
12. Liquid Chromatography Detectors from on-line Chrom-Ed Series by Scott RPW www.chromatography-online.org/5/contents.html
13. McMahon, G.P. and Kelly, M.T. (1998) Determination of aspirin and salicylic acid in human plasma by column-switching liquid chromatography using on-line solid-phase extraction. *Anal Chem*, **70 (2)**, 409–414.
14. McMahon, G. (2000) Design of a column-switching LC system with on-line solid phase extraction for the determination of polar drugs. *Am Biotech Lab*, **18 (3)**, 52–4.
15. Delgado-Andrade, C., Rufian-Henares, J.A., Jimenez-Perez, S. and Morales, F.J. (2006) Tryptophan determination in milk-based ingredients and dried sport supplements by liquid chromatography with fluorescence detection. *Food Chem*, **98 (3)**, 580–585.
16. Vittal, S., Shitut, N.R., Kumar, T.R. *et al.* (2006) Simultaneous quantitation of rosuvastatin and gemfibrozil in human plasma by high-performance liquid chromatography and its application to a pharmacokinetic study. *Biomed Chrom*, **20 (11)**, 1252–1259.
17. Venkatesh, P., Harisudhan, T., Choudhury, H. *et al.* (2006) Simultaneous estimation of six anti-diabetic drugs-glibenclamide, gliclazide, glipizide, pioglitazone, repaglinide and rosiglitazone: development of a novel HPLC method for use in the analysis of pharmaceutical formulations and its application to human plasma assay. *Biomed Chrom*, **20 (10)**, 1043–1048.
18. Hvattum, E., Uran, S., Sandbaek, A.G. *et al.* (2006) Quantification of phosphatidylserine, phosphatidic acid and free fatty acids in an ultrasound contrast agent by normal-phase high-performance liquid chromatography with evaporative light scattering detection. *J Pharm Biomed Anal*, **42 (4)**, 506–512.
19. Collins, P.J., McMahon, G. and O'Brien, P., O'Connor, B. (2004) Purification, Identification and Characterisation of Seprase From Bovine Serum. *Int J Biochem Cell Biol*, **36 (11)**, 2320–33.
20. Kinderman, M.W. (2003) A new-generation integrated ion chromatography system. *Am Lab*, 26–31.
21. Haddad, P.R. (2004) IC: an eye on the future. *Today's Chemist at Work*, 38–44.

22. Li, G.X., Shi, Q.H., Guo, Y.Y. *et al.* (2006) Determination of cations and anions in the sea water by ion chromatography with suppressor conductivity detection. *Fenxi Kexue Xuebao*, **22 (2)**, 153–156.
23. Wang, H.B. and Shooter, D. (2001) Water-soluble ions of atmospheric aerosols in three New Zealand cities: seasonal changes and sources. *Atmos Environ*, **35 (34)**, 6031–6040.
24. Yang, P., Shi, W.B., Zhou, H.Y. *et al.* (2004) Determination of sodium monofluoroacetate in human blood and food samples by ion chromatography. *Sepu*, **22 (2)**, 177–180.
25. Santos, B., Simonet, B.M., Rios, A. and Valcarcel, M. (2006) Automatic sample preparation in commercial capillary-electrophoresis equipment. *TRAC*, **25 (10)**, 968–976.
26. Huang, Y.F., Huang, C.C., Hu, C.C. and Chang, H.T. (2006) Capillary electrophoresis-based separation techniques for the analysis of proteins. *Electrophoresis*, **27 (18)**, 3503–3522.
27. Carvalho, A.Z., Pauwels, J., de Greef, B. *et al.* (2006) Capillary electrophoresis method development for determination of impurities in sodium cysteamine phosphate samples. *J Pharm Biomed Anal*, **42 (1)**, 120–125.
28. Ptolemy, A.S. and Britz-McKibbin, P. (2006) Sample preconcentration with chemical derivatization in capillary electrophoresis: capillary as preconcentrator, microreactor and chiral selector for high-throughput metabolite screening. *J Chrom A*, **1106 (1–2)**, 7–18.
29. Santilio, A., D'Amato, M., Cataldi, L. *et al.* (2006) Chiral separation of metalaxyl by capillary zone electrophoresis using cyclodextrins. *J Cap Electrophor Microchip Tech*, **9 (5/6)**, 79–84.
30. Zhou, S.S., Ouyang, J., Baeyens, W.R.G. *et al.* (2006) Chiral separation of four fluoroquinolone compounds using capillary electrophoresis with hydroxypropyl-βcyclodextrin as chiral selector. *J Chrom A*, **1130 (2)**, 296–301.
31. Suarez-Luque, S., Mato, I., Huidobro, J.F. and Simal-Lozano, J. (2006) Rapid capillary zone electrophoresis method for the determination of metal cations in beverages. *Talanta*, **68 (4)**, 1143–1147.
32. Gurel, A., Hizal, J., Oztekin, N. and Erim, F.B. (2006) CE determination of carbohydrates using a dipeptide as separation electrolyte. *Chromatographia*, **64 (5/6)**, 321–324.
33. Carabias-Martinez, R., Rodriguez-Gonzalo, E., Miranda-Cruz, E. *et al.* (2006) Comparison of a non-aqueous capillary electrophoresis method with high performance liquid chromatography for the determination of herbicides and metabolites in water samples. *J Chrom A*, **1122 (1–2)**, 194–201.
34. Clark, J. (2004) Pushing the frontiers of olefin determination in gasoline. *Petrol Industry News*, June/July 34–35. www.petroindustrynews.com/pdf/pin/2004/06/pin200406_034_035.pdf
35. Wang, Z.Y., Li, S.M., Jonca, M. *et al.* (2006) Comparison of supercritical fluid chromatography and liquid chromatography for the separation of urinary metabolites of nobiletin with chiral and non-chiral stationary phases. *Biomed Chrom*, **20 (11)**, 1206–1215.
36. Mukherjee, P.S. and Cook, S.E. (2006) A feasibility study on direct assay of an aqueous formulation by chiral supercritical fluid chromatography. *J Pharm Biomed Anal*, **41 (4)**, 1287–1292.
37. Maurer, H., Hyphenated mass spectrometric techniques – indispensable tools in clinical and forensic toxicology and in doping control. *J Mass Spectrom*, **41 (11)**, 1399–1413.
38. Gas Chromatography–Tandem Techniques from on-line Chrom-Ed Series by Scott RPW. http://www.chromatography-online.org/13/contents.html
39. Surowiec, I., Baena, J.R., Frank, J. *et al.* (2005) Flow-through microdispenser for interfacing μ-HPLC to Raman and mid-IR spectroscopic detection. *J Chrom A*, **1080 (2)**, 132–139.
40. Niedzielski, P. (2005) The new concept of hyphenated analytical system: Simultaneous determination of inorganic arsenic(III), arsenic(V), selenium(IV) and selenium(VI) by high performance liquid chromatography–hydride generation–(fast sequential) atomic absorption spectrometry during single analysis. *Anal Chim Acta*, **551 (1–2)**, 199–206.
41. Niessen, W.M.A. (2003) Progress in liquid chromatography-mass spectrometry instrumentation and its impact on high-throughput screening, *J Chrom A*, **1000**, 413–436.
42. McMahon, G., Collins, P. and O'Connor, B. (2003) Characterisation of the active site of a newly-discovered and potentially significant post-proline cleaving endopeptidase called ZIP using LC–UV–MS. *Analyst*, **128 (6)**, 670–5.
43. Ruth, D., McMahon, G. and O'Fagain, C. (2006) Peptide synthesis by recombinant Fasciola hepatica cathepsin. *Biochimie*, **88 (1)**, 117–120.

44. McMahon, G., Wall, R., Nolan, K. and Diamond, D. (2002) Characterisation of the ester-substituted products of the reaction of p-t-butyl calix[4]arene and ethyl bromoacetate using LC–UV–MS and LC–DAD. *Talanta*, **57**, 1119–1132.
45. Schazmann, B., McMahon, G. and Diamond, D. (2005) Identification and recovery of an asymmetric calix[4]arene tetranitrile derivative using liquid chromatography and mass spectrometry. *Supramol Chem*, **77 (5)**, 393–399.
46. Gennaro, L.A., Salas-Solano, O. and Ma, S. (2006) Capillary electrophoresis–mass spectrometry as a characterization tool for therapeutic proteins. *Anal Biochem*, **355 (2)**, 249–258.
47. Garza, S. and Moini, M. (2006) Analysis of complex protein mixtures with improved sequence coverage using (CE–MS/MS)^n. *Anal Chem*, **78 (20)**, 7309–7316.
48. Stutz, H. (2005) Advances in the analysis of proteins and peptides by capillary electrophoresis with matrix-assisted laser desorption/ionization and electrospray-mass spectrometry detection. *Electrophoresis*, **26 (7–8)**, 1254–1290.
49. Balaguer, E. and Neusuess, C. (2006) Glycoprotein characterization combining intact protein and glycan analysis by capillary electrophoresis–electrospray ionization–mass spectrometry. *Anal Chem*, **78 (15)**, 5384–5393.
50. Choudhary, G., Chakel, J., Hancock, W. *et al.* (1999) Investigation of the potential of capillary electrophoresis with off-line matrix-assisted laser desorption/ionization time-of-flight mass spectrometry for clinical analysis: examination of a glycoprotein. *Anal Chem*, **71 (4)**, 855–9.

4

Imaging Instruments

Microscopy uses radiation and optics to obtain a magnified image of an object. The resolution of the imaging is limited by the minimum focus of the radiation due to diffraction. For light microscopy the diffraction limit is approximately 1 μm. In the 1930s, electron microscopes were developed; these used an electron beam, rather than light rays, to give a theoretical resolution limit of about 4 nm. There are two common types – the scanning electron microscope (SEM) and the transmission electron microscope (TEM). Then, in the 1980s, scanning probe microscopes (SPMs) were developed which finally allowed atomic resolution. In 1982, the scanning tunnelling microscope (STM) was invented and, in 1986, the atomic force microscope (AFM).

4.1 Optical Microscopy

Principle

Light microscopy uses electromagnetic radiation in the UV or visible wavelength range to obtain a magnified image of an object. The resolution of the imaging is limited by the diffraction limit, which is approximately 1 μm, and the ultimate theoretical resolution is about 200 nm. Imaging objects smaller than 1 μm is not possible using light. The simplest light microscope consists of an objective lens and an eyepiece. Light is focused on the specimen and the objective lens magnifies the image before the light travels through the microscope column. The light finally passes through the eyepiece and into the viewer's eye which interprets the image. Microscope objectives and eyepieces usually contain complex lens systems of two or more lenses to correct for lens aberrations.

Instrument

A schematic of the light microscope is given in Figure 4.1.

Analytical Instrumentation: A Guide to Laboratory, Portable and Miniaturized Instruments G. McMahon

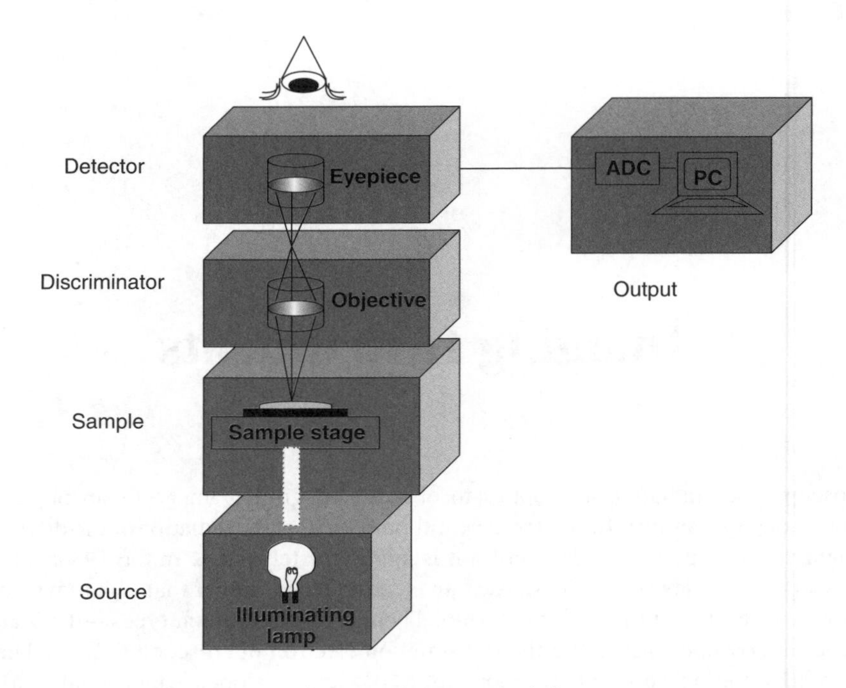

Figure 4.1 *Schematic diagram of a light microscope.*

Source
The source in the light microscope is usually a white light source lamp (illuminator). A mirror focuses this light. High magnification requires very bright illumination of the sample and a condenser lens is usually placed between the light source and the sample stage to focus light onto the sample.

Sample
The sample is placed on a microscope slide, prepared as needed and mounted on a stage.

Discriminator
The objective lens is the discriminator. It forms a real intermediate image that is then greatly magnified by the eyepiece. The objective lens and eyepiece are maintained at a fixed distance and focusing is achieved by moving the whole assembly up and down in relation to the sample.

Detector
The detector is the eyepiece that magnifies the image, and of course the human eye.

Output
Sometimes the images can be displayed digitally on a PC and manipulated there as required.

Information Obtained

The information gleaned from examining a sample underneath a light microscope is mainly qualitative in nature, i.e. what is present and what does it look like? However, there are a number of semiquantitative techniques available based on counting – manually or digitally by use of grids. The counting technique might be used to estimate the number of white blood cells per millilitre of a patient plasma sample, for example.

Developments/Specialist Techniques

This diffraction limit prevents optical imaging beyond the micrometre scale. However, near-field optical microscopy (NFOM) avoids these limits by using ultra-miniaturised components. For example, it employs a light source of less than 100 nm diameter to overcome the diffraction limit of conventional optical microscopy. The main concept of NFOM is to place the sample very close to the light source, i.e. in the near field, so that the imaging resolution is determined by the narrow diameter of the light source.

The near-field light source is made from a glass capillary that is heated with a carbon dioxide laser and pulled to an atomically-sharp tip. The outside of the capillary is coated with silver for reflectivity except at the very end of the tip. Laser light is focused into the glass capillary and propagates to the tip by internal reflection (as in optical fibres). A small amount of the light leaks out of the tip via the evanescent wave. The near-field source is brought very close to a sample, and the sample is rastered back and forth to produce an image. NFOM can use either absorption or fluorescence as the optical signal. Scanning near-field optical microscopy (SNOM) is an emerging technique in surface analysis for observing specimens with lateral resolution of even better than 100 nm[1].

Applications

Applications for light microscopy exist mainly in the biological arena. However, in recent years, especially with the huge interest in material science, these microscopes are increasingly used in analytical laboratories to investigate surfaces, e.g. electrode or chip surfaces that have been coated.

4.2 Confocal Microscopy

Principle

This technique generates the image in a completely different way than the normal visual bright field microscope. Confocal microscopes are light microcopes that use lasers and very small apertures to produce images of thin layers of specimens. The technique gives slightly higher resolution but, most importantly, it provides optical sectioning without out-of-focus light degrading the image. The layers can be reconstructed to give a 3-D image of the specimen. It is often used in conjunction with fluorescence microscopy.

Instrument

A confocal microscope consists of the modules shown in Figure 4.2.

Source
The source in the confocal microscope is a laser, which uses the objective lens in the discriminator to focus it onto different planes on the sample.

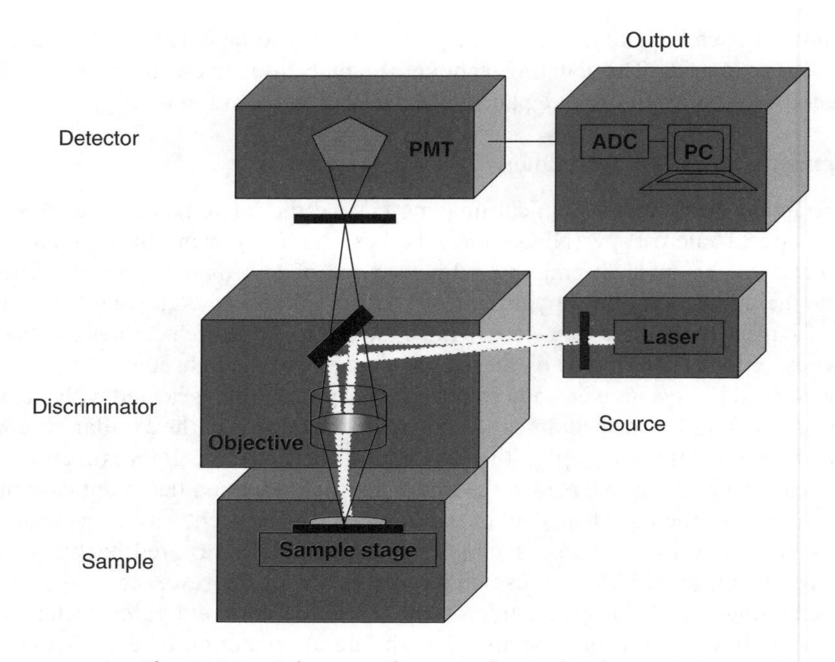

Figure 4.2 Schematic diagram of a confocal microscope.

Sample
The sample is prepared as required, placed on a slide and mounted on a stage.

Discriminator
The objective lens is the discriminator. It forms a real intermediate image that is then detected by the photomultiplier tube (PMT).

Detector
The detector is the PMT, which picks up the image and transfers the information to the computer for magnification.

Output
The images are displayed digitally on a PC and can be manipulated there as required.

Another diagram of the confocal microscope is shown in Figure 4.3.

Information Obtained

Confocal microscopy gives the following qualitative information: topography (the surface features of the object and their texture), morphology (the shape, size and arrangement of the particles making up the object that are lying on the surface of the sample) and composition (the elements and compounds the sample is composed of and their relative ratios) if so equipped. All of these features are in the micrometre region of size. Quantitative information may be possible using counting methods based on grids.

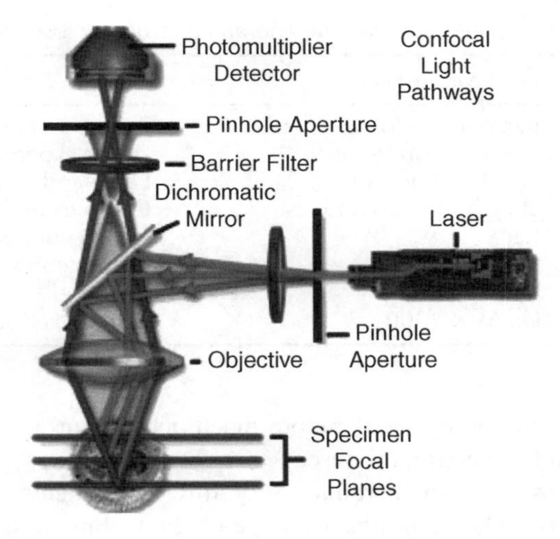

Figure 4.3 *Schematic diagram of a specific confocal microscope (Courtesy of the Olympus Fluoview website www.olympusfluoview.com).*

Developments/Specialist Techniques

One of the developments in the area of confocal microscopy is the use of fluorescence to highlight or visualise certain features that either fluoresce naturally or are derivatised with probes prior to microscopic examination. Another new development is two-photon microscopes, which allow operation in the UV spectrum and, hence, deeper penetration of structures such as living cells, skin and other biological samples.

Applications

The laser scanning confocal microscope enables visualisation deep within both living and fixed cells and tissues, is 3-D with high-resolution and is nondestructive. Confocal microscopy offers several advantages over conventional optical microscopy, including the ability to control depth of field (using the z-axis), elimination or reduction of blurring and the capability of collecting serial optical sections from thick specimens. The confocal approach can generate images of relatively thin sections within a thick sample, avoiding the need for fine physical sectioning or specimen compression. Confocal microscopy is very popular due to the extremely high-quality images that can be obtained from specimens prepared for conventional fluorescence microscopy, and the growing number of applications in cell biology that rely on imaging both fixed and living cells and tissues[2].

4.3 Electron Microscopy

Electron microscopy is an imaging technique that uses an electron beam to probe a material. Since the wavelength of electrons is much smaller than the wavelength of visible light, diffraction effects occur at much smaller physical dimensions. The imaging

Table 4.1 Comparison of a light microscope and an electron microscope.

Criterion	Light Microscope	Electron Microscope
Cost	Inexpensive to buy (< €2000) and cheap to run	Expensive to buy (> €1 000 000) and expensive to run
Size	Small and portable	Large and cumbersome
Sample preparation	Straightforward and fast	Complex and slow
Artefacts	Rare	Can occur due to sample preparation
Vacuum	Not required	Required
Magnification	Up to × 2000	Up to × 1 000 000 or more

resolution in electron microscopy is therefore much better than in light microscopy. In Table 4.1 the light and the electron microscope are compared.

The electron microscope can reproduce very tiny components in an image called an electron micrograph. The main disadvantage of the technique is that samples have to be viewed in an air lock to maintain internal vacuum conditions. This means that no *living* material can be studied. Also, samples have to be specially prepared to give proper detail, e.g. freezing, fixation or dehydration, which may result in artefacts being produced. This can give rise to the problem of distinguishing artefacts from the material of interest, particularly in biological samples. There are two main types of electron microscopy – scanning electron microscopy (SEM) and transmission electron microscopy (TEM).

4.3.1 Scanning Electron Microscopy

Principles

In brief, the scanning electron microscope (SEM) images the electrons that are *reflected from a sample*. These images are useful for studying surface morphology or measuring particle sizes. The SEM produces the images by detecting secondary electrons that are emitted from the surface due to excitation by the primary electron beam. Generally, SEM resolution is about an order of magnitude less than TEM resolution, but because the SEM image relies on surface processes rather than transmission it is able to image bulk samples and has a much greater depth of view, and so can produce images that are a good representation of the 3-D structure of the sample. The SEM has two major advantages over a conventional microscope: higher magnification and greater depth of field. At higher magnification more details may be evident and the great depth of field makes the examination of specimens easier.

Instrument

An SEM instrument can be depicted as illustrated in Figure 4.4.

Source

An electron gun delivers the electron beam, which is produced by applying a high voltage to a hot tungsten filament, and accelerates the emitted electrons through a high electric field, 10–50 kV. The electron beam is then focused with magnetic field lenses to a spot of 100 nm or less on the sample. The microscope column is under high vacuum.

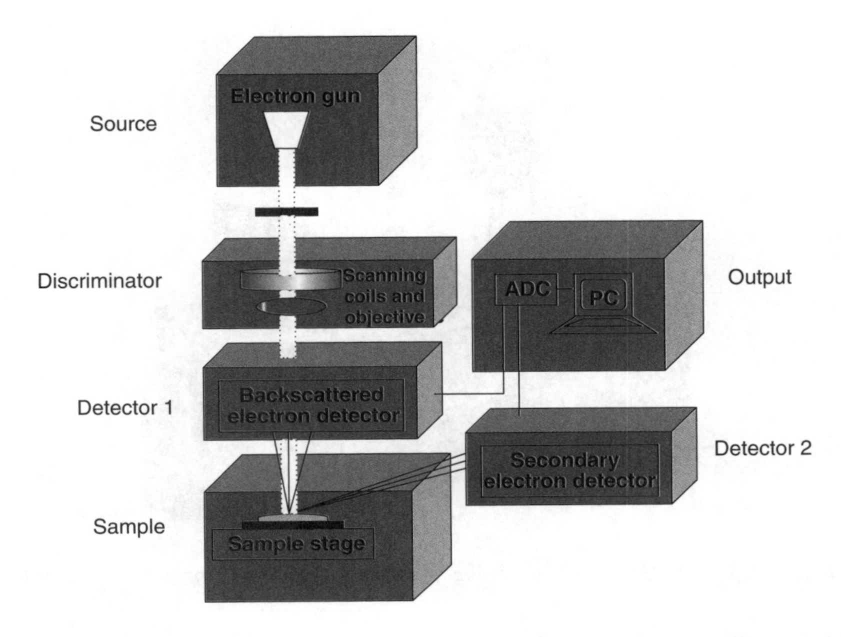

Figure 4.4 *Schematic diagram of a scanning electron microscope.*

Sample
When using an SEM, biological samples are normally first coated with a metal that readily reflects electrons, e.g. gold. This coating also provides a conducting surface for electrons to avoid charging of the sample. The SEM has a high voltage electron emitter that sends a beam of electrons down the column, which is at very high vacuum, onto the sample where they bounce off and are used to form the image. The observer therefore sees a picture of the surface of the sample, without any internal information. The sample, after being prepared, is mounted on the compustage/sample grid.

Discriminator
The electron beam is rastered across the sample via the scanning coils by ramping voltages on the x- and y-deflection plates through which the electron beam passes (the z-axis is the electron-beam direction).

Detectors
The sample can be viewed by detecting back-scattered electrons, secondary electrons or even X-rays emitted by the sample. Each type of detection can give different information about the sample.

Output
The computer controls the scanning coils and also manipulates and displays the data received by the detectors.

Information Obtained

Scanning electron microscopy gives the following qualitative information: topography (the surface features of an object and their texture), morphology (the shape, size and

Figure 4.5 The JSM-7700F scanning electron microscope (Image courtesy of Jeol).

arrangement of the particles making up the object that are lying on the surface of the sample) and composition (the elements and compounds the sample is composed of and their relative ratios) if so equipped. All of these features are in the nanometre region of size. Crystallographic information is also possible in SEM, i.e. the arrangement of atoms in the specimen and their degree of order (only useful on single crystal particles $>20\,\mu m$). An example of an SEM is shown in Figure 4.5.

Developments/Specialist Techniques

A recent innovation in electron microscopy is environmental SEM, which allows samples to be studied at pressures and humidities that approach ambient conditions. To achieve this, several stages of differential pumping between the electron gun and the sample are used, and the sample itself is placed in a vacuum of a few hundred Pascals. Environmental SEM enables many materials to be examined without pretreatment, unlike conventional SEM, in which specimens must be solid, dry and usually electrically conductive. This now makes possible studies of the natural, unadulterated surfaces of specimens such as polymers, biological tissues and cells, food and drugs and forensic materials.

A new environmental secondary electron detector (ESED) has been produced by Hitachi for its SEM. It provides an alternative to the traditional back-scattered imaging and closely mimics a conventional secondary electron detector to yield good surface information. The new ESED picks up ions as well as electrons, creates an avalanche effect with the ions and produces a better quality image.

SEMs are now often equipped with an energy-dispersive X-ray (EDX) analysis system for the elemental analysis of certain specimens. For example, the standard forensic laboratory test for identifying gunshot residue is based on the use of an electron microscope equipped with an EDX analyser.

Applications

SEM has been used to image DNA molecules tagged with gold nanoparticles[3]. This technique allowed counting and hence quantitative measurement of the molecules of interest. SEM with EDX has also been employed to measure aerosol particles in workroom air[4]. The size, morphology and chemical composition of over 2000 particles were determined. SEM is useful too for the study of geological materials, grains and crystal structure.

4.3.2 Transmission Electron Microscopy

Principles

The transmission electron microscope (TEM) images the electrons that *pass through a sample*. Since electrons interact strongly with matter, they are attenuated as they pass through a solid; this requires that the samples are prepared in very thin sections. The image of the sample is observed on a phosphor screen below the sample and can be recorded with film. Generally, the TEM resolution (0.5 nm) is about an order of magnitude better than the SEM resolution (10 nm). Further information on TEM is available elsewhere[5].

Instrument

A TEM instrument can be depicted as shown in Figure 4.6.

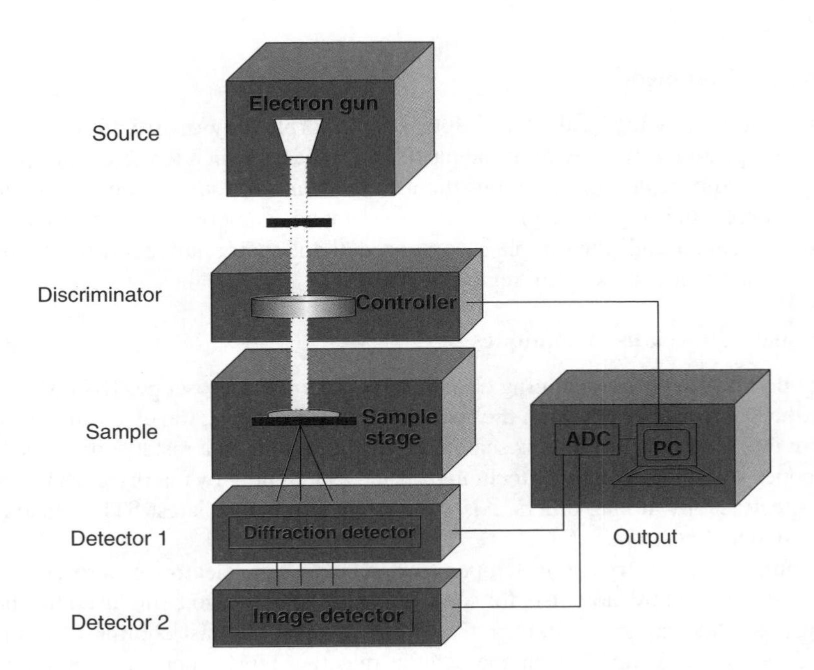

Figure 4.6 *Schematic diagram of a transmission electron microscope.*

Source
An electron gun delivers the electron beam, as in SEM, and the microscope column is under high vacuum also.

Sample
A TEM produces an image that is a projection of the entire object, including the surface and the internal structures. The incoming electron beam interacts with the sample as it *passes through* the entire thickness of the sample (like a slide projector). The TEM column is at a very high vacuum, as with SEM. Objects with different internal structures can be differentiated because they give different projections. However, the image is two-dimensional and depth information in structures is lost. Furthermore, the samples need to be thin ($\leq 0.1\,\mu m$), or they will absorb too much of the electron beam. The prepared sample is mounted on the compustage/sample grid.

Discriminator
The controller is used to scan through the sample.

Detectors
The image strikes the phosphor image screen and light is generated, allowing the user to see the image. Diffraction information is also imaged on a separate platform. TEM microscopes can image individual atoms and their relative placement, and can give compositional information over an area of interest.

Output
The computer controls the scanning coils and also manipulates the data received by the detectors.

Information Obtained

TEM gives the following qualitative information: morphology or structural information (the size, shape and arrangement of the particles or phases which make up the specimen/sample), crystallographic information (the arrangement of atoms in the specimen, their degree of order and any defects) and, if so equipped, compositional information (the elements and compounds the sample is composed of and their relative ratios). All of these features are in the low nanometre region of size.

Developments/Specialist Techniques

It is now possible to obtain scanning transmission electron microscopes (STEMs). With a STEM, the electrons pass through the specimen but, as in SEM, the electron optics focus the beam into a narrow spot that is scanned over the sample in a raster. This makes these microscopes suitable for analysis techniques such as mapping by energy dispersive X-ray (EDX) spectroscopy among others. Also, the resolution on the latest STEM instruments is less than one Angstrom.

A technique called cryogenic temperature transmission electron microscopy (cryo-TEM) has been used by biologists for some years but is now also being used by chemists to examine a wide range of samples. CryoTEM involves the fast cooling of a thin film of liquid sample and transferring the sample into the TEM where it is maintained at cryogenic temperature. It can be used to look at polymers, lipids, surfactants and C-60 suspensions[6,7,8].

Applications

Because the electron beam goes through the sample, TEM reveals the interior of a specimen/sample. It is often used for the study of physical samples, such as minerals and rocks in geology, and new materials. It is also suitable for many types of biological, chemical and environmental samples, the main limitation being the ability to prepare sample sections thin enough to allow the electron beam to pass through. One environmental study used TEM to identify and quantify inorganic particles in a colloidal size range by applying a straightforward method of sample preparation (direct centrifugation of the samples on transmission electron microscopy grids) in conjunction with particle analysis using TEM and EDX[9]. Another environmental study examined ultrafine ($<$100 nm) ash particles in three coal fly ashes[10]. Crystalline phases down to 10 nm in size were identified.

4.4 Scanning Probe Microscopy

Both the scanning tunnelling microscope (STM) and the atomic force microscope (AFM) fall under the umbrella of scanning probe microscopy instruments. The heart of any scanning probe microscope is the ultra sharp electrified probe tip itself that is scanned over the sample surface to build up some form of image. The probe tip traces a map as it moves across the sample's surface. The sample height information (topography) usually forms one aspect of the image but the images can yield other important information about the sample, such as crystal structure and hardness.

4.4.1 Scanning Tunnelling Microscopy

Principle

The scanning tunnelling microscope (STM) is used to obtain images of conductive surfaces on an atomic scale. It is a nonoptical microscope that exploits the quantum mechanical tunnelling effect to determine the distance between the probe and a surface. It requires the sample to be conductive. It is an electroanalytical instrument because it moves a tiny electrode over the surface and uses electric current measurements to discern the atomic-scale features of the surface. High quality STMs can reach sufficient resolution to show single atoms. The STM is able to get within a few nanometres distance of what it is observing. It can also be used to alter the observed material by manipulating individual atoms, triggering chemical reactions, and creating ions by removing individual electrons from atoms and then reverting them to atoms by replacing the electrons. The STM produces 3-D images. The STM has higher resolution (0.2 nm) than the atomic force microscope (AFM) but requires a sample to be conductive whereas AFM does not and so has wider applicability.

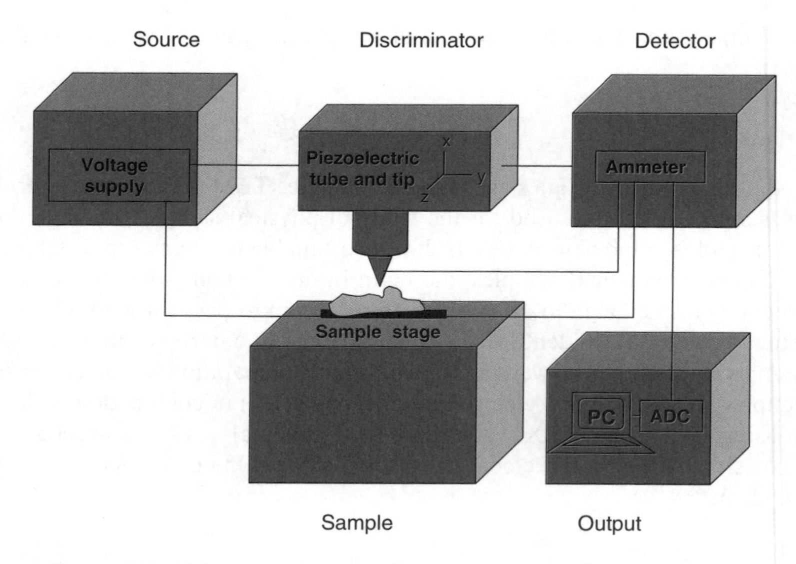

Figure 4.7 *Schematic diagram of a scanning tunnelling microscope.*

Instrument

An STM instrument is shown in Figure 4.7.

Sample
The sample is prepared appropriately for STM and mounted on the sample stage. Obviously, for a current to occur the substrate being scanned must be conductive. Insulators cannot be scanned through the STM.

Source
A voltage supply is supplied to the piezoelectric tube and tip. This voltage is applied to keep the preset tunnelling current constant.

Discriminator
A sharp tip is mounted on a piezoelectric tube, which allows tiny movements by applying a voltage at its electrodes. This is the discriminator. This fine sharpened tip, which is presumed to have a single atom at the apex, is scanned at approximately 1 nm above the surface of the sample. When the tip and sample are connected with a voltage source, electrons will 'tunnel' or jump from the tip to the surface (or vice versa depending on the polarity), resulting in a weak electric current; this is the tunnelling current. This current can be measured, its size being exponentially dependent on the distance between probe and the surface. When the tip reaches a step in the sample, the tunnelling current increases and the feedback loop, which keeps the tunnelling current constant, retracts the piezoelectric tube (constant current mode) to allow it to climb the step. By scanning the tip over the surface and measuring the height, which is directly related to the voltage applied to the piezo element, the surface structure of the material under study can be reconstructed.

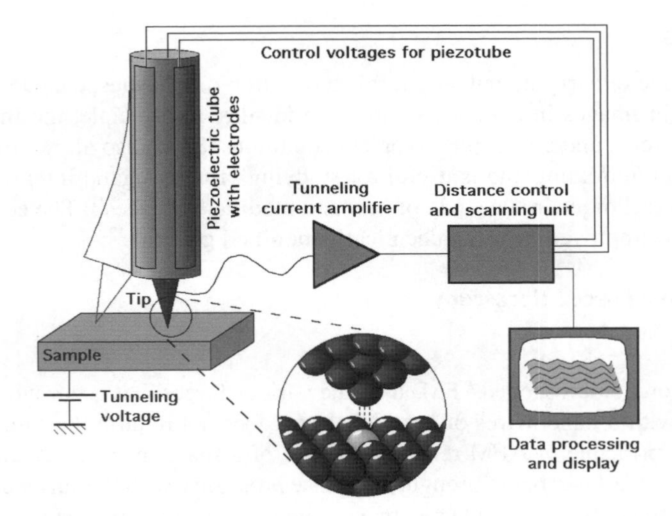

Figure 4.8 *Schematic diagram of how the STM works (Courtesy of Michael Schmid under the Creative Commons Attribution Sharealike 2.0 Austria License. From website: www.iap.tuwien.ac.at/www/surface/STM_Gallery/stm_schematic.html.).*

Detector
The variation in tunnelling current as it passes over the specimen is then recorded and a three-dimensional 'map' of the surface can be obtained.

Output
The computer manipulates the data received and generates the images.

Another means of representing the STM is shown in Figure 4.8.

Information Obtained

STM gives a 3-D profile of the surface, which is very useful for characterising surface roughness, observing surface defects and determining the size and conformation of molecules and aggregates on the surface. Other possible measurements based on electron density include topography, which shows the total density of electrons on the surface as a function of position, and mapping, which shows the density of electrons at a particular energy as a function of position on the surface. STM can also perform a 'line cut', which is a series of spectra taken at equally spaced spatial points along a line drawn across the surface, and, finally, a 'spectrum', which shows the density of states as a function of energy at a particular spatial position. With STM, atoms can also be manipulated and moved around using the tip.

Developments/Specialist Techniques

Electrochemical STM is available; this enables the analyst to study electrochemical reactions as they occur. Potential profiling and mapping as well as topographic imaging of the electrode itself are all possible with this range of instruments.

Applications

STM is capable of carrying out structural studies of biomolecules, e.g. DNA, and can be used to obtain images in aqueous solutions, allowing, in principle, the investigation of biological systems under near-physiological conditions. STM also allows investigation of protein folding/unfolding and is useful for studying protein–ligand interactions. STM is an important technique in the study of new and modified surfaces[11]. The contrast of STM images can be improved by using chemically modified gold tips[12].

4.4.2 Atomic Force Microscopy

Principle

The atomic force microscope (AFM) uses the various forces that occur when two objects are brought within nanometres of each other and does not require the sample to be conducting. The principle of AFM relies on the use of a sharp, pyramidal, ultra-fine probe mounted on a cantilever being brought into close proximity with the surface where it feels a chemical attraction or repulsion and moves up or down on its supporting cantilever. The probe movement is monitored using a laser beam that is reflected or diffracted from the back of the cantilever. AFM topographic images of the surface are obtained by recording the cantilever deflections as the sample is scanned.

Like STM, AFM produces 3-D images. Unlike STM, AFM does not require the sample to be conducting and so has wider applicability. Unlike optical microscopes, AFM does not use a lens, so the resolution of the technique is limited by probe size (and strength of interaction between surface and probe tip) rather than by diffraction effects. The key to the sensitivity of AFM is in carefully monitoring the movement of the probe tip.

Instrument

An AFM instrument consists of the modules shown in Figure 4.9.

Sample
As in all SPM techniques, sample preparation is key. An AFM can work either when the probe is in contact with a surface, or when it is a few nanometres away. AFM can also be carried out on liquids.

Source
A laser is shone onto the probe tip at the end of the cantilever.

Discriminator
The discriminating element in the AFM is the cantilever and tip. When the tip moves over the surface, the cantilever moves and so too does the laser. The negative feedback loop moves the sample up and down via a piezoelectric scanner so as to maintain the interactive force. Typical AFM probe tips are made of silicon or silicon nitride, which provide lateral resolutions of 5–10 nm and 10–40 nm, respectively. There are three main modes of AFM and all depend on this interaction between the tip and the sample:

- Contact mode – with this mode, ionic repulsion forces are responsible for creating the image.

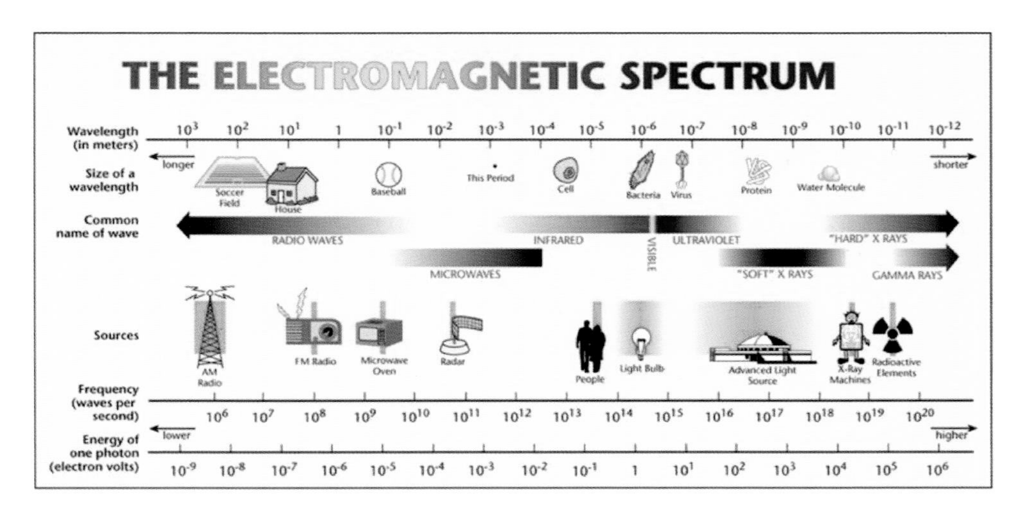

Figure 2.1 *The electromagnetic spectrum (Courtesy of the Advanced Light Source, Berkeley Laboratory).*

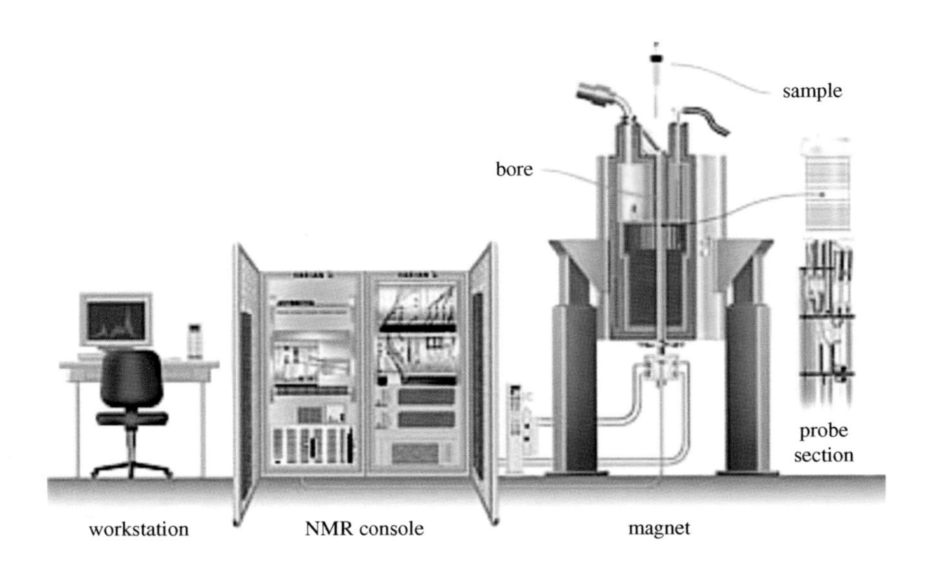

Figure 2.27 *Components of an NMR instrument (Reproduced with permission from Varian Inc.).*

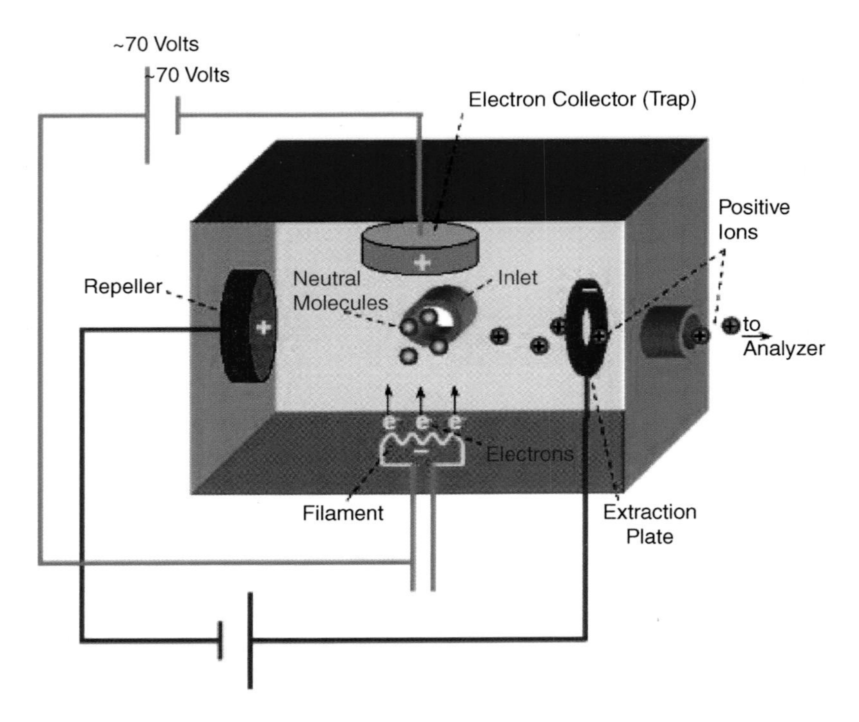

Figure 2.31 *Schematic diagram of an electron impact ion source (Figure used by kind permission of Jim Roe, Department of Chemistry, University of Washington, Seattle, USA).*

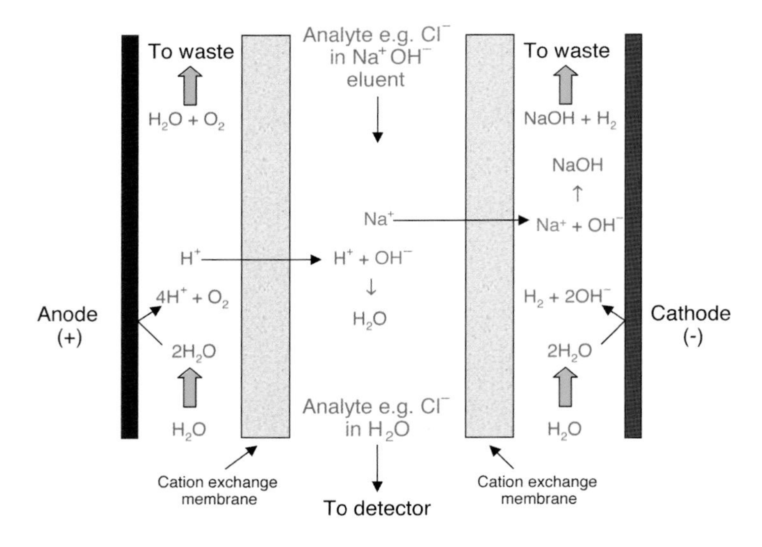

Figure 3.25 *Schematic diagram of a membrane suppressor for use in anion-exchange chromatography (H^+ ions replace the Na^+ ions yielding water as the background electrolyte which has low conductivity).*

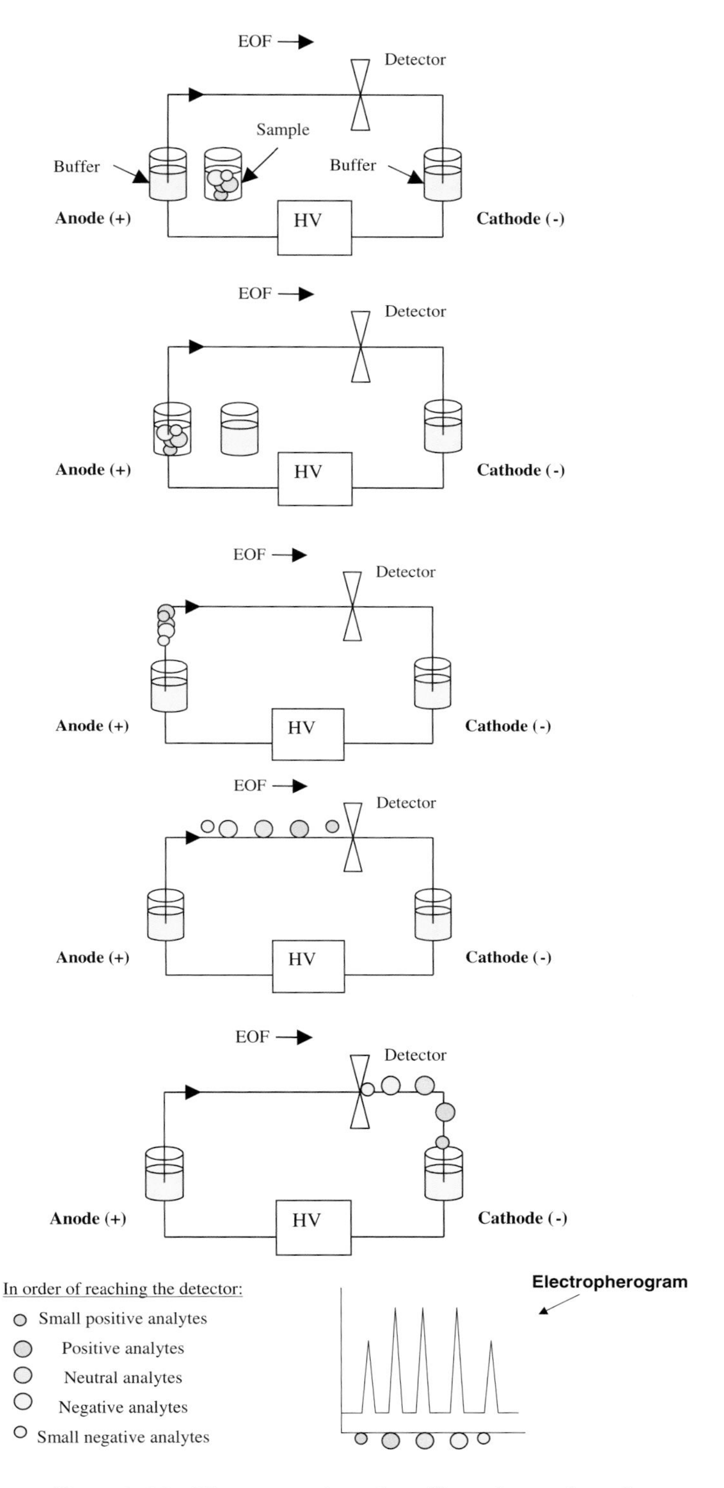

Figure 3.26 *The progression of capillary electrophoresis.*

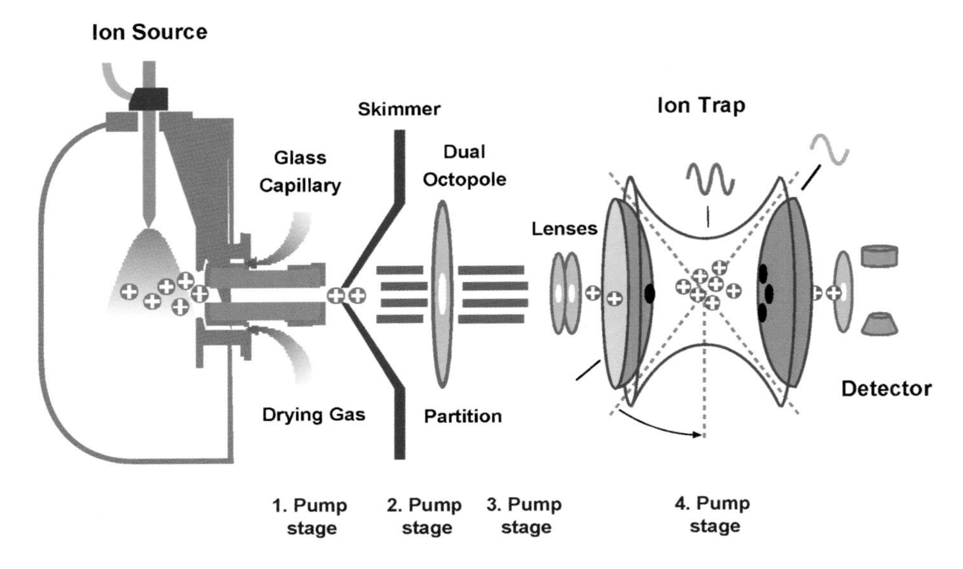

Figure 3.42 *Schematic of ion trap MS system with orthogonal ionisation source (ESI or APCI) attached (Courtesy of Bruker Daltonics).*

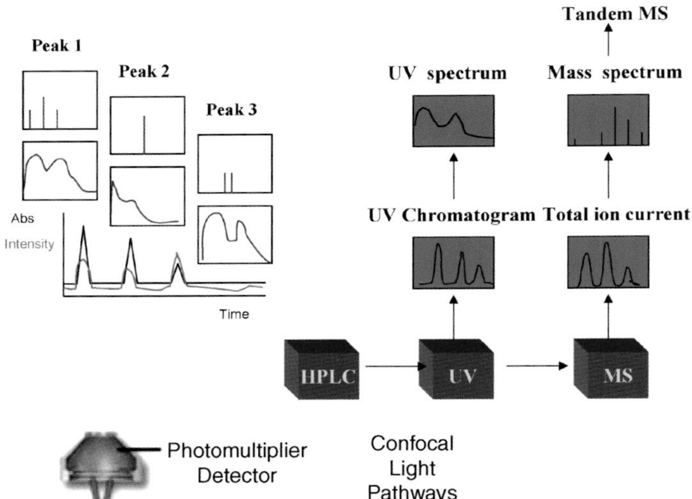

Figure 3.44 *Schematic of LC–UV–MS illustrating the 3-D nature of the data obtained (Eluent from the HPLC flows first into the UV detector yielding a UV chromatogram and, if the detector is a PDA, UV spectral data for each component. The eluent then flows into the mass spectrometer and yields a total ion current, mass spectral data for each component and tandem mass spectral data if so equipped.).*

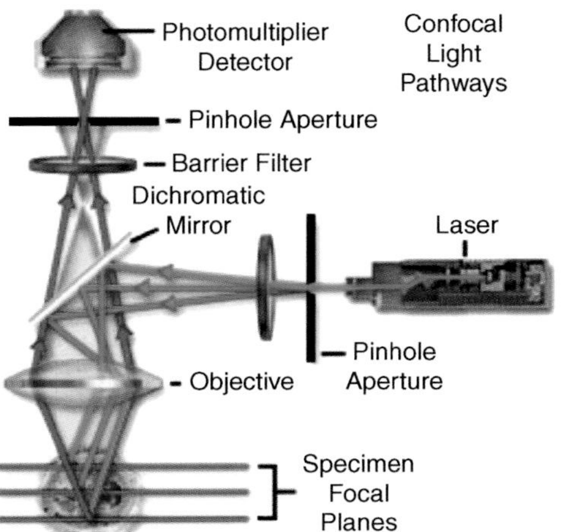

Figure 4.3 *Schematic diagram of a specific confocal microscope (Courtesy of the Olympus Fluoview website www.olympus-fluoview.com).*

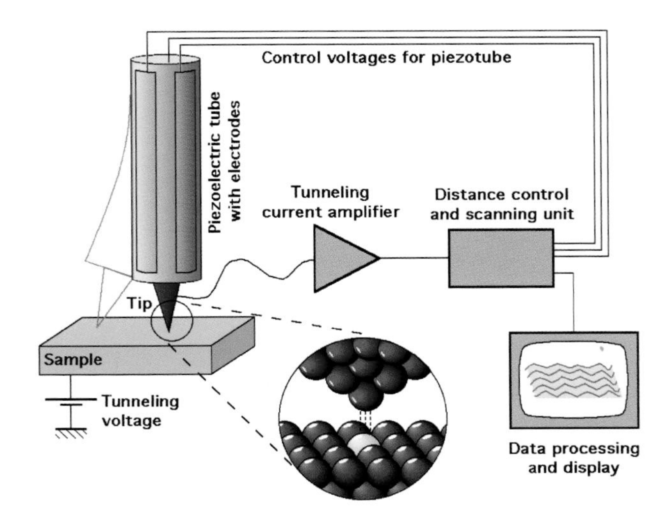

Figure 4.8 Schematic diagram of how the STM works (Courtesy of Michael Schmid under the Creative Commons Attribution Sharealike 2.0 Austria License. From website: www.iap.tuwien.ac.at/www/surface/STM_Gallery/stm_schematic.html.).

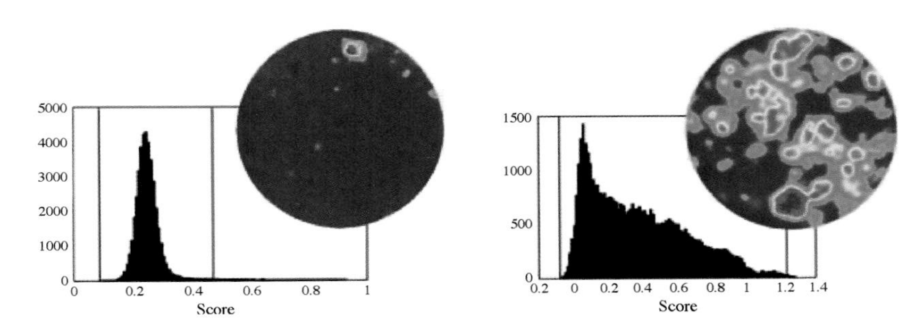

Figure 4.12 Spectral images showing the differences in active ingredient distribution (Reproduced, with permission, from Lyon, R. C., Lester, D. S., Lewis, E. N. et al. (2002), 'Near infrared spectral imaging for quality assurance of pharmaceutical products: analysis of tablets to assess powder blend homogeneity'. AAPS Pharm Sci Tech, **3** (3), 1. Website: www.malvern.com/LabEng/products/sdi/nir_chemical_imaging_range.htm.).

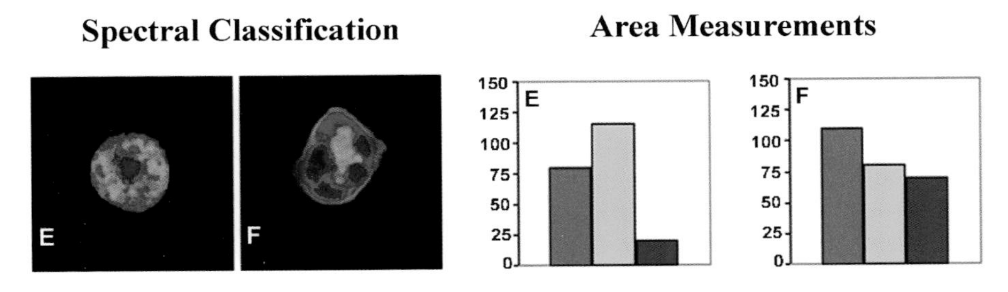

Figure 4.14 Spectral images of a normal (E) and cancerous (F) human liver cell and the respective area measurements of both images (Reproduced by kind permission of Applied Spectral Imaging).

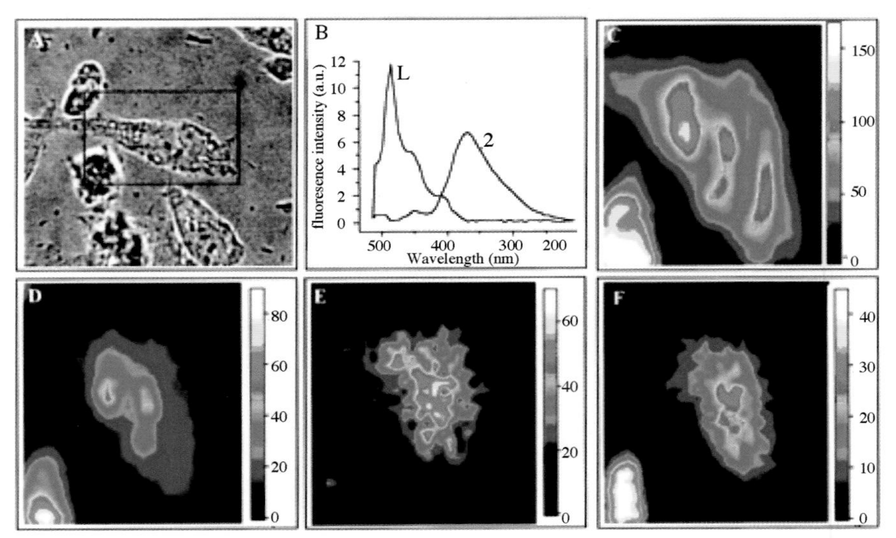

(A) Photonic microscopic image, (B) model spectra of the molecular probe (1: non-specific fluorescence emission and 2: fluorescence emission in Golgi apparatus). (C) contribution of non-specific fluorescence of the probe, (D) contribution of fluorescence in the Golgi apparatus, (E) fluorescence emission of the anthracycline drug in a resistant-like environment and (F) fluorescence of the anthracycline drug in a sensitive-like environment.

Figure 4.15 *Confocal fluorescence imaging showing drug uptake and distribution in tumour cells (Reproduced, with permission, from Belhoussine, R., Morjani, H., Millot, J.M. et al. (1998), 'Confocal scanning microspectrofluorometry reveals specific anthracycline accumulation in cytoplasmic organelles of multi-drug resistant cancer cells'. Journal of Histochemistry & Cytochemistry,* **46** *(12), 1369–1376[19].).*

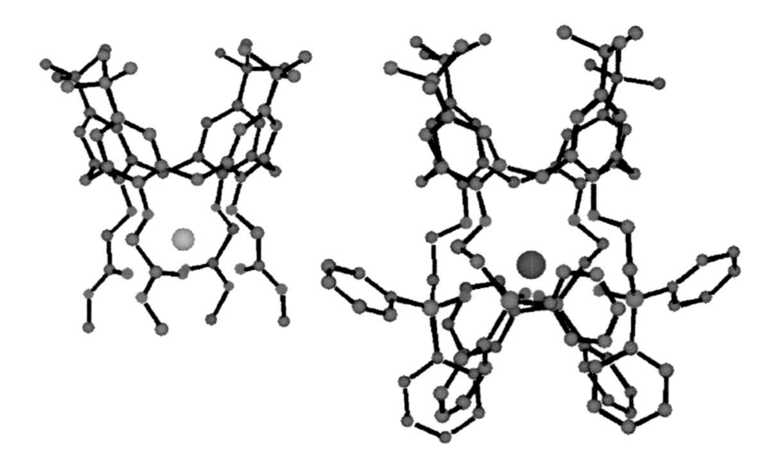

Figure 5.3 *Tetraethylestercalix[4]arene and tetraphosphine oxide calix[4]arene (The tetraethylester cavity is selective for sodium while the larger tetraphosphine oxide cavity is selective for calcium. Images kindly provided by Prof. Dermot Diamond, Dublin City University, Ireland.).*

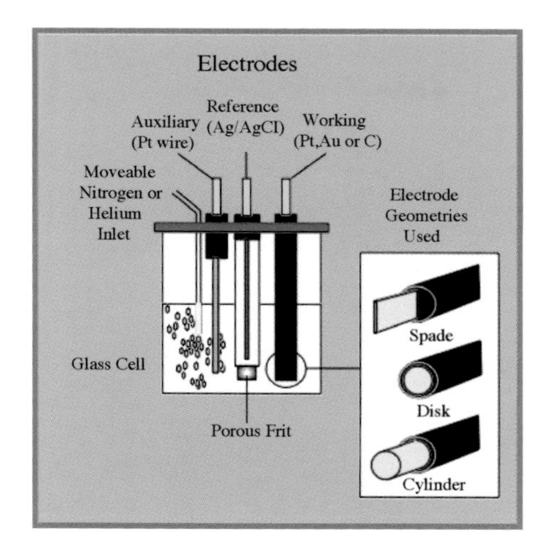

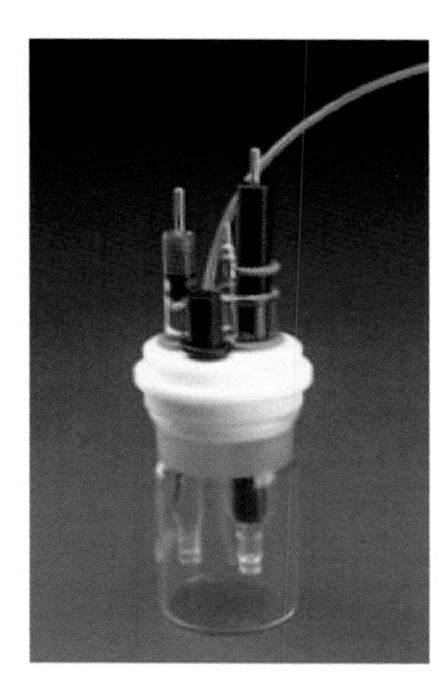

Figure 5.6 *Schematic diagram and photograph of a three-electrode voltammetric cell (Figure used by kind permission of Dr. Marco Cardosi, Dept of Biological Sciences, University of Paisley, UK; photograph courtesy of BAS Inc., USA.).*

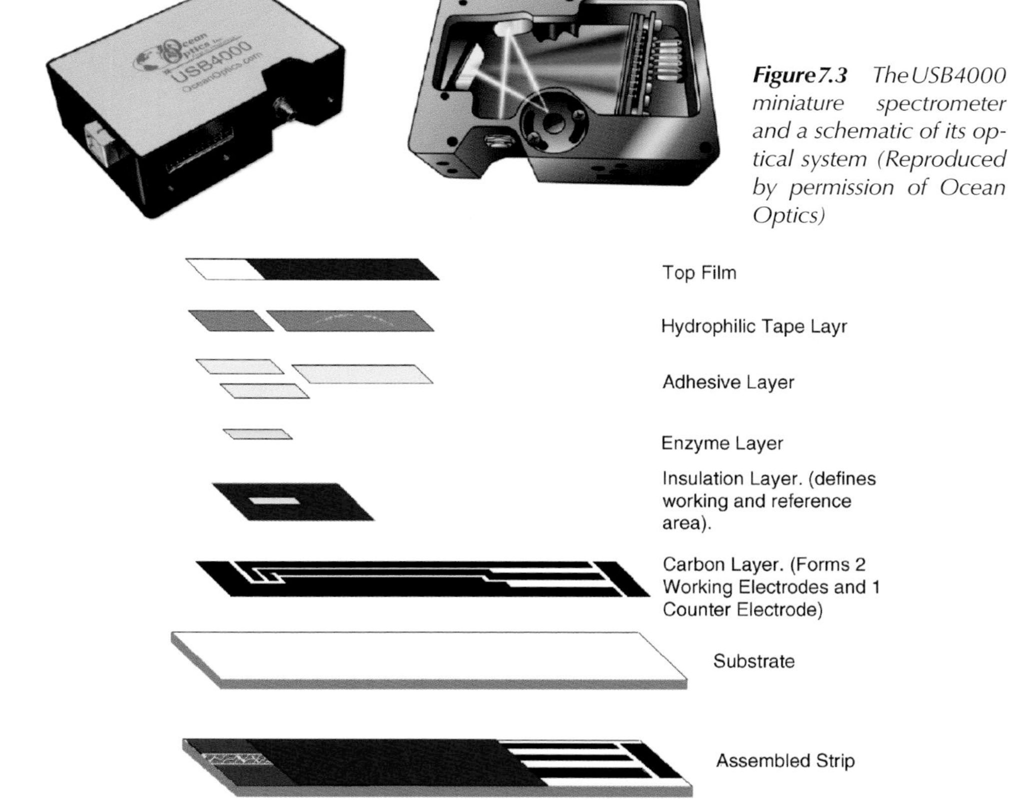

Figure 7.3 *The USB4000 miniature spectrometer and a schematic of its optical system (Reproduced by permission of Ocean Optics)*

Figure 8.3 *An exploded view of the OneTouch® Ultra glucose test strip (Reproduced by permission of Lifescan).*

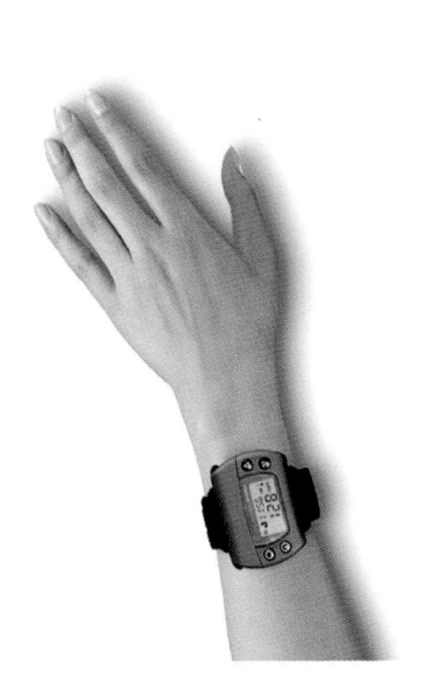

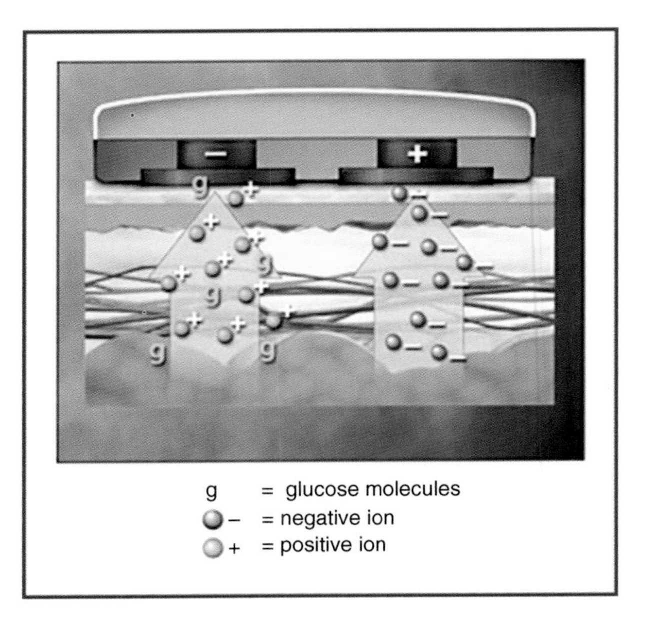

Figure 8.4 *The Glucowatch Biographer from Cygnus/Animas and a schematic illustrating how the glucose is collected and measured (Reproduced by permission of Animas).*

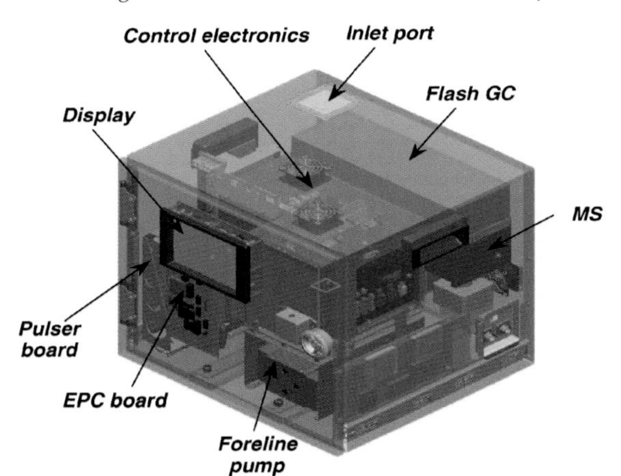

Figure 8.12 *The FieldMate prototype system (Image courtesy of Syagen).*

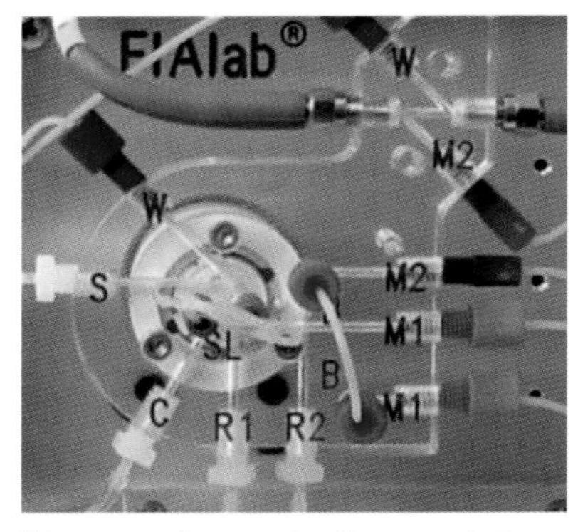

W = waste, S = sample, SL = sample loop, B = bridge, C = carrier buffer/solvent, R = reagent, M = mixing coil.

Figure 10.6 *The FIAlab Lab-On-Valve® integrates all connections, sample loop and flow cell into one simple manifold (Image provided by kind permission of FIAlab Instruments).*

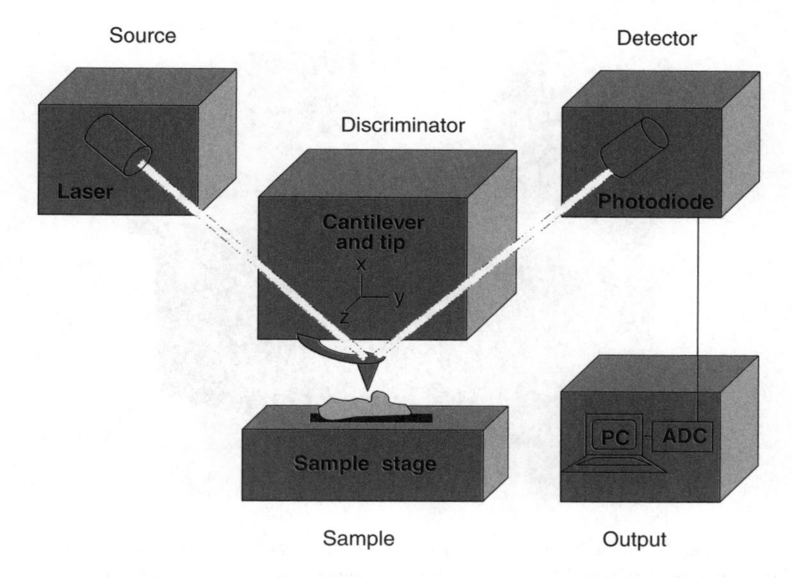

Figure 4.9 *Schematic diagram of an atomic force microscope.*

- Non contact mode – with distances greater than 10 Å between the tip and the sample surface, Van der Waals, electrostatic, magnetic or capillary forces produce the surface images. This generally provides lower resolution than contact mode.
- Tapping mode – this was developed as a method of achieving high resolution on soft, fragile samples without inducing destructive frictional forces.

Detector
Any deflection in the laser is picked up by the photodiode detector and is converted into topographical information about the sample.

Output
The detector sends a signal to the computer for analysis and image generation.

Information Obtained

Atomic force microscopy gives topographical information which is generally displayed by using a colour map for height. AFMs can measure force and sample elasticity by pressing the tip into the sample and measuring the resulting cantilever deflection. AFM can also measure the hardness or softness of a specimen or part of that specimen; information on crystallinity can also be obtained. AFM has recently been used to study dynamic interactions between emulsion droplets[13]. An example of an AFM is shown in Figure 4.10.

Developments/Specialist Techniques

A version of AFM is chemical force microscopy (CFM), where an AFM tip is coated with a thin chemical layer that interacts selectively with the different functional groups exposed on the sample surface. For example, gold-coated tips functionalised with a thiol

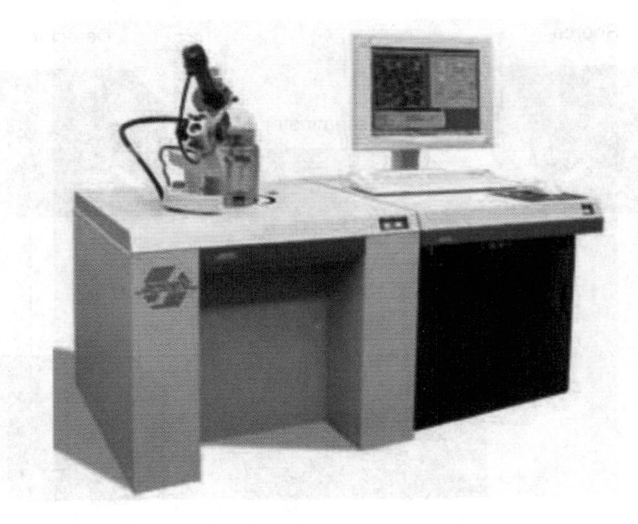

Figure 4.10 The JSPM-5200 scanning probe microscope can be configured as either an atomic force microscope (AFM) or scanning tunnelling microscope (STM) by changing the tip (Image courtesy of Jeol).

monolayer terminated with carboxyl (COOH) groups interact strongly with a substrate also bearing protruding carboxyl groups, due to hydrogen bonding, but interact weakly with a substrate containing pendant methyl (CH_3) groups. Pharmaceutical companies use CFM to measure the cohesive and adhesive forces between particles to predict how drug molecules interact with each other and to investigate how well a drug can be processed. Chemically modifying tips in this way can provide detailed local information on the chemical nature of the sample surface. Some researchers are adding tiny electrodes into the probe tips to enable high resolution electrochemical imaging.

Much work has been invested in AFM tips in order to improve resolution further. Single-wall carbon nanotubes offer higher resolution than conventional tips as they combine extremely small size with low adhesion to the sample surface. They also buckle rather than break when put under stress and are very durable. However, one hurdle to their use is the expense and difficulty involved in fabricating them, which has made their manufacture a new area of research[14].

High speed AFM is another area of development for the technique. Hasma *et al.* have achieved 30 frames per second, which is 500 times faster than conventional AFM[15]. Applications of this work include the ability to measure real-time biological interactions.

Applications

AFM is particularly suited to more fragile samples that could be damaged by the tunnelling current in STM. These include organic materials, biological macromolecules, polymers, ceramics and glasses, and they can be imaged in different environments, such as in liquid solution, under vacuum and at low temperatures. AFM is very useful in biology and is capable of carrying out structural studies of biomolecules such as DNA[16, 17]. Because it is amenable to aqueous solutions, the investigation of biological systems under near-physiological conditions is possible. AFM also allows investigation of protein folding and unfolding and is useful for studying protein–ligand interactions.

An interesting article by Anderson has described the potential of AFM in chromatography and microfluidics as it could provide shear-driven pumping of fluid, a mechanism for injecting samples, imaging of the liquid surfaces in the microchannels and, finally, removal of samples for further spectral analysis[18].

4.5 Spectral Imaging

Principle

A new departure in spectroscopy is spectral imaging, which enables the spectral and spatial identification of chemical species in a sample. Spectral imaging is also sometimes called imaging spectroscopy or chemical imaging. The final image can be considered to be a chemical map. The original idea of spectral microscopes was to allow spectra to be taken of very small samples but it soon became clear that collecting multiple spectra from across a larger sample showed the distribution of components or properties across its surface. Imaging spectrometers for UV, Vis, NIR, IR, Raman and luminescence have come onto the market in the last few decades but can still be expensive. These instruments often look more like microscopes and collect images at a single wavelength or at a number or range of wavelengths.

Every compound is formed by chemical bonds and therefore has the potential to be detected with spectroscopy. A camera is used to measure the spatial image at each wavelength, which yields a number of images of the same specimen; contrast is based on chemical differences in the sample. The technique involves the application of reflectance/emission spectroscopy to every pixel in the spatial image. There are two main approaches. The first is taking a grid pattern of, for example, 20 by 20 spectra, over a 1 mm^2 section, which would lead to a total of 400 independent spectra. Images can then be constructed by plotting the intensity of a key band or bands over the grid. When using more than one band, different colours are assigned. The second approach is to use an array detector (perhaps with 128×128 elements) that collects all spectra simultaneously. Detection is dependent on the spectral range covered, the spectral resolution, the signal-to-noise ratio of the spectrometer part of the instrument and the strength of absorption for the sample in the wavelength region measured.

Confocal microscopy already allows the mapping of slices of a sample by spectral imaging at different depths. By combining traditional spectroscopy with microscopic and macroscopic imaging capabilities, complex heterogeneous samples can be analysed. In summary, the technique allows both the physical characteristics e.g. particle size and distribution, and chemical characteristics, e.g. the structural properties, of a sample to be investigated simultaneously.

Instrument

Dedicated spectral imaging instruments often look more like elaborate microscopes attached to computers. However, another approach is the interfacing of spectroscopic equipment to a microscope. Examples of both are shown in Figure 4.11.

Figure 4.11 The SpectraView® spectral imaging instrument from Applied Spectral Imaging and the FTIR spectral imaging system from Varian (Reproduced by permission of Applied Spectral Imaging and Varian Inc.).

Information Obtained

The spectral ranges for these instruments vary from close to 0 nm up to 15 000 nm. Bulk images or specific pixel-based images can be obtained by zooming in on a certain region of the sample. The spectroscopy can be used to detect individual absorption or emission features due to specific components.

Developments/Specialist Techniques

There are three main types of spectral imaging: multispectral, hyperspectral and ultraspectral. Multispectral means there are a number of images, each corresponding to a discrete band. This is most useful when the bands of interest are known. Hyperspectral imaging gives discrete bands over a continuous range of the electromagnetic spectrum and so can produce a 'spectrum' of the specimen. This is useful when the bands of interest are unknown. Ultraspectral means that the images are taken with a very fine spectral resolution but spatial resolution may be low. In all cases, there is often a trade-off between spatial and spectral resolution.

Applications

UV microscopy
This type of spectral imaging has most of its applications in space exploration – for studying solar radiation, comets and planet atmospheres.

Visible microscopy
All natural and man-made features (e.g. rocks, water, vegetation etc.) absorb somewhere in the electromagnetic spectrum and, in return, reflect a distinguishable electromagnetic 'signature'. Vegetation, for example, is highly reflective from approximately 750–1300 nm (visible and NIR region). Hence, mapping of specific materials from the air can be carried out over large areas of land. Spectral imaging is, therefore, an excellent tool for environmental assessments, mineral mapping and exploration and vegetation studies.

Near infrared microscopy
Based on infrared array detector technology, NIR chemical imaging techniques are the most popular in the market. Spatial, chemical, structural and functional information can

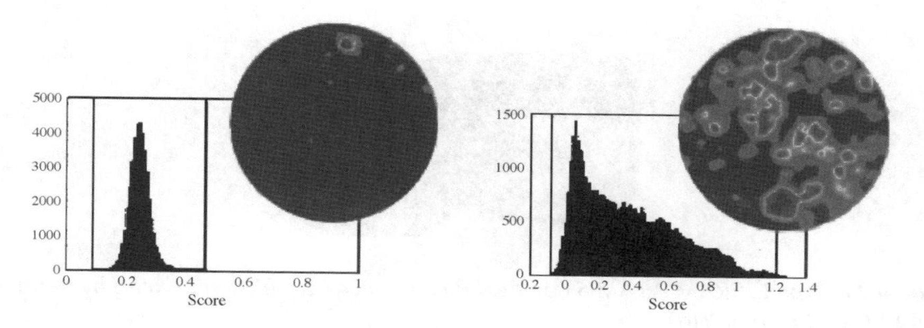

Figure 4.12 *Spectral images showing the differences in active ingredient distribution (Reproduced, with permission, from Lyon, R. C., Lester, D. S., Lewis, E. N. et al. (2002), 'Near infrared spectral imaging for quality assurance of pharmaceutical products: analysis of tablets to assess powder blend homogeneity'. AAPS Pharm Sci Tech,* **3** *(3), 1. Website:* www.malvern.com/LabEng/products/sdi/nir_chemical_imaging_range.htm.).

be derived from a single sample very quickly. Near infrared spectra are collected by an indium-antimony (InSb) camera for each pixel of the sample, creating a 3-D cube consisting of both spatially resolved spectra and wavelength dependent images. A liquid crystal tunable filter can be used as the wavelength selector, which allows for rapid scanning. The image cube can be seen as a series of wavelength resolved images or, alternatively, as a series of spatially resolved spectra, one for each point on the image.

Chemical imaging is a particularly useful technique in the pharmaceutical industry. NIR for chemical imaging of blend uniformity in tablets is now an important in-process check. The example in Figure 4.12 illustrates how the two tablet images are identical in their overall chemical composition while their NIR spectral images clearly show that in one tablet the active ingredient (in yellow and red) is more uniformly blended than in the other. The histograms portray the same blend information quantitatively over slices or sections of the tablet. Other problems with solid samples, such as polymorph distribution and impurity location, can also be investigated.

Infrared microscopy

Since the introduction of FTIR, small samples could be imaged by infrared spectroscopy. Correct placement of the sample in the beam is facilitated by using the microscope to visually magnify the sample itself. Once the sample is fixed in place, the visible illumination is turned off and the IR beam is switched on. Both transmission and reflectance modes can be used, as well as attenuated total reflectance (ATR) techniques. IR microscopy has many important medical applications and sending the video images to a computer for further analysis renders the technique very powerful. Spectra can be taken of each of the various components in living cells, e.g. lipids (cell walls), nucleic acids (DNA, RNA) and proteins. Each of these major classes of cellular components has distinct IR markers. Hence, FTIR microspectroscopy of malignant tumours has been carried out and can be very useful for diagnostic purposes.

Raman microscopy

Raman microscopy can be applied to many areas of scientific analysis, such as surface and materials science, forensic research and industry. The technique is noninvasive and

Figure 4.13 The LabRAM ARAMIS confocal Raman microscope (Reproduced by permission of HORIBA Jobin Yvon Ltd).

nondestructive and requires minimal sample preparation. Samples can be subjected to various temperatures and analysis within sealed systems is also possible. Normally, a laser source is used and, if fluorescence from the sample is an issue, excitation by a near IR laser at 785 or 1064 nm to minimise this is possible. Because of the excitation wavelengths used, the technique has better resolving power when compared to IR. Confocal Raman spectroscopy is an important technique for getting spectral images at different sample depths, e.g. in a tissue sample. An example of such an instrument is shown in Figure 4.13. Combination Raman–FTIR microscopes are also commercially available.

Fluorescence microscopy
This technique also has important applications in medicine. For example, it can be used for the spectral classification of a normal human liver cell versus a cancerous liver cell as shown in Figure 4.14. Both the normal human liver cell (E) and the cancerous liver cell (F) contain three dominant types of spectra, each of which is displayed as a distinct colour. However, when the two cells are compared quantitatively, as shown in the histogram area measurements, the abnormalities in the cancerous cell can be quantified objectively. Another example of the use of fluorescence microscopy in medicine is illustrated in Figure 4.15. Using confocal fluorescence imaging, the uptake and distribution of drug (in this case an anthracycline) can be profiled in tumour cells.

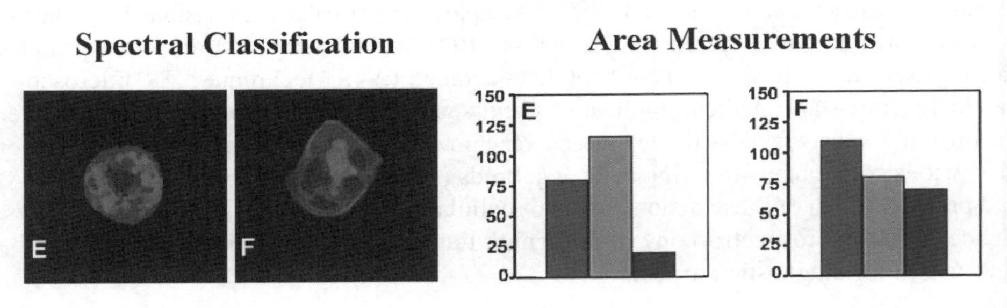

Figure 4.14 Spectral images of a normal (E) and cancerous (F) human liver cell and the respective area measurements of both images (Reproduced by kind permission of Applied Spectral Imaging).

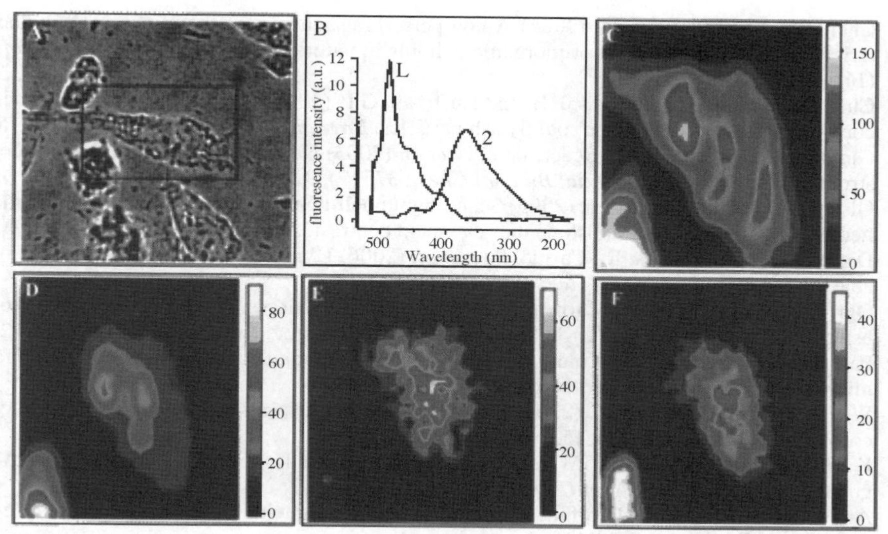

(A) Photonic microscopic image, (B) model spectra of the molecular probe (1: non-specific fluorescence emission and 2: fluorescence emission in Golgi apparatus). (C) contribution of non-specific fluorescence of the probe, (D) contribution of fluorescence in the Golgi apparatus, (E) fluorescence emission of the anthracycline drug in a resistant-like environment and (F) fluorescence of the anthracycline drug in a sensitive-like environment.

Figure 4.15 *Confocal fluorescence imaging showing drug uptake and distribution in tumour cells (Reproduced, with permission, from Belhoussine, R., Morjani, H., Millot, J.M. et al. (1998), 'Confocal scanning microspectrofluorometry reveals specific anthracycline accumulation in cytoplasmic organelles of multi-drug resistant cancer cells'. Journal of Histochemistry & Cytochemistry,* **46** *(12), 1369–1376*[19]*.).*

References

1. Rasmussen, A. and Deckert, V. (2005) New dimension in nano-imaging: breaking through the diffraction limit with scanning near-field optical microscopy. *Anal Bioanal Chem*, **381 (1)**, 165–172.
2. Navratil, M., Mabbott, G.A. and Arriaga E.A. (2006) Chemical microscopy applied to biological systems. *Anal Chem*, **78 (12)**, 4005–4019.
3. Nie, B., Shortreed, M.R. and Smith, L.M. (2006) Quantitative detection of individual cleaved DNA molecules on surfaces using gold nanoparticles and scanning electron microscope imaging. *Anal Chem*, **78 (5)**, 1528–1534.
4. Hoeflich, B.L.W., Weinbruch, S., Theissmann, R., *et al.* (2005) Characterization of individual aerosol particles in workroom air of aluminium smelter potrooms. *J Env Mon*, **7 (5)**, 419–424.
5. TEM Basics at www.matter.org.uk/tem/ Goodhew P, 2000. Based on textbook *Transmission Electron Microscopy – Basics* by Williams DB and Carter CB.
6. Davies, E. (2006) Colloids in the cold. *Chemistry World*, **3 (2)**, 46–50.
7. Fortner, J.D., Lyon, D.Y., Sayes, C.M. *et al.* (2005) C_{60} in water: Nanocrystal formation and microbial response. *Environ Sci Technol*, **39 (11)**, 4307–4316.
8. Kesselman, E., Talmon, Y., Bang, J. *et al.* (2005) Cryogenic transmission electron microscopy imaging of vesicles formed by a polystyrene–polyisoprene diblock copolymer. *Macromolecules*, **38 (16)**, 6779–6781.

9. Chanudet, V. and Filella, M. (2006) A non-perturbing scheme for the mineralogical characterization and quantification of inorganic colloids in natural waters. *Environ Sci Technol*, **40 (16)**, 5045–5051.
10. Chen, Y.Z., Shah, N., Huggins, F.E. and Huffman, G.P. (2005) Transmission electron microscopy investigation of ultrafine coal fly ash particles. *Environ Sci Technol*, **39 (4)**, 1144–1151.
11. Gunhold, A., Goemann, K., Beuermann, L. *et al.* (2003) Nanostructures on lanthanum-doped strontium titanate surfaces. *Anal Bioanal Chem*, **375 (7)**, 924–928.
12. Olson, J.A. and Buehlmann, P. (2003) Scanning tunnelling microscopy with chemically modified gold tips: *in situ* reestablishment of chemical contrast. *Anal Chem*, **75 (5)**, 1089–1093.
13. Dagastine, R.R., Manica, R., Carnie, S.L. *et al.* (2006) Dynamic forces between two deformable oil droplets in water. *Science*, **313 (5784)**, 210–213.
14. Ogrin, D., Colorado, R. Jr., Maruyama. B. *et al.* (2006) Single-walled carbon nanotube growth using $[Fe_3(\mu_3\text{-}O)(\mu\text{-}O_2CR)_6(L)_3]^{n+}$ complexes as catalyst precursors. *Dalton Trans*, 229–236.
15. Fantner, G.E., Schitter, G., Kindt, J.H. *et al.* (2006) Components for high speed atomic force microscopy. *Ultramicroscopy*, **106 (8–9)**, 2006, 881–887.
16. Wu, A.G., Wei, G. and Li, Z. (2004) Biological samples investigated by atomic force microscopy. *Fenxi Huaxue*, **32 (11)**, 1538–1543.
17. Kim, J.M., Jung, H.S., Park, J.W., *et al.* (2004) AFM phase lag mapping for protein-DNA oligonucleotide complexes. *Anal Chim Acta*, **525 (2)**, 151–157.
18. Anderson, M.S. (2005) Microfluidics and chromatography with an atomic force microscope. *Anal Chem*, **77 (9)**, 2907–2911.
19. www.jhc.org/cgi/content/full/46/12/1369/F4

5

Electrochemical Instruments

Electrochemical analytical techniques are some of the oldest in chemistry and can be divided into potentiometry, voltammetry and conductimetry. They are most important as detectors after chromatographic separations and as chemical and biological sensors. They generally involve the use of electrodes that are housed in electrochemical cells. All electrochemical cells contain two electrodes but some have three. The first electrode is the actual working electrode (also called a sensing or indicator electrode) and the second is a combined reference electrode and auxiliary (counter) electrode. If there are three electrodes, the reference and counter electrodes are separate.

There are two types of electrochemical cells: voltaic (galvanic), which *produce* energy from a chemical reaction, and electrolytic (voltammetric), which *require* or use up energy. In voltaic cells, a spontaneous chemical reaction produces electricity. These cells are important in potentiometry. In electrolytic cells, electrical energy is used to force a chemical reaction to take place such as in voltammetry. In summary:

- Potentiometric instruments measure potential (voltage)
- Voltammetric instruments measure amps (current)
- Conductimetric instruments measure conductivity.

5.1 Potentiometry

Principle

Owing to its simplicity and flexibility, potentiometry is probably the most widely used analytical technique. It is most commonly used for measuring pH and for the selective determination of analyte concentrations in a wide variety of sample solutions. Potentiometry is based on the measurement of the potential difference between the reference and working electrodes in a voltaic cell. In this type of cell, as mentioned above, a spontaneous redox chemical reaction occurs due to one reagent being oxidised (losing electrons) at the anode

Analytical Instrumentation: A Guide to Laboratory, Portable and Miniaturized Instruments G. McMahon

and one reagent being reduced (gaining electrons) at the cathode. The reagents are not physically in contact with each another, so the electrons are forced to flow through an external circuit to move between the reagents and hence potential can be measured along this path by a voltmeter or similar device.

Potentiometry requires a reference electrode, a working electrode and a potential-measuring instrument, e.g. voltmeter, otherwise known as a potentiometer. The test solution must be in direct contact with the working electrode, which is sometimes referred to as the chemical sensor as it is 'sensing' the output of a chemical reaction. The reference electrode can also be placed in the test solution or can be brought into contact with the test solution via a salt bridge. The measured potential can be related to the concentration of the species being measured and this approach is called direct potentiometry.

The most common type of potentiometric instruments are ion-selective electrodes (ISEs) which act like probes. An ISE is a complete electrochemical cell that is selective for a particular ion and can measure that ion almost to the exclusion of others when dipped into a test solution. However, unlike typical voltaic cells, ISEs are not based on redox processes. Instead, they operate based on the fact that analyte ions of interest will migrate across a selective membrane from a region of higher concentration to a region of lower concentration. This causes a potential difference (charge build-up) across the membrane, the magnitude of which can indicate the concentration of analyte on both sides of the membrane. The best known ISE is the pH glass electrode, which can measure the concentration of hydrogen ions in almost any aqueous solution.

Most ISEs have high resistance and must be used with a direct reading instrument, e.g. pH meter. ISEs measure only the free ion in solution and sometimes a special buffer must be added before measurement. They have many advantages, such as the fact that they function in turbid or coloured solutions, they have a log response and hence a wide dynamic range (four to six orders of magnitude), they have a rapid response, can be used in flowing streams and also, importantly, the sample is not destroyed. However, ISEs can be subject to fouling, especially in biological samples, and are susceptible to interference from other ions to lesser or greater extents. Further information on ISEs is available elsewhere[1].

Instrument

A schematic diagram of an ISE is illustrated in Figure 5.1.

Source

The source in this instrumental set-up is the reference electrode, the most popular examples of which are the silver/silver chloride electrode or the mercury/mercury chloride (saturated calomel) electrode. The important requirements of a reference electrode are reversibility, reproducibility and stability in time. Often these are incorporated into the ISE housing so that it contains the working and reference electrodes together and this combination electrode is attached to the voltmeter.

Discriminator

The discriminator is the working electrode, which is the ion-selective membrane of the ISE itself and is where the selectivity for the analyte is based. It is often called an electrochemical

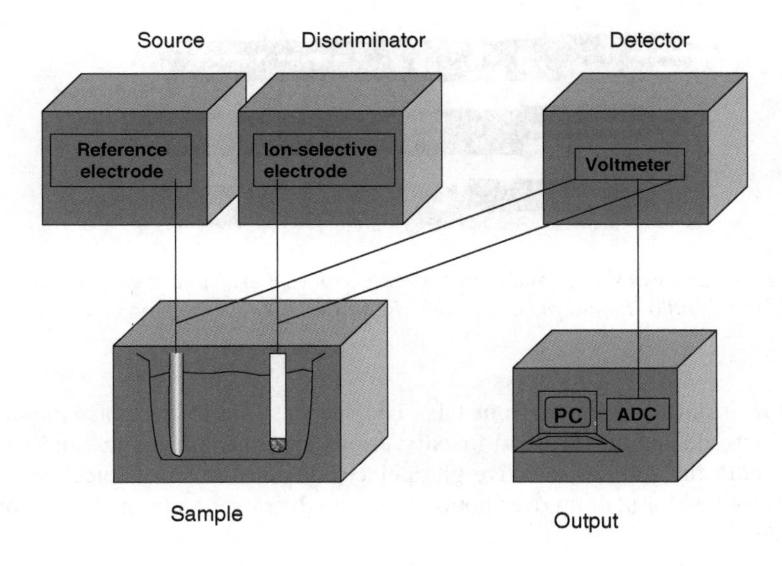

Figure 5.1 *Schematic diagram of an ion-selective electrode.*

sensor. There are many types of working electrodes available commercially though many more versions can be manufactured by the analyst, often quite easily:

- Glass membrane electrodes
- Crystalline electrodes
- Liquid membrane electrodes
- Gas-sensing membrane electrodes
- Enzyme-based electrodes
- Semiconductor electrodes.

Glass membrane electrodes were the first ISEs to be discovered and are widely used for the determination of pH, ions and for in-process monitoring. This type of ISE has good selectivity, but only for several single-charged cations, mainly hydrogen (H^+), sodium (Na^+) and silver (Ag^+), and in some cases double-charged metal ions, such as lead (Pb^{2+}), and cadmium(Cd^{2+}). Glass membranes are usually made from an ion-exchange type of glass that has excellent chemical durability and can work in very aggressive media, e.g. liquid ammonia and biological matrices. The most common glass electrode is the hydrogen ion ISE (the pH electrode). In this electrode there is a reference electrode and the working (sensing) electrode, which is a glass membrane. The special glass membrane is highly selective for hydrogen ions as long as it remains hydrated. The potential varies logarithmically with hydrogen ion concentration at the sample interface. A pH meter is usually attached to the pH probe (ISE) to accurately record the values.

It is possible to get all sorts of shapes and sizes of pH electrode, (see Figure 5.2 for example). It is also possible to obtain electrodes that are application specific, for example pH probes for the life sciences which allow immersion in very small samples and have a protein resistant junction or pH probes for food samples where the

Figure 5.2 *A variety of shapes and sizes of pH electrodes is available, e.g. epoxy body (A), flat (B) or round (C) bulb shaped pH electrodes (Reproduced by permission of Weiss Research).*

probe is resistant to clogging from fats. The composition of the glass membrane in the pH electrode can be varied to favour various ions such as sodium and silver. For example, both the sodium-selective glass electrode and the silver-selective electrode have selectivities for sodium over potassium and silver over sodium of three orders of magnitude.

Crystalline electrodes use an ionically conducting salt, sparingly soluble in water that either forms the membrane, e.g. a pure crystal of the measured species, or is incorporated into the membrane using an inert material. Crystalline membranes are made from mono- or polycrystallites of a single substance. They have good selectivity because only ions which can introduce themselves into the crystal structure can affect the electrode's response. Selectivity of crystalline membranes can be for both the cation and anion of the membrane-forming substance. The most well known of these crystalline electrodes is the fluoride electrode which is based on lanthanum (III) fluoride crystals (doped with Europium(II) to increase membrane conductivity). It is sensitive to fluoride concentrations at the parts per billion level. The electrode has a 1000-fold selectivity for fluorine ions (F^-) over among other ions chlorine (Cl^-), bromine (Br^-) and iodine (I^-). The main interferent is the hydroxy ion, through its effect can be minimised using a special buffer called TISAB (total ionic strength adjustment buffer).

Liquid membrane electrodes are composed of a water-immiscible liquid ion carrier held in place by a porous hydrophobic membrane that allows contact between the test solution and the ion carrier but minimises actual mixing. The ion carrier can be electrically charged (ion exchange resin) or electrically neutral. These ion carriers are compounds with ion-binding sites and are usually incorporated into the soft hydrophobic plastic membrane, e.g. polyvinyl chloride (PVC) matrices. The plastic membranes are easy to make and pliable enough that they can be mounted on the end of an electrode body. The ionophore is lipophilic (so it does not leach out of the membrane in aqueous solutions) and in the membrane it selectively complexes with the ion of interest.

The two most well known liquid membrane electrodes are cation-selective: the potassium and calcium electrodes. For the potassium ISE, the soft plastic membrane has the neutral hydrophobic ionophore *valinomycin* immobilised in it. This electrode has a 10 000-fold selectivity for potassium ions over sodium ions. For the calcium ISE, the liquid ion exchanger membrane is a water-immiscible calcium chelator. Calcium ions are

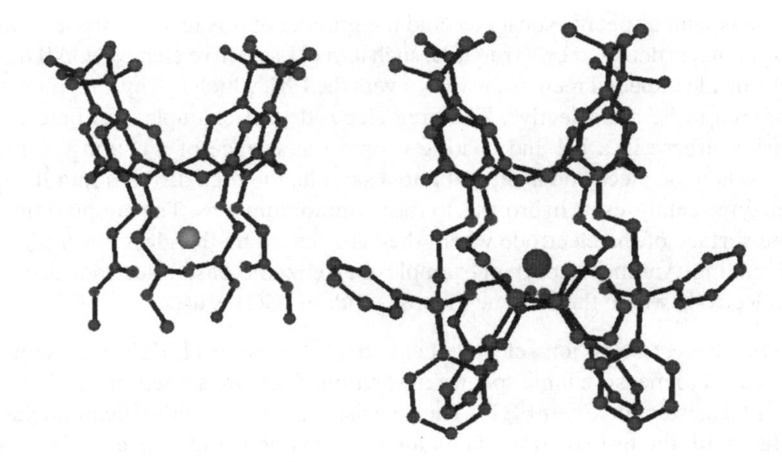

Figure 5.3 *Tetraethylestercalix[4]arene and tetraphosphine oxide calix[4]arene (The tetraethylester cavity is selective for sodium while the larger tetraphosphine oxide cavity is selective for calcium. Images kindly provided by Prof. Dermot Diamond, Dublin City University, Ireland.).*

selectively transported across the membrane to establish a voltage related to the difference in concentration between the sample and the internal filling solution. This electrode has a 3000-fold selectivity for calcium ions over sodium and potassium ions. Liquid membrane ISEs are also the most widespread electrodes with anionic selectivity and an important example is the nitrate-selective electrode.

New ionophores have become available in recent years to act as ion carriers in ion-selective electrodes, such as crown ethers and calix[4]arenes, macrocyclic molecules with polar cavities to hold the analyte ion and a nonpolar (neutral) outer shell to prevent mixing with or leaching into the aqueous sample. The development of some of these tailored ionophores has led to the availability of very selective ISEs for a number of metal ions, such as the sodium and potassium electrodes (see Figure 5.3)[2, 3].

ISEs can be prepared by simply coating a bare electrode wire with a selective PVC membrane such as those described above. These coated wire electrodes are very easy to prepare and use. A THF solution of the membrane ingredients is simply coated onto the wire and allowed to evaporate.

Gas-sensing membrane electrodes use a gas-permeable membrane that allows measured species (as a dissolved gas) to pass through and be measured within the electrode. These probes are very selective and sensitive for gases such as ammonia and carbon dioxide. For carbon dioxide, the working electrode is usually a modified glass pH electrode covered in a teflon membrane. Carbon dioxide in solution forms carbonic acid which lowers the pH.

Enzyme-based electrodes: ISEs can be used in conjunction with immobilised enzymes to make them selective for enzyme substrates. These electrodes are not true ion-selective electrodes but are based upon them. Such an electrode has a two-step mechanism: an enzyme reacts with a specific substance and the product of this reaction (usually hydrogen or hydroxyl ions) is detected by a true ISE, such as a pH-selective electrode. All these reactions occur inside a special membrane that covers the ISE, which is why enzyme electrodes are considered to be ion-selective. The urea electrode, for example, can be prepared by immobilising urease in a gel and coating it onto the surface of a cation-sensitive glass electrode. When the electrode is dipped into a sample, the urea diffuses into the gel layer and the enzyme catalyses its hydrolysis to form ammonium ions. The ammonium ions diffuse to the surface of the electrode where they are sensed by the glass electrode to give a potential reading. Another important example of an enzyme-based electrode is the glucose selective electrode where the enzyme glucose oxidase (GO) is used.

Semiconductor electrodes: ion-selective field effect transistors (ISFETs) are semiconductor devices used to measure ionic species in solution. They are sometimes called chemical field-effect transistors or ChemFETs. The transistor is coated with silicon nitride, which is in contact with the test solution via an analyte-sensing membrane and also connected to a reference electrode. A variation in the concentration of the analyte ions changes proportionally with the voltage of the ISFET. ISFETs are rugged, have a faster response time than membrane electrodes and can be stored dry.

Sample

The electrochemical cell in an ISE has two electrodes, the working and reference electrodes. Often these are combined into a single probe. Measurement is usually by direct immersion of the ISE probe into a liquid sample. ISEs respond to the activity of analyte ion as opposed to concentration. As chemists wish to measure concentration, TISAB can be added to bring up the ionic strength of the samples and standards. Under such conditions of high, constant ionic strength, the measured potential gives a concentration result. If the matrix composition is not known, the standard addition method can be used instead of direct immersion. This involves measuring the potential of the unknown solution first and then adding known amounts of standard to aliquots of the unknown solution for measurement. A standard curve can then be plotted and the intercept of the curve yields the background (unknown) concentration.

Detector

The voltmeter (potentiometer) measures the difference in potential between the two electrodes (reference and working). It indicates a positive voltage when the electrons flow into the negative terminal and a negative voltage when they flow in the opposite direction. When a pH meter is used as a voltmeter, the positive terminal is the one to which the working electrode is attached and the negative terminal the one to which the reference electrode is attached. The measured potential is proportional to activity (concentration) of the analyte ion.

Output

The signal is converted into a useful form and displayed at the PC.

Information Obtained

ISEs are usually used quantitatively for measuring the concentration of an analyte species in solution. They are calibrated in standard solutions first before determining the potential (concentration) of the unknown sample. Commercial ISEs display the analyte concentration on a visual display on the voltmeter.

Developments/Specialist Techniques

Micro ISEs can now be manufactured and can be used in the human body and inside cells and animals. Examples include microelectrodes for dopamine in cerebrospinal fluid[4] and in cells[5] and glucose in blood[6]. In terms of instrumentation, there are now ISEs available which (via an interface) can be directly attached to a laptop or PC. This eliminates the need for a conventional pH/mV/ion meter and permits the use of more sophisticated and efficient procedures for data acquisition, processing, recording and display (see Figure 5.4).

Applications

Ion-selective electrodes are used for many applications but most commonly for the measurement of pH and metal ions such as tin[7], lead[8], silver[9] and nickel[10] as well as various other analytes such as surfactants[11]. The samples investigated range from surface waters, e.g. lakes and streams, ground waters, rain, to soil and food. ISEs for anion detection are not as prevalent due to solubility issues but more are becoming available, such as one for iodide with a detection limit of 5.3×10^{-7} M,[12] and with the help of ionic liquids another has been developed for sulfate detection[13].

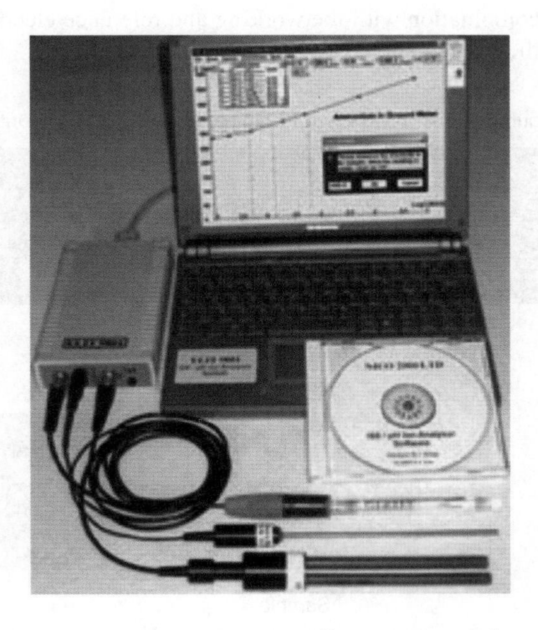

Figure 5.4 *The ELIT ion/pH analyser system (Different ISEs can be attached as required. Reproduced by permission of Nico 2000 Ltd.).*

5.2 Voltammetry

Principle

Voltammetry is the measurement of current–voltage and current–time relationships at a working electrode immersed in a solution containing an electroactive species. Such species are capable of being oxidised or reduced. Voltammetry is electrolysis on a very small scale where the working electrode is a microelectrode. The current generated in voltammetric measurements is like the absorbance observed when obtaining a UV scan. With voltammetry, potentials are scanned to give a current response yielding a voltammogram while with UV, wavelengths are scanned to give an absorbance response called a spectrum. The measured changes in current are proportional to concentration. Because the microelectrode keeps the current very small (microamps), and because the test solution volume is relatively large and the analysis time short, the concentration of the test solution is not significantly changed during analysis. This means that repeated measurements can be carried out in the same solution and that the sample can be recovered almost unchanged at the end of the experiment.

Instrument

The components of a voltammetric instrument are shown in Figure 5.5.

Source
The source in this instrumental setup is the power supply.

Discriminator
The potentiostat in combination with the working and reference electrodes form the discriminating part of the instrument.

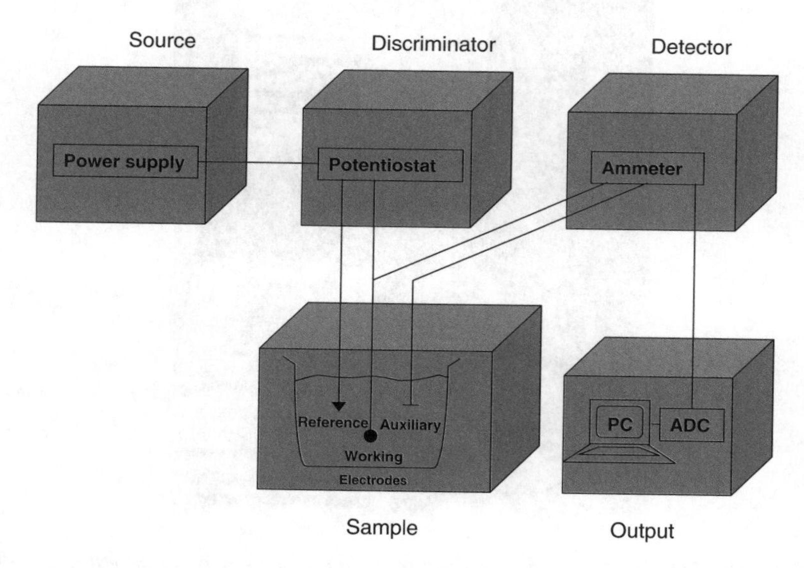

Figure 5.5 *Schematic diagram of a voltammetric instrument.*

The *potentiostat* provides and controls the voltage between the reference and working electrodes according to a pre-selected voltage–time program and includes a potentiostatic control circuit and a current-to-voltage converter. Generally, the potential applied to working electrode is swept (linearly) or pulsed with time and the resulting current measured, but other programs are possible. The potential scan rates are typically between 0.001 and 1000 V/s. At a certain potential during the experiment, the analyte will be oxidised or reduced.

In voltammetric analysis, the *working electrode* can be one of a few types of electrode: dropping mercury electrode (DME), the static mercury drop electrode (SMDE) or a solid electrode. Voltammetric techniques employing a mercury electrode are called *polarography*. In this technique, the DME or SMDE is used as a micro working electrode. In polarography, current is measured between the DME and the auxiliary electrode and voltage is measured between the DME and the reference electrode. The mercury drop is formed with highly reproducible diameter and lifetime. One advantage of this technique is that a mercury electrode can tolerate high negative potentials for the reduction of analytes that could not be reduced at other electrodes. The other advantage is that there is a new electrode surface generated each time (with each new drop), which eliminates contamination of the surface and gives more reproducible current–potential data.

The SMDE is based on the DME but the drop size and drop time are controlled even more precisely and it has a constant surface area, so it is often employed in pulsed polarography techniques. This results in smoother polarogram traces. Solid electrodes are used for the determination of oxidisable analytes. The most popular types are graphite, carbon paste and glassy carbon electrodes.

The *reference electrode* is usually a saturated calomel electrode (SCE) and its potential remains constant throughout the experiment, so the applied potential at the working microelectrode is always known. In aqueous solutions, a simpler two-electrode instrument can be used where the working and reference electrodes are incorporated into one.

Sample

The electrochemical cell in voltammetry has three electrodes: working, reference and auxiliary (counter) (Figure 5.6). The sample is placed in this voltammetric cell, the voltage or voltage ramp applied as required and the resulting current measured.

Detector

The potentiostat is connected to the auxiliary electrode to measure current between it and the working electrode. The auxiliary electrode is the current-supporting partner of the working electrode. Significant current flows between the working and auxiliary electrodes.

Output

The signal is converted into a useful form and displayed at the PC.

Information Obtained

Both qualitative and quantitative information is obtained in voltammetry. The half wave potential is characteristic for a particular compound in a particular medium, so the

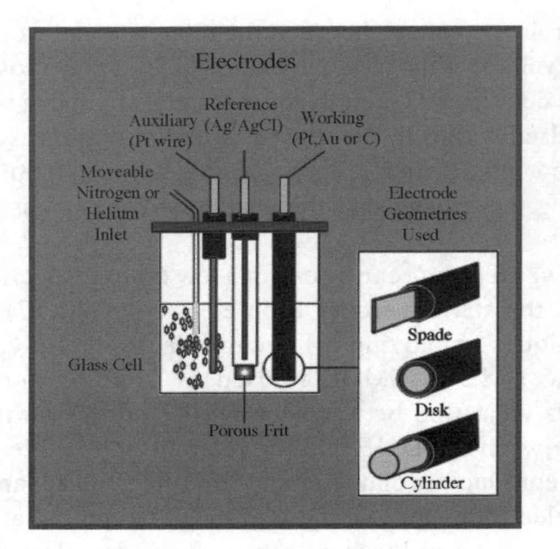

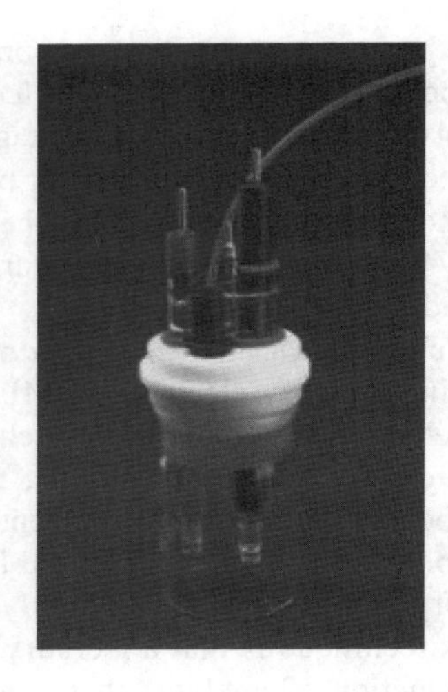

Figure 5.6 *Schematic diagram and photograph of a three-electrode voltammetric cell (Figure used by kind permission of Dr. Marco Cardosi, Dept of Biological Sciences, University of Paisley, UK; photograph courtesy of BAS Inc., USA.).*

identity of a compound can be established. The measured current is proportional to the concentration of the electroactive species as long as there is sufficient concentration of supporting electrolyte. For quantitative work, external standards can be run and a calibration curve used, the standard addition method can be employed or internal standards can be used. Some examples of typical instruments used in voltammetry are shown in Figure 5.7.

Developments/Specialist Techniques

There are a number of techniques that can be used in voltammetry to increase sensitivity; these include the following:

- Linear sweep voltammetry
- Cyclic voltammetry
- Differential pulsed polarography
- Adsorptive stripping voltammetry
- Amperometry.

Linear sweep voltammetry is based on the potential being ramped up between the working and auxiliary electrodes as current is measured. The working electrode is usually a SMDE nowadays, in which case this technique would be called linear sweep polarography. In this set-up, the auxiliary electrode is a mercury pool electrode and may also serve as the reference electrode. The resultant current–potential recording (the polarogram) can yield much information which can be used to qualitatively identify the species and the medium in which it is determined as well as calculate concentrations. Analysis of mixtures is also possible. The detection limit is of the order of 10^{-5} M.

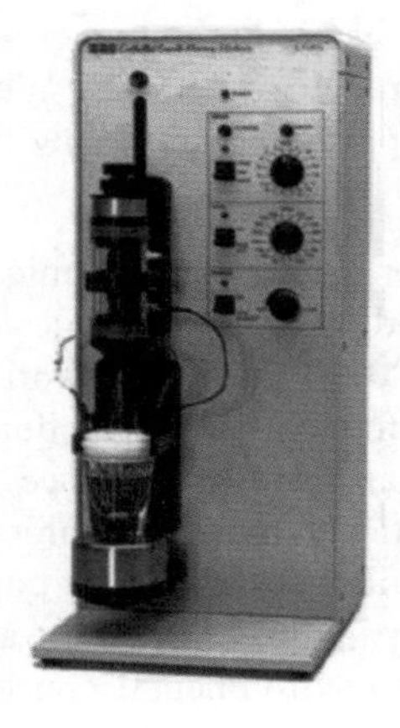

Figure 5.7 *Controlled growth mercury electrode for polarography and C-3 cell stand for voltammetry (Reproduced by permission of BAS Inc.).*

If the direction of the potential ramp is reversed at the end of the sweep back to the starting potential, the technique is called *cyclic voltammetry*. In this case, one scan is due to the reduction of a species (cathodic current) and the other is due to oxidation of that species (anodic current). The two current plots combined make a voltammogram (Figure 5.8). Anodic and cathodic currents are always opposite in sign and sigmoidal in shape. Cyclic voltammetry allows variations in scan rate to study kinetics and can also characterise the oxidation and reduction behaviour of various electroactive compounds.

Differential pulsed voltammetry (DPV) is a technique in which potential pulses of fixed but small amplitudes are superimposed periodically on a linear voltage ramp. The most commonly used working electrode is the SMDE and one pulse is applied for each drop. The mV pulse is applied near the end of the life of the mercury drop. Current is measured once before the pulse and after the pulse. The difference between the currents is plotted against potential (Figure 5.8). The resultant peak-shaped current–voltage signal, which

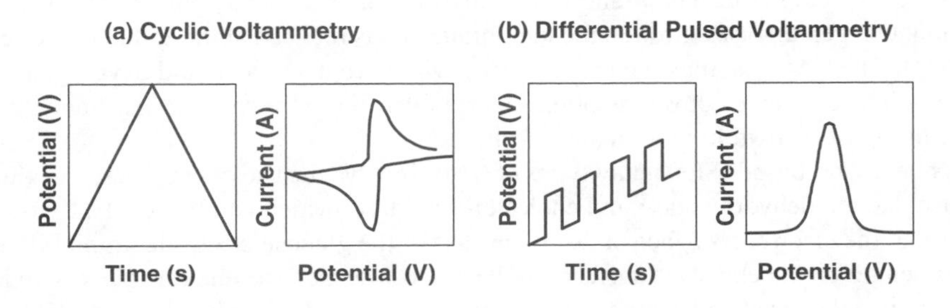

Figure 5.8 *The applied voltage ramp profile and the resulting voltammogram for two different types of voltammetry experiments.*

is like the derivative of the direct current voltammogram, has a higher selectivity and sensitivity (10^{-7}M) than the aforementioned voltammetric methods. With the use of the SMDE, the technique can be called differential pulsed polarography (DPP) and the current–voltage signal a polarogram.

Adsorptive stripping voltammetry (ASV) is another specialised technique where the SMDE electrode is used for reducible species and carbon paste electrodes for oxidisable ones. This allows enrichment (by factors of 100–1000) of ions at the working electrode before 'stripping' them off for measurement; this improves the detection limits. This technique is rapid, sensitive (10^{-11}M), economical and simple for trace analysis. The basic instrumentation for stripping analysis is a potentiostat (with voltammetric analyser), electrode and recorder. While voltammetry is generally very useful for compounds that do not have a chromophore or fluorophore, stripping analysis is the best analytical tool for direct, simultaneous determination of metals of environmental concern, e.g. lead, cadmium, zinc and copper in sea water.

Amperometry is when voltammetry is used at a fixed potential and the current alone is followed. This current is proportional to the concentration of a species in a stirred or flowing solution. It is, therefore, a very suitable detection method for use in flow injection analysis systems and in chromatographic separations; its use as a detector for separation techniques is discussed in that section. The current is the result of the electrochemical oxidation or reduction of the analyte after application of a potential pulse across the working and auxiliary electrodes.

Electrochemical sensors based on amperometric detection are popular small instruments that can be used to directly probe samples. The electrodes in these sensors do not always have to be made of metal, e.g. platinum or mercury, and do not always have to be bare. A commonly employed working electrode is the rotating disk electrode, which is preferred over the DME for easily reduced species and anodic reactions. It works by convection mixing of the solution so that fresh sample is constantly passed over the surface.

Chemically modified electrodes (CMEs) can provide new dimensions in amperometric sensing. The most common CMEs are coated chemically with molecules or coated with a layer of polymer film in which selected molecules are embedded. Both types have huge applications in sensors. One well-known type of CME is the oxygen electrode, where the working electrode is usually platinum over which is stretched a thin oxygen permeable membrane of Teflon. The membrane allows diffusion of gases but is impermeable to ions in solution. Oxygen diffuses through the membrane and is reduced at the cathode, producing a current. The electrode must be precalibrated with both de-aerated and oxygen saturated samples. The oxygen electrode is often incorporated into clinical multigas analysers for determining oxygen within blood and serum.

As in enzyme-based ISEs, many types of CME involve the use of enzymes, e.g. glucose oxidase, lactate dehydrogenase and cholesterol oxidase, which can be covalently bonded directly to the electrode surface or, as in the case of the glucose electrode, immobilised in a polymer gel on the electrode surface. In the glucose sensor, the enzyme glucose oxidase is immobilised in a gel and coated onto the surface of a platinum wire electrode. This type of electrode can also be called a conducting polymer electrode. The enzyme catalyses the aerobic oxidation of glucose to give gluconic acid and hydrogen peroxide. The hydrogen peroxide diffuses to the platinum electrode where it is oxidised to give a current in proportion to the glucose concentration in the test sample.

Amperometric electrodes that are extremely thin (<1 μm diameter) are called ultramicroelectrodes and have a number of advantages over conventional electrodes. Being narrower than the diffusion rate thickness, mass transport is enhanced, the signal-to-noise ratio is improved and the measurements can be made in resistive matrices such as nonaqueous solvents. These have huge applications in medicine as they can fit inside a living cell. Carbon fibre electrodes are coated in insulating polymer and plated with a thin layer of metal at the exposed tip to prevent fouling of the carbon itself. These can then be used to measure analytes of interest in various cells and membranes of the human body.

Applications

Amperometric techniques are very useful for detecting analytes that have been separated by chromatographic means but have no chromophores or other easy means of detection. Adsorptive stripping voltammetry (ASV) can be used for the direct sensitive analysis of metals in many types of sample matrix. For example, ASV has been used to determine cadmium, lead and zinc in urine[14], copper and bismuth in human hair[15, 16], tin in fruit juice[17], zinc and copper in fish[18] and lead in gunshot residue[19]. Stripping analysis can also be used for other applications such as determining flavanols in wine[20], inorganic compounds such as cyanide[21] and pharmaceuticals[22].

5.3 Conductimetry

Principle

Conductimetric instruments use electrodes to measure the total ion current in a solution. Conductimetry actually measures resistance between electrodes according to Ohm's Law and the inverse of resistance is conductance. Conductivity is conductance with the cross-sectional area and length between electrodes taken into account. No oxidation or reduction takes place in conductivity detection. It is a bulk (non-specific) measurement and all solutions have some degree of conductivity. Electrodes are often used in flow-through systems to measure changes in conductivity, i.e. the presence, absence, increase or decrease in ions over time.

Conductimetry depends on a number of factors:

- surface area of electrodes
- shape of electrodes
- relative position of electrodes
- type of solution
- concentrations
- temperature.

Therefore, conditions for conductivity experiments need to be controlled. There are two techniques for conductivity measurement: the use of contacting electrodes, where the contacting cells can have two or four electrodes, and the use of noncontacting electrodes, where conductance measurements can be made without physical contact between the solution and any metallic conductors by using electromagnetic induction, e.g. radiofrequency (RF).

Instrument

Commercial instruments for both contacting and noncontacting electrode types come in the form of meters, probes, sensors and flow-through cells. Top-of-the-line instruments will have temperature compensation (a thermistor) built in. It is a simple instrument with the following components: a conductivity cell, electrodes and a voltage supply. The conductivity cell consists of two platinum plates with an AC voltage applied between them. The eluent flows through the cell and the current, which depends on the concentration and type of ions, is measured. A conductivity detector responds to all ions but not to molecular compounds like ethanol and water.

The most common forms of the conductivity instrument are small meters or probes for direct measurement of the conductivity of samples in the field or as conductivity detectors following separation techniques such as high performance liquid chromatography (and ion chromatography in particular). In the detector situation, the reading will be a difference measurement between the background eluent and the eluent with analyte(s) of interest. Conductivity instruments will be discussed further in each of these relevant sections.

Information Obtained

A conductivity measurement is normally quantitative in terms of relating increasing or decreasing conductivity to the concentration of the analyte of interest. An example of a conductivity device is shown in Figure 5.9.

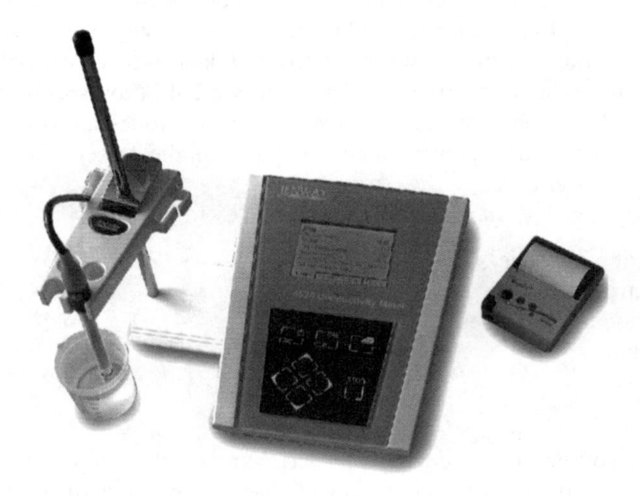

Figure 5.9 Laboratory conductivity meter (Courtesy of Jenway).

Developments/Specialist Techniques

Benchtop conductivity meters are not as prevalent now, as portable and handheld devices have become commercially available. These are discussed further in the sections on portable instruments and process instruments. The conductivity meter is often combined with other electrochemical measurements on the same device, e.g. pH or dissolved oxygen.

Applications

There are many applications for conductimetric measurements, some of which are very specific. For example, in a medical application, a conductivity sensor has been used to determine levels of ammonia in human breath[23]. In the same article, the authors also demonstrated a second sensor for breath analysis of carbon dioxide. In environmental analysis, a samarium iron oxide ($SmFeO_3$) gas sensor has been developed for the detection of ozone based on conductance at sub-ppm levels[24].

References

1. Rundle, C.C. A beginners guide to ion-selective electrode measurements. *Nico*, www.nico2000.net/Book/Guide1.html (2005).
2. Diamond, D. and Nolan, K. (2001) Calixarenes: designer ligands for chemical sensors. *Anal Chem*, **73 (1)**, 22A–29A.
3. Lynch, A., Eckhard, K., McMahon, G. *et al.* (2002) Cation binding selectivity of partially substituted calix[4]arene esters. *Electroanalysis*, **14 (19–20)**, 1397–1404.
4. Sun, Y.H. and Zheng, X.X. (2000) Study of ion-selective microelectrode sensitive to the neurotransmitter dopamine. *Fenxi Huaxue*, **28 (8)**, 993–996.
5. Cui, H.F., Ye, J.S., Chen, Y. *et al.* (2006) *In situ* temporal detection of dopamine exocytosis from L-dopa-incubated MN9D cells using microelectrode array-integrated biochip. *Sens Actuat B*, **B115 (2)**, 634–641.
6. Zhu, J.Z., Xie, J.F., Lu, D.R. *et al.* (2000) Micromachined glucose sensor and potassium ion ISE based on containment array. *Sens Actuat B*, **B65 (1–3)**, 157–159.
7. Arvand, M., Moghimi, A.M., Afshari, A. and Mahmoodi, N. (2006) Potentiometric membrane sensor based on 6-(4-nitrophenyl)-2,4-diphenyl-3,5-diaza-bicyclo{3.1.0}hex-2-ene for detection of Sn(II) in real samples. *Anal Chim Acta*, **579 (1)**, 102–108.
8. Ardakani, M.M., Kashani, M.K., Salavati-Niasari, M. and Ensafi, A.A. (2005) Lead ion-selective electrode prepared by sol-gel and PVC membrane techniques. *Sens Actuat B*, **107 (1)**, 438–445.
9. Yan, Z.N., Lu, Y.Q. and Li, X. (2006) The preparation of silver ion selective PVC membrane electrode based on neutral ionophore tetramethylene bis(N-ethyl-N-phenyldithiolcarbamate). *Fenxi Ceshi Xuebao*, **25 (5)**, 7–10.
10. Kumar, K.G., Poduval, R., Augustine, P. *et al.* (2006) A PVC plasticized sensor for Ni(II) ion based on a simple ethylenediamine derivative. *Anal Sci*, **22 (10)**, 1333–1337.
11. Wei, F.X. and Hao, L.L. (2006) Preparation of flow-through tubular ion-selective electrode for the determination of anionic surfactants. *Fenxi Ceshi Xuebao*, **25 (6)**, 53–55.
12. Xu, W.J., Chai, Y.Q., Yuan, R. *et al.* (2006) Highly selective iodide electrode based on the copper(II)-*N,N'*-bis(salicylidene)-1,2-bis(*p*-aminophenoxy)ethane tetradentate complex. *Anal Sci*, **22 (10)**, 1345–1350.
13. Coll, C., Labrador, R.H., Manez, R.M., *et al.* (2005) Ionic liquids promote selective responses towards the highly hydrophilic anion sulphate in PVC membrane ion-selective electrodes. *Chem Comm*, **24**, 3033–3035.

14. Kefala, G. and Economou, A. (2006) Polymer-coated bismuth film electrodes for the determination of trace metals by sequential-injection analysis/anodic stripping voltammetry. *Anal Chim Acta*, **576 (2)**, 283–289.
15. Qi, J.R., Li, Z. and Xu, H.D. (2005) Stripping voltammetric determiantion of copper (II) with tetrahydroxyanthraquione-activated carbon paste electrode. *Fenxi Huaxue*, **33 (12)**, 1740–1742.
16. Guo, H.S., Li, Y.H., Xiao, P.F. and He, N.Y. (2005) Determination of trace amount of bismuth(III) by adsorptive anodic stripping voltammetry at carbon paste electrode. *Anal Chim Acta*, **534 (1)**, 143–147.
17. Prior, C. and Walker, G.S. (2006) The use of the bismuth film electrode for the anodic stripping voltammetric determination of tin. *Electroanalysis*, **18 (8)**, 823–829.
18. Celik, U. and Oehlenschlaeger, J. (2004) Determination of zinc and copper in fish samples collected from Northeast Atlantic by DPSAV. *Food Chem*, **87 (3)**, 343–347.
19. de Donato, A. and Gutz, I.G.R. (2005) Fast mapping of gunshot residues by batch injection analysis with anodic stripping voltammetry of lead at the hanging mercury drop electrode. *Electroanalysis*, **17 (2)**, 105–112.
20. Volikakis, G.J. and Efstathiou, C.E. (2005) Fast screening of total flavonols in wines, tea-infusions and tomato juice by flow injection/adsorptive stripping voltammetry. *Anal Chim Acta*, **551 (1–2)**, 124–131.
21. Safavi, A., Maleki, N. and Shahbaazi, H.R. (2004) Indirect determination of cyanide ion and hydrogen cyanide by adsorptive stripping voltammetry at a mercury electrode. *Anal Chim Acta*, **503 (2)**, 213–221.
22. El-Shahawi, M.S., Bashammakh, A.S. and El-Mogy, T. (2006) Determination of trace levels of diosmin in a pharmaceutical preparation by adsorptive stripping voltammetry at a glassy carbon electrode. *Anal Sci*, **22 (10)**, 1351–1354.
23. Toda, K., Li, J.Z. and Dasgupta, P.K. (2006) Measurement of ammonia in human breath with a liquid-film conductivity sensor. *Anal Chem*, **78 (20)**, 7284–7291.
24. Hosoya, Y., Itagaki, Y., Aono, H. and Sadaoka, Y. (2005) Ozone detection in air using $SmFeO_3$ gas sensor. *Sens Actuat B*, **108 (1–2)**, 198–201.

6
Other Instruments

There are a host of other instruments that could be included in this section but the two types that will be discussed here are thermoanalytical and X-ray diffraction instruments. Thermal analysis involves monitoring a specified physical property as a function of temperature. It finds most of its applications in materials research and quality control. The purpose of making thermal analysis measurements is to study the chemical and physical changes that occur in compounds (usually solids) during a temperature program. The plot of the property against temperature is called a thermal analysis curve. The property and the temperature are measured automatically and continuously. The temperature of the sample is altered in a predetermined manner. Thermoanalytical methods are versatile as they can be combined with each other, e.g. thermogravimetric analysis and differential thermal analysis, to provide more information from the same sample during a single measurement.

X-ray diffraction experiments enable the structures of unknown crystals, as well as the orientation and perfection of large single crystals and their lattice parameters, to be determined. Many solids from small molecules to large proteins can be profiled using this analytical technique.

6.1 Thermogravimetric Analysis

Principle

Thermogravimetric analysis (TGA) is a technique in which the mass of the sample is measured against time or temperature while the temperature is changed under a certain atmosphere. This method is useful for determining sample purity and water content (adsorbed versus absorbed), carbonate and organic content and for studying decomposition reactions. The graph of the resultant change in mass with respect to temperature is called a thermogravimetric (TG) curve. The rate at which the property changes with temperature can also be measured, e.g. kinetics. This is derivative thermogravimetry (DTG), the curve of which can be extremely useful in interpreting the TGA curves by resolving overlapping thermal reactions.

Analytical Instrumentation: A Guide to Laboratory, Portable and Miniaturized Instruments G. McMahon

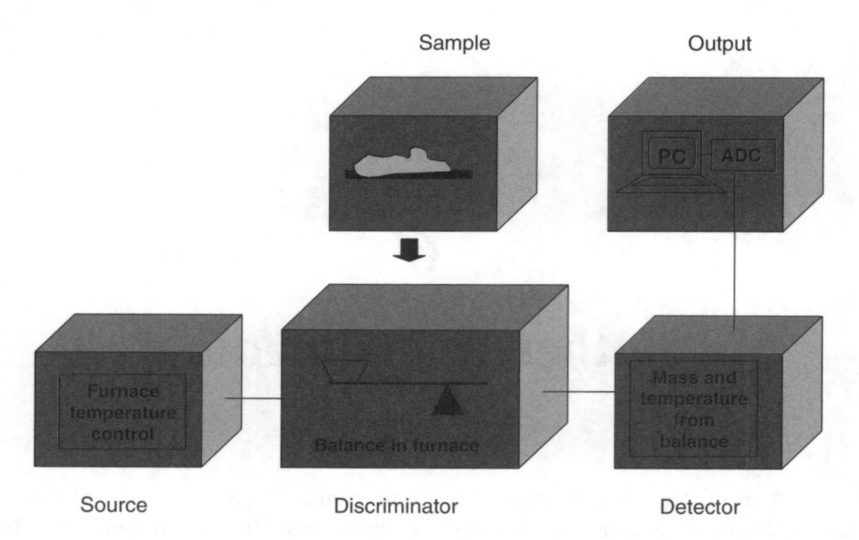

Figure 6.1 *Schematic diagram of a thermobalance.*

Instrument

The instrument used is called a thermobalance or thermogravimetric analyser. The two most important parts are the furnace and the balance. A schematic diagram is shown in Figure 6.1.

Source
The source is the furnace temperature (and environment) controller. The required temperature programs, heating and/or cooling, come from here.

Sample
A sample is placed into a tared TGA sample pan attached to a sensitive microbalance assembly. The sample holder portion of the TGA balance assembly is subsequently placed into the high temperature furnace. There are a number of different set-ups for how the sample is placed in the balance, each having its own advantages and disadvantages. The first is where the sample is placed horizontally relative to the balance and furnace, the second is where the sample is placed in the pan by top-loading and the third is where the sample is suspended in the furnace.

Discriminator
The discriminator is the balance within the furnace. The balance allows the measurement of any changes in the weight of the sample under the influence of temperature and is a crucial part of the thermobalance instrument. Sensitivities of the order of 0.1 μg or less are determined. A null-point weighing mechanism is favoured as, in this case, the sample itself does not move in the furnace. Electromagnetism is often used as the basis for the measurement as it is very sensitive and has high thermal and vibrational stability. For the furnace, the construction depends on the temperature range required. If very high, e.g. 1500–1700 °C, alumina or mullite may need to be used. All commercial thermobalances offer the use of controlled, inert (nitrogen or argon) or oxidative (air or oxygen) atmospheres for the measurements within the furnace. Some of the major factors affecting TGA measurements include sample size,

heating rate, buoyancy, electrostatic effects, gas flow and sample holder. These should all be controlled to minimise error and increase reproducibility.

Detector
The balance assembly measures the initial sample weight at room temperature and then continuously monitors changes in sample weight (losses or gains) as heat is applied to the sample. The balance and furnace data are collected during the experiment and sent to the PC for manipulation.

Output
A computer carries out all the manipulations, calculations and plotting of TG graphs required. It also calculates the derivative thermogravimetry (DTG) curve if required.

Information Obtained

TGA tests may be run in a gradient temperature program or isothermally. Typical weight loss profiles are analysed for the amount of or percentage weight loss at any given temperature and the amount of or percentage noncombusted residue at some final temperature. Hence the sort of information that can be obtained for materials is how much loss there is of moisture (loss on drying), volatiles and other gases, as well as information about the decomposition of the sample itself. TGA is often used to show the decomposition of a compound, as when a material is heated or cooled there is a change in its structure or composition. These changes are connected with heat exchange. Each downward step in the TGA trace corresponds to a weight loss due to the loss of various gases and liquids present in the sample, e.g. water, carbon monoxide and carbon dioxide (Figure 6.2)

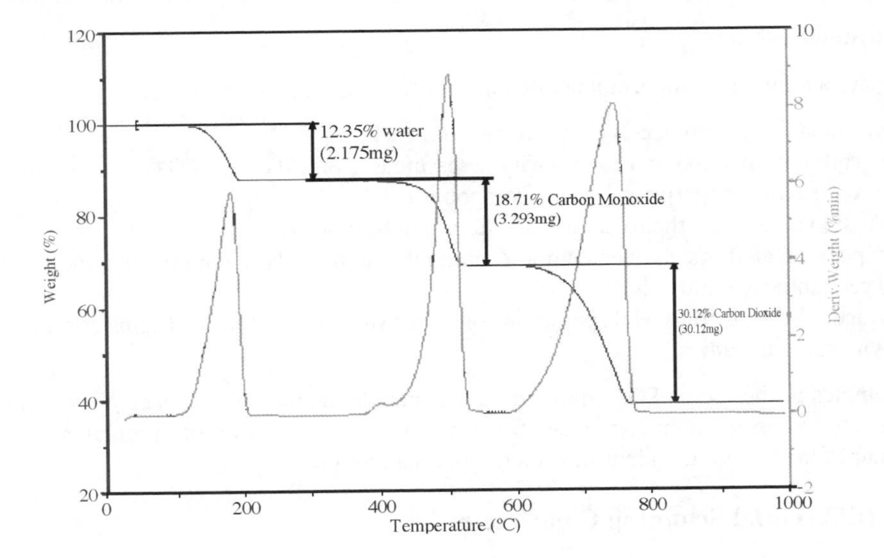

Figure 6.2 *A TG/DTG curve for calcium oxalate monohydrate (This material gives three distinct weight losses over a wide temperature range. The plot also shows the derivative of the TG curve (DTG curve), which can reveal extra detail such as the small event around 400°C that would not have been seen on the TG curve itself. Courtesy of Anasys Thermal Methods Consultancy Ltd, UK.).*

Figure 6.3 *The STA PT1600 thermobalance can be coupled to FTIR or MS (Reproduced by permission of Linseis.).*

Developments/Specialist Techniques

It is now possible to get a thermobalance combined with either FTIR or MS[1] (Figure 6.3). The TGA–FTIR or TGA–MS allow real-time identification and monitoring of gases as they evolve from a sample. The combined TGA–FTIR system measures the change in weight of a sample as function of temperature or time in a controlled atmosphere and the evolved gases pass into the flow cell of the interface where the infrared spectra are collected. This instrument can aid the determination of sample characteristics such as decomposition pathways, thermal stability or sample integrity. Thermogravimetry can also be coupled with GC–MS[2].

Applications

There are a number of important applications of the technique such as:

- Thermal stability and decomposition mechanisms.
- Material identification and even purity assessment, as the TG and DTG curves can be used to provide 'fingerprints' of the compounds.
- Corrosion studies, as the oxidation reactions can be studied.
- Composition analysis, as even complex materials can have their components selectively and sequentially removed.
- Characterisation of drugs while processing or during simulated/accelerated aging conditions.
- Polymorphism studies.

Examples of the use of TGA in the literature include monitoring the decomposition of potassium and ammonium tetraphenylborates[3], the characterisation of chromatographic stationary phases[4] and the identification of polymers in forensic cases.

6.2 Differential Scanning Calorimetry

Principle

Differential scanning calorimetry (DSC) is a thermal analysis technique where energy changes in a sample are investigated with temperature, as distinct from mass changes in the technique of TGA. In DSC, the temperature of a sample of the substance in question is raised in increments via a temperature program, while a reference is heated at the same rate. The heat flow to or from the sample and to or from the reference is monitored as a function of temperature or time. Such measurements provide qualitative and quantitative information about physical and chemical changes that involve endothermic and exothermic processes, or changes in heat capacity, i.e. thermal transitions. Differential thermal analysis (DTA) is a closely related technique in which the measurement of the difference in temperature between a sample and a reference is recorded as heat is applied to the system. DSC is the measurement of the difference in heating power required to keep a sample and a reference at the same temperature. DSC is useful for making the same measurements as DTA with the added capability that is measures heat capacities.

Instrument

A DSC instrument is very similar to a standard thermobalance with an extra pan for a reference sample (Figure 6.4). For quantitative measurements, the instrument has to be calibrated using standard reference materials.

Source

The source is the furnace temperature (and environment) controller. The required temperature programs, as with the TGA come from here.

Sample

A sample is placed into a tared TGA sample pan attached to a sensitive microbalance assembly. A reference sample (or nothing) is placed into a tared reference pan, which is also attached to the microbalance assembly. Both pans are located in the high temperature furnace, which is often cylindrical in shape.

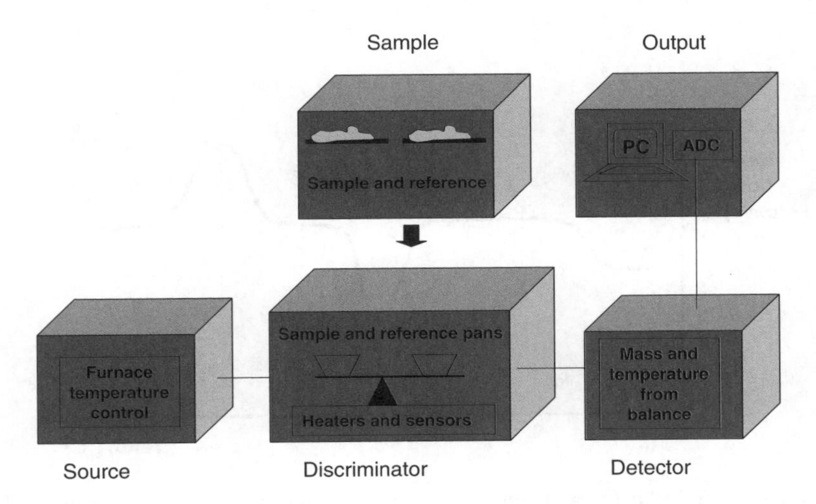

Figure 6.4 *Schematic diagram of a differential scanning calorimeter.*

Discriminator

The discriminator is the balance within the furnace. The balance allows the measurement of any temperature changes between the sample and reference. For a DSC experiment, the individual heaters under both pans work to keep the temperatures the same (or the differential temperature equal to zero) and the power drawn to maintain this is measured. For a DTA experiment, the difference in temperature between the sample and the reference is recorded as a function of furnace temperature. The environment in the furnace is controlled.

Detector

The temperatures, temperature differences and/or power are recorded as the heating program proceeds. The balance and furnace data are collected during the experiment and sent to the PC for manipulation.

Output

The data is recorded and can be plotted at the computer. From these DSC plots, thermal events such as melting points, phase change temperatures, chemical reaction temperatures and glass transition temperature of polymers can be determined. An endothermic peak is plotted in the upward direction and an exothermic peak in the downward direction

Information Obtained

DSC is a more qualitative technique which detects any reaction that has a change in heat capacity. It is a very useful fingerprint technique for qualitative analysis, e.g. analysis of plastic waste containing different polymers, each recognised on the basis of their melting or transition temperatures. The DSC plot for a polymer can contain the glass transition temperature (T_g), the crystallisation temperature (T_c) and the melting temperature (T_m) (Figure 6.5); T_c and T_m will only show up for polymers that can form crystals. An example of a DSC instrument is shown in Figure 6.6. TGA is a quantitative technique and responds only to reactions that incur a mass change.

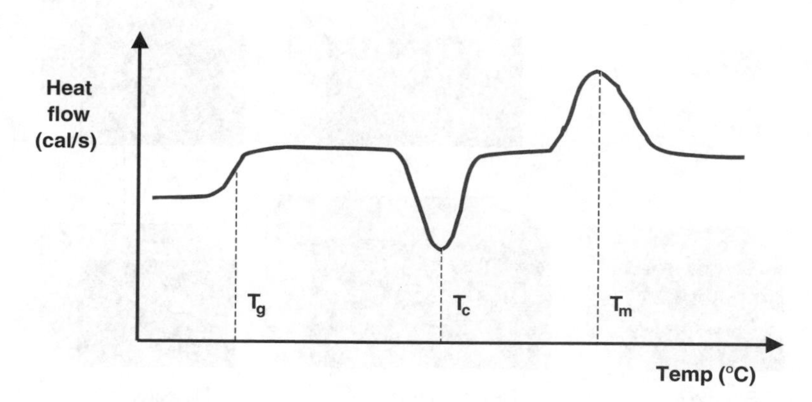

Figure 6.5 *A DSC plot for a polymer showing the glass transition temperature (T_g), the crystallisation temperature (T_c) and the melting temperature (T_m).*

Figure 6.6 *The Jade DSC (Reproduced by permission of Perkin Elmer Inc.).*

Developments/Specialist Techniques

Modulated DSC (MDSC) is now possible as well as traditional DSC. MSDC applies two simultaneous heating profiles to the sample and reference: the conventional underlying linear rate and a sinusoidally modulated rate. The linear rate provides the same information as traditional DSC, while the modulated rate provides unique information about the sample's heat capacity. Benefits of MDSC include increased sensitivity and resolution and the ability to directly measure heat capacity.

Both TGA and DSC can now be coupled together in one instrument which is capable of the resolution of TGA (of 1 mg), uses samples of 5–100 mg and can give the sensitive and quantitative performance of DSC.

Applications

Thermal analysis is an extremely important analytical tool for the pharmaceutical industry. All transitions in materials involve the flow of heat (either into the sample during an endothermic event or out of the sample during an exothermic event) and DSC is the universal detector for measuring a wide variety of transitions in pharmaceutical materials. These include measurement of amorphous structure, crystallinity (and polymorphs), drug–excipient interaction and many other applications.

In one study, DSC was compared with X-ray powder diffraction for quantifying amorphous nifedipine in mixtures of crystalline nifedipine and compared with FTIR for quantifying polymorphs in pharmaceutical formulations[5]. The DSC gave a limit of determination of 0.06 % compared to 5 % for the X-ray technique and the DSC gave a limit of determination of 0.02 % compared to 7 % by FTIR.

In terms of medical applications, DSC has been used to examine intervertebral disc degeneration in humans[6]. The results suggested that definitive differences exist between the stages of disc degeneration in calorimetric measures.

In food chemistry, DSC testing has been carried out on fillets stored at −20 °C; it showed a shift to a lower transition temperature and a decrease in enthalpy compared with control fillets stored at −30 °C[7]. This change was enhanced when fillets were stored for a longer period, confirming protein denaturation and the formation of aggregates in the meat.

6.3 X-Ray Diffraction

Principle

X-ray diffraction (XRD) is a nondestructive technique that operates on the nanometre scale based on the elastic scattering of X-rays from structures that have long range order (i.e. an organised structure of some sort, e.g. periodicity, such as in a crystal or polymer). It can be used to identify and characterise a diverse range of materials, such as metals, minerals, polymers, catalysts, plastics, pharmaceuticals, proteins, thin-film coatings, ceramics and semiconductors. The two main types of XRD are X-ray crystallography and X-ray powder diffraction.

X-ray crystallography, also known as single crystal diffraction, is a technique that is used to examine the whole structure of a crystal. The crystal is hit with X-rays and, in a typical experiment, the intensity of the X-rays diffracted from the sample is recorded as a function of angular movement of both the detector and the sample. The diffraction pattern of intensity versus angle can also be converted into its more useful form of probability distribution versus distance. The diffraction pattern produced can be analysed to reveal crystal details such as the spacing in the crystal lattice, bond lengths and angles. It can be difficult to obtain a pure crystal but, if achieved, the data obtained with this method can be very informative. Many compounds can be subjected to X-ray crystallography, such as macromolecules, small inorganic materials, biological compounds such as proteins and even small pharmaceuticals.

When a single pure crystal cannot be obtained, X-ray powder diffraction can be used instead. It can still yield important information about the crystalline structure, such as crystal size, purity and texture, but the data set may not be as complete as X-ray crystallography. The sample under investigation is usually ground down to a fine microcrystalline powder first. Sometimes the sample must be rotated to obtain the optimal diffraction pattern.

Instrument

The instrument used is called a diffractometer. A schematic diagram is shown in Figure 6.7.

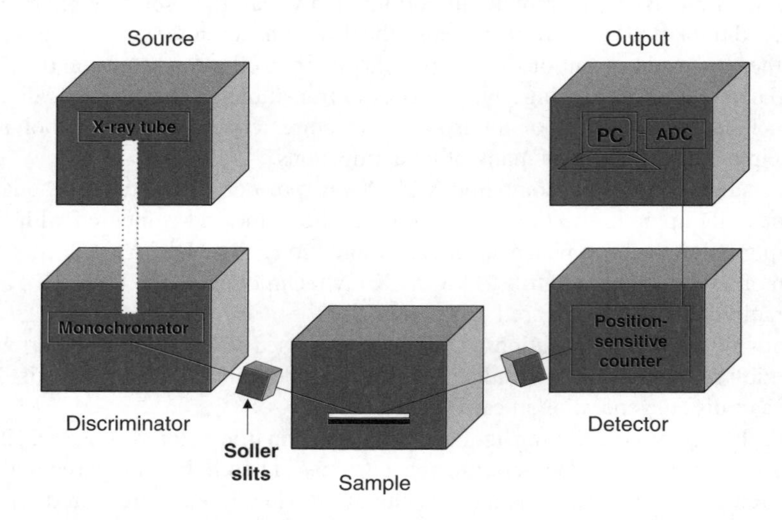

Figure 6.7 Schematic diagram of an X-ray diffraction instrument.

Source
The source is the sealed X-ray tube or a synchrotron (with much higher photon flux).

Sample
A single crystal can be mounted in a thin glass tube or on a glass fibre using grease or glue to hold it in place. The crystals are often cooled to reduce radiation damage and thermal motion during the experiment. The solid sample can be rotated about an axis during exposure to the X-rays to increase the chances of all orientations of the crystals in a powder sample being detected. The crystals act as 3-D diffraction gratings. Sample stages on most modern instruments allow for many different sample environments. These range from cryostats to room temperature to high temperature and humidity conditions.

Discriminator
The discriminator is a crystal monochromator such as graphite. Soller slits after the monochromator keep divergence of the beam to a minimum.

Detector
The position-sensitive detector registers the diffraction pattern of the sample by moving around the sample. The detector is usually a scintillation counter or more recently an array of X-ray detectors (CCD), which allows more data to be collected simultaneously.

Output
The diffraction data is recorded, manipulated and can be plotted at the computer in the form required. The diffraction pattern can be compared to a library of patterns (International Centre for Diffraction Data) and, therefore, a positive identification made.

Information Obtained

Information such as spacing in the crystal lattice, bond lengths and angles, crystal size, purity and texture can all be obtained using XRD. Information about thermal motion can also be obtained. Overall, a picture of the molecules, unit cells and the crystal can be built up.

Developments/Specialist Techniques

There are now instruments that combine high-temperature XRD and DSC in one. In this method, any phase changes in the sample can be detected while the sample is subjected to a changing temperature program.

Applications

Areas of application include qualitative and quantitative phase analysis, chemical crystallography, texture and residual stress investigations, investigation of semiconductors and nanomaterials, and high-throughput polymorph screening. In medicine, XRD is the most important technique for determining the chemical composition and phases of urinary stones[8]. XRD is also very useful for identifying drugs, especially in forensic applications and in the case of illegal or confiscated drugs[9]. XRD is also widely employed in proteomics[10].

References

1. Basalik, T. TGA with evolved gas analysis. *Am Lab*, 24–27 (Jan 2005).
2. Raffin, G. and Grenier-Loustalot, M.-F. (2004) Coupling thermogravimetry with cold trap/gas chromatography and mass spectrometry. *Spectra Analyse*, **33 (241)**, 36–43.
3. Adamczyk, D. and Cyganski, A. (2006) Thermogravimetric method for simultaneous determination of potassium and ammonium involving thermal decomposition of tetraphenylborates. *Chemia Analityczna*, **51 (2)**, 241–249.
4. Lumley, B., Khong, T.M. and Perrett, D. (2004) The characterisation of chemically bonded chromatographic stationary phases by thermogravimetry. *Chromatographia*, **60 (1–2)**, 59–62.
5. Song, M., Liebenberg, W. and de Villers, M.M. (2006) Comparison of high sensitivity micro differential scanning calorimetry with X-ray powder diffractometry and FTIR spectroscopy for the characterization of pharmaceutically relevant non-crystalline materials. Die *Pharmazie*, **61 (4)**, 336–340.
6. Doman, I. and Illes, T. (2004) Thermal analysis of the human intervertebral disc. *J Biochem Biophys Meth*, **61 (1–2)**, 207–214.
7. Saeed, S. and Howell, N.K. (2004) Rheological and differential scanning calorimetry studies on structural and textural changes in frozen Atlantic mackerel (*Scomber scombrus*). *J Sci Food Agric*, **84 (10)**, 1216–1222.
8. Ouyang, J.M. (2006) The application of X-ray diffraction in the study of urinary stones. *Guangpuxue Yu Guangpu Fenxi*, **26 (1)**, 170–174.
9. Thangadurai, S., Abraham, J.T., Srivastava, A.K. *et al.* (2005) X-ray powder diffraction patterns for certain β-lactam, tetracycline and macrolide antibiotic drugs. *Anal Sci*, **21 (7)**, 833–838.
10. Liu, H.L. and Hsu, J.P. (2005) Recent developments in structural proteomics for protein structure determination. *Proteomics*, **5 (8)**, 2056–2068.

Section II

Portable Analytical Instrumentation

The Drive for Portable Analytical Instruments

Many benchtop instruments are now available in a compact version either for portability or simply to save on space. The recent availability of portable versions of analytical equipment brings many advantages to analytical science, some of which are:

- They can be brought to the sample instead of vice versa.
- They can be brought to remote locations.
- They are smaller and lighter, therefore easily transported.
- They take up less space in a laboratory.
- Many of them can work off batteries and/or mains power.
- They often cost less than conventional benchtop instruments.
- They require less sample and reagents and produce less waste.
- They are often simpler to use.

The drive to reduce the size of conventional instrumentation has come about from a need to take equipment out of the laboratory to the sampling site. This requires an instrument that can be transported easily and that can work in the field, ideally from batteries for at least a few hours. The batteries may be disposable or rechargeable. Some portable devices have the option that they can work from mains electricity; consequently, they can also be used back in the laboratory, where they save space due to their smaller footprint and often cost less to run and maintain than their benchtop counterparts. Where equipment may need to be moved from one location to another, even for coupling to another instrument, a smaller footprint and portability is again advantageous. Compact instruments usually require less sample and reagent volumes, which in turn reduces waste. Often, portable equipment is more user-friendly than the benchtop version as it is designed for use by nonscientists as well as scientists.

However, in reducing the size, there can be some disadvantages too. Trade-offs are usually made when optimising for weight, power and performance. This is application-specific but may, for example, result in the portable version of an instrument being less sensitive than the benchtop version. However, if the limits of detection are adequate for the field measurements, this may not be an issue.

Many portable instruments have been made possible by the use of clever designs incorporating smaller components, for example optical fibres, light emitting diodes (LEDs), liquid crystal displays (LCDs) and miniature gratings for use in spectrometers, each of which are discussed below. These developments, amongst others, have contributed immensely to the development of the whole area of portable analytical instruments. They have seen particular use in portable electrochemical and spectrometric instruments.

Optical fibres are one of the most important advances that have enabled the design and development of portable analytical instrumentation. These fibres are capable of transmitting light from one place to another along a flexible waveguide or light pipe. They were originally developed for the communications industry to replace electric wires and can transmit UV, Vis, NIR and IR radiation. Optical fibres are not as susceptible to electrical noise, they can send data more rapidly and can carry more information than wires. Silica fibres are most common and are useful from the UV to the NIR regions of the spectrum, while special materials such as fluoride glass allow transmission of radiation in the IR region. In the visible region of the electromagnetic spectrum, silica, glass or plastic fibres can be used. In conventional absorbance and fluorescence spectrophotometers, where both source radiation and emitted radiation are required, a bifurcated cable consisting of two cables combined in the one casing is often used to achieve this.

Light emitting diodes are finding widespread use as light sources and detectors; they are now available for use over a wide region of the electromagnetic spectrum from UV through to IR. They act like tiny lightbulbs and can be used as sources of radiation or as detectors of radiation (as photodiodes).

Liquid crystal displays are very important to the portable instrument market. They are small, slim, lightweight and consume little power. They are now capable of displaying ever more complex graphical and digital information with speed. Due to the pressure from users of portable equipment, modern LCDs are also rugged, shock-proof and have long backlight life. A new transflective LCD is now available, which means that the display screen can be read in any environment, from a darkened room to full sunlight. Especially in the medical device sector, touch screen interactive technology is prevalent, even in small devices.

Miniature gratings are now available owing to the development of laser technologies. These small gratings make it possible to reduce the size of spectrometers.

In designing portable instruments, one of the considerations that must be taken into account is that they should be very user-friendly, as they need to be used in the field and should work quickly. Ease of use is important as nonscientists need to be able to use these instruments in certain situations, and the speed of obtaining the result is important from a battery life perspective. Portable devices should ideally give no false positive results but, in the real world, this means they should be as accurate as possible as the resultant data are often used there and then to make important decisions. Finally, they should be robust and easy to maintain and they should have enough memory to store the pertinent data obtained, at least until the user returns to a base station.

Although some instruments are more amenable to being reduced in size than others, all types of equipment are feeling the squeeze. Of particular note are spectrometric, imaging, separations and electrochemical instruments.

7

Portable Instruments in the Laboratory

7.1 Spectrometric Instruments

UV–Vis and UV–Vis–NIR Spectrophotometers

UV–Vis spectrometers have become much smaller in recent years due, in part, to the availability of fibre optics and very small grating monochromators. Miniature spectrometers have been available commercially since the 1990s. These spectrometers are often devised so that they can be interfaced to the sample and source via optical fibres. This enables difficult locations to be accessed, analytes of interest to be measured in the field and in-process and biological measurements to be taken. Some of these mini UV–Vis–NIR spectrometers can be used both in the laboratory and outside the laboratory for field testing.

The SMART Spectro from LaMotte (Figure 7.1) is an example of a portable spectrophotometer that is easy to use and compact in size. It has automatic wavelength selection and pre-programmed tests. It employs a 1200 lines/mm grating that provides for an extended range (350–1000 nm) for which the accuracy is ±2 nm and the resolution 1 nm. It is precalibrated for over 40 tests but can be calibrated by the user for more. It weighs 4.65 kg.

Another example of a portable spectrophotometer is the 2800 UV–Vis spectrophotometer from Hach. It has over 240 pre-installed analytical methods. It has a very small footprint and can run from direct electrical power or a battery. It also has an optional pour-through cell for high volume or trace analysis and a USB port. The wavelength range is 340–900 nm, accuracy is ±1.5 nm and resolution 1 nm. The instrument weighs 4.38 kg including battery. Once a cuvette containing sample is inserted, measurements are made and the result shown immediately in mg/L. If the backlighting is switched off in the field, the batteries allow up to 40 hours of operation.

Analytical Instrumentation: A Guide to Laboratory, Portable and Miniaturized Instruments G. McMahon

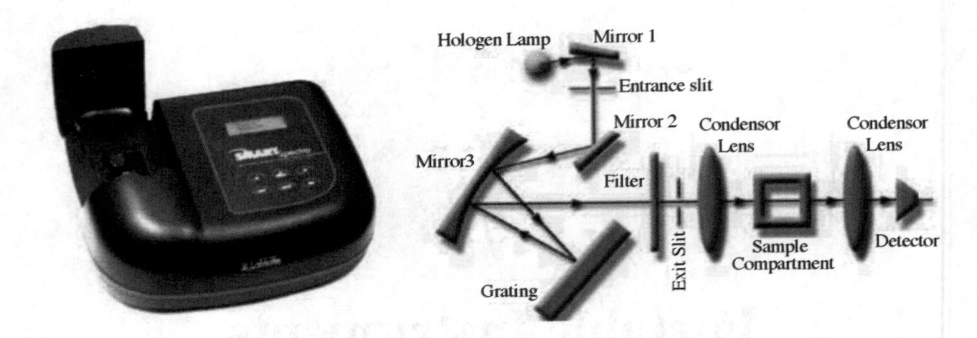

Figure 7.1 The SMART Spectro portable spectrometer and a schematic of its optical system (Reproduced by permission of La Motte).

Even smaller in size is the OSM-400 from Newport (Figure 7.2), a self-contained, portable spectrometer with a choice of spectral ranges to cover the ultraviolet through near-infrared spectral regions, e.g. the 250–850 nm range has an optical resolution of 1 nm. Optical input is via a fibre optic cable, or directly though the input slit. Data can be stored and transported on removable memory cards or on the built-in 4 MB memory. Data may also be downloaded via an RS-232 link or optional ethernet port. The OSM-400 can be configured for various types of optical spectroscopy, including fluorescence, atomic emission, near IR absorption and Raman. It weighs 2.7 kg and its battery provides up to four hours of continuous portable use.

The Ocean Optics USB4000 miniature spectrometer is small enough to fit into a hand (Figure 7.3) and can be configured for UV, Vis and NIR applications from 200–1100 nm. The actual spectrometer part is very small (the size of a circuit board) and weighs only 190 g. Fourteen different gratings and a number of CCD array detectors can be fitted depending on the spectral region of choice. Optical resolution ranges from 0.3–10 nm and integration time is a mere 10 μs. It can be interfaced to samples and sources using optical fibres. The batteries last 8–12 hours depending on the light source used.

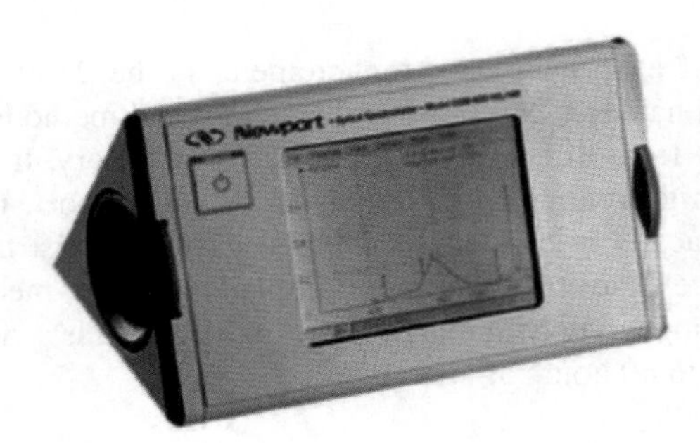

Figure 7.2 The OSM-400 portable spectrophotometer (Reproduced by permission of Newport Spectra-Physics Ltd).

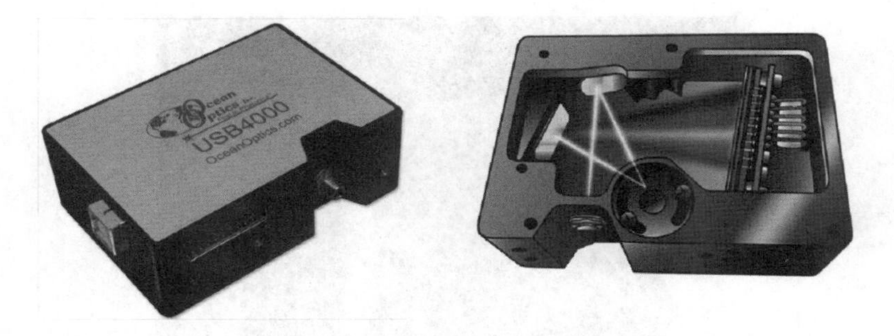

Figure 7.3 The USB4000 miniature spectrometer and a schematic of its optical system (Reproduced by permission of Ocean Optics).

Another mini-spectrometer is the VS140 from HORIBA Jobin Yvon (Figure 7.4), a portable linear array spectrometer that can be interfaced with various devices for a variety of applications. Depending on the version, the spectral range is between 190–1100 nm. It is available with a choice of either a linear CCD or high-dynamic-range PDA array detector. Resolution is between 1.4–3.0 nm. The device is 178 × 123 × 58 mm and weighs only 0.6 kg.

NIR Spectrometers

Near infrared instruments have been reduced in size to a considerable extent in recent years. Axsun Technologies has produced the NIR Analyzer XL (Figure 7.5); it is based on the IntegraSpec microspectrometer platform, which is itself only 14 mm long. The detector assembly enables both direct-to-detector and fibre-to-detector set-ups. A number of accessories, designed to work with the system, are also available, including a diffuse reflectance probe for solid phase analysis and a liquid transmission probe. Gases can also be analysed. The battery-operated IntegraSpec minispectrometer platform contains the source, wavelength selector and detector. There are six different models of the NIR instrument available, each with its own wavelength range, the most extensive one being 1350–1970 nm. The main advantage of this device is that it is a laboratory-grade instrument with high specifications, such as wavelength accuracy of ±0.025 nm and a spectral resolution of 3 cm^{-1}. The total platform is only 25 × 15 × 7.5 cm in size.

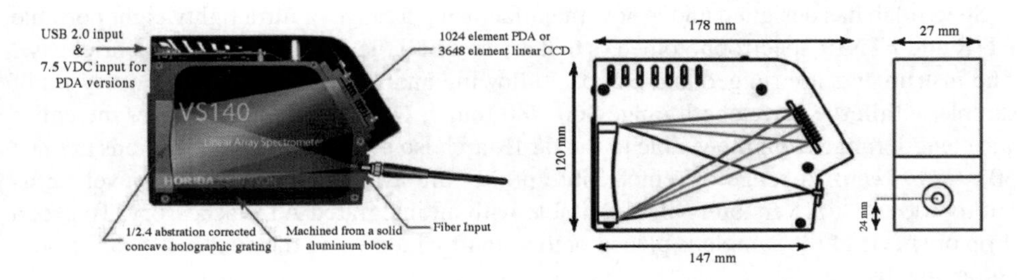

Figure 7.4 The VS140 miniature spectrometer and a schematic of its optical system (Reproduced by permission of HORIBA Jobin Yvon).

Figure 7.5 The NIR Analyzer XL (Photograph reproduced by permission of Axsun Technologies).

Polychromix has introduced the Phazir, the first handheld NIR digital transform spectrometer (DTS) analyser. It combines the DTS engine (with InGaAs detector), light source, reflectance probe, rechargeable batteries and an on-board computer with LCD display. It weighs only 1.7 kg and contains no moving parts. The spectral range is either 1000–1800 nm or 1600–2400 nm, the optical resolution is 11 nm and the batteries last at least five hours.

Portable NIR analysers have been used to measure protein, cholesterol and glucose in whole blood and plasma samples in their collection tubes with no sample preparation required[1].

IR Spectrometers

FTIR has become more portable in recent years due in part to the miniaturisation of the interferometer. However, FTIR instruments have only recently been reduced in size to the same extent as UV–Vis or NIR devices. D&P Instruments has developed the Model 102 portable interferometer-based FTIR spectrometer (Figure 7.6) for use in remote sensing applications. Weighing less than 7 kg, it can run off batteries or a mains supply. A PC is built into the case along with the FTIR module. The spectral range is 625–5000 cm^{-1} with a resolution of 4 cm^{-1}.

Spectrolab has designed and is now manufacturing a range of ultra lightweight portable FTIR and FTNIR spectrophotometers that are so small they can be carried in a brief case. The instruments are rugged and compact allowing analysis to be made on most types of sample within the wavelength range 360–7800 cm^{-1}. Other versions that cover the entire wavelength range from the visible to the far IR are also available. The dimensions are just $30 \times 22 \times 12$ cm. A range of remote fibre probes are available covering all wavelengths out to 600 cm^{-1}. A version is also available with an integrated ATR accessory. To use, a drop or smear of the sample is placed onto a small window and the analysis is carried out automatically.

The MIRAN (miniature IR analyser) series of gas analysers from Thermo Electron Corp. is a versatile gas detection system which allows accurate and fast wavelength

Figure 7.6 The Model 102 portable FTIR spectrometer and a close-up of the small interferometer used in the Model 102 (Photographs reproduced by permission of D&P Instruments).

selection. It is available with gas calibrations for up to 100 gases and sub-ppm detection. It has the ability to identify unknown airborne compounds using special software. It can be set up for flow-through analysis using sample probes based on transmission or reflectance. The instrument weighs 10 kg and the battery lasts four hours.

Bruker Optics has launched the ALPHA, a very small FTIR designed to save space in the laboratory. It weighs less than 7 kg and has a footprint of only 22 × 30 cm. Solid, liquid and gas samples can be analysed and transmission, attenuated total reflection (ATR) and diffuse reflection experiments can be carried out.

Raman Spectrometers

Raman spectrometers have been miniaturised using diode laser sources and other small components. An example of a small Raman spectrometer is the InPhototeTM portable Raman system from InPhotonicsTM (Figure 7.7), which is used to identify chemicals (solids, liquids or slurries) in many different applications. Only the fibre optic probe needs to be close to the sample and analysis is nondestructive. The probe is compact and lightweight for handheld use, and can be mounted onto remote-controlled vehicles to identify potentially hazardous materials at a safe distance. Probe extension cables (up to 200 m) can be used to extend the sampling reach of the unit. The InPhotote comprises a 785 nm diode laser, spectrograph and CCD detector integrated into the case. Set-up time is minimal and measurement is simple with the probe. Optical components can withstand elevated ambient temperatures and are shock-mounted for routine shipping and use. Two

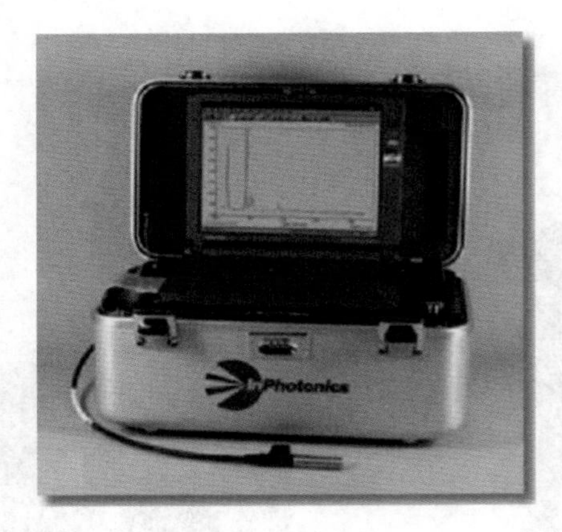

Figure 7.7 The InPhote portable Raman spectrometer (Reproduced by permission of InPhotonics).

spectral ranges are available: 250–1800 cm^{-1} (resolution 4–6 cm^{-1}) and 250–2300 cm^{-1} (resolution 8–10 cm^{-1}). It weighs 10 kg.

Even smaller is the RSLPlus from Raman Systems (Figure 7.8), with laptop and inbuilt battery, which is designed specifically for quick material identification and verification in various on-site settings. The laser is at 785 nm and it has a linear array CCD detector, fibre optic sampling probe and sample compartment with holder for vials and cells. The spectral range is 200–2700 cm^{-1} with a spectral resolution of better than 10 cm^{-1}. The dimensions are $30 \times 25 \times 7$ cm and it weighs only 1.6 kg. Batteries allow two hours of continuous operation.

Fluorescence Spectrometers

The reduction in size of fluorescence spectrometers has mainly taken place in biological laboratories where fluorescence is used for analysing DNA, proteins, peptides and other

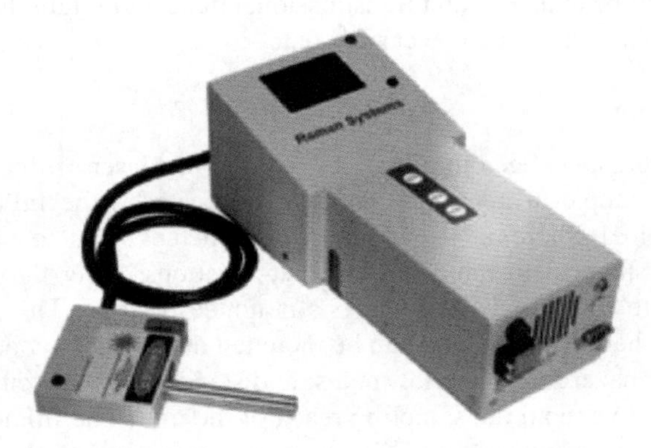

Figure 7.8 The RSL Plus portable Raman spectrometer (Reproduced by permission of Raman Systems).

Figure 7.9 *The CP20 miniature spectrograph (Reproduced by permission of HORIBA Jobin Yvon).*

important molecules of biological significance. Turner Designs Inc. has developed the Trilogy™ laboratory fluorometer for making fluorescence, absorbance and turbidity measurements in one unit by using snap-in application modules. Calibration can be done with as few as one or up to five standards. The source and detector are an LED and photodiode respectively. The instrument needs a mains supply and weighs 3.65 kg.

The USB4000 from Ocean Optics (Figure 7.3) can be configured for fluorescence as the USB4000-FL. This unit includes a 200 μm slit and a grating with a range of 350–1000 nm.

The CP20 20 mm focal length spectrograph from HORIBA Jobin Yvon (Figure 7.9) can also be configured for fluorescence. This device is a very compact $5 \times 5 \times 5$ cm in size with a spectral range of 380–760 nm or 350–900 nm depending on the array type used.

Designed more for the life sciences, the mini and pico fluorometers from Turner Biosystems show that these instruments can be made very small indeed. The TBS-380 is a fluorometer designed for dedicated assays such as measurement of DNA and other proteins. A new minicell adaptor enables the measurement of samples with an assay volume as low as 50 μL. Results are based on single point calibration and averaged data and the readouts are given to ng/ml concentrations. The dimensions are only $18.4 \times 6.9 \times 14$ cm and weight 0.7 kg. The Picofluor™ is an even smaller handheld fluorometer, configured for the fluorescent probes commonly used the quantitative measurement for nucleic acid and proteins. Like the TBS-380, it has two channels to enable quick analysis of samples for two different assays. The four AA batteries allow over 1000 readings to be taken. It weighs only 400 g.

A portable fluorescence system has been developed to measure beryllium in workplace air[2]. Another design of fluorescence probe has been used to determine carbofuran and paraoxon in fruit and vegetable samples[3].

NMR Spectrometers

Gone are the days when a nuclear magnetic resonance (NMR) instrument took up a whole room. In recent years, they have been considerably reduced in size when compared to traditional laboratory-based equipment but are still not available as a portable instruments, at least not commercially. However, a compact version of an NMR machine

Figure 7.10 *The NMR Minispec (Reproduced by permission of Bruker Optics).*

is available from Bruker Optics. The Bruker Minispec (Figure 7.10) is a benchtop instrument suited to a full range of low resolution NMR measurements. It has a multinuclear and variable frequency (2–65 MHz) range and also has a number of magnet fields and probe heads (absolute, ratio, combined, temperature controlled, gradients) available. It is generally configured for relaxation experiments, in particular applications such as droplet size, fluorine or percentage hydrogen analysis.

Nuclear magnetic resonance instruments have proved very difficult to miniaturise due to inhomogeneity of the static magnetic field. However, a new 'single-sided sensor' has been developed where a sample is simply placed on the surface of the RF probe. In 1996 a prototype called the NMR-MOUSE™ (mobile universal surface explorer) was developed by Professor Blumich's group in Germany[4]. Bruker has developed a handheld NMR instrument called the minispec ProFiler for relaxation measurements in the first few millimetres below the surface of any sized samples. This low-cost single-sided magnet instrument reduces the spatial restrictions of the sample size required in both conventional NMR experiments and in the Minispec. As a result, the surface of large objects, including ferromagnetic samples, may be examined. It can also measure through packaging. In the past two years, researchers at the Lawrence Berkeley National Laboratory in the USA, in conjunction with Professor Blumich's group at the Institute for Technical Chemistry and Macromolecular Chemistry in Germany, has been working on the development of a field-portable NMR sensor. The current prototype uses single-sided sensing that enables high-resolution NMR measurements to be taken in the field for samples of any size[5, 6].

Much research is being done on making magnetic resonance imaging (MRI) more portable and even handheld. Savukov and Romalis in Princeton University have developed a device called an atomic magnetometer that does not require the giant magnets and cooling systems currently required for MRI experiments[7, 8]. The magnetometer is composed of a 4 cm wide glass container filled with hot vaporised potassium atoms suspended in a gas. By observing how the potassium atoms move, with the help of lasers, it is possible to work out the magnetic signals of nearby water samples. However, at the moment, the magnetometer and samples have to be protected from noise using large shields but the researchers envisage a handheld device that will be able to take an MRI of the body as easily and as quickly a digital camera takes a photo.

Mass Spectrometers

Portable, robust MS instruments are now available despite the fact that the components of a mass spectrometer, e.g. the vacuum pump, have been challenging to reduce in

Figure 7.11 *The miniTOF II mass spectrometer (Photograph reproduced by permission of Comstock).*

size. A successful example of a portable MS is the miniTOF® II mass spectrometer from Comstock (Figure 7.11). It is based on electron impact (EI) ionisation and a linear time-of-flight (TOF) analyser and has a heated ion source, vacuum chamber and pumps. Applications include fast GC–MS with storage of up to 100 averaged spectra per second, gas and headspace monitoring. Specifications include a mass resolution of better than 1 amu for the mass range 0–600 amu and low ppm level sensitivity. The system dimensions are approximately 53 × 33 × 56 cm excluding pump with a footprint of 53 × 22.9 × 50.8 cm which includes the spectrometer, power supply and turbo pump.

Another example is the MS-200 from Kore Technology (Figure 7.12), a portable, battery-powered mass spectrometer for gas analysis entirely contained in a single case and weighing only 20 kg including battery. In analysis mode, the battery lasts 6.6 hours.

Figure 7.12 *The MS-200 portable mass spectrometer (Reproduced by permission of Kore Technologies).*

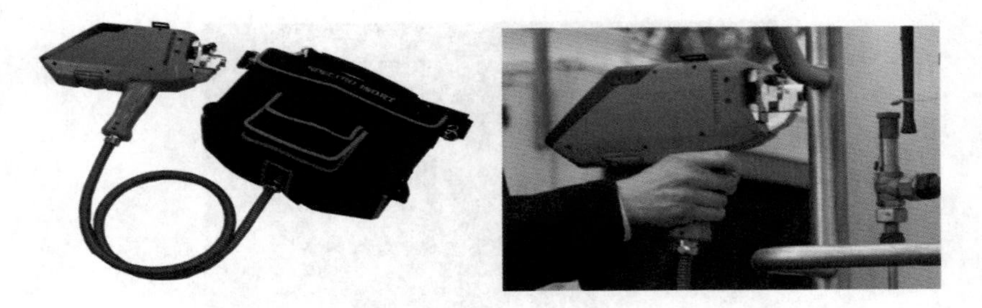

Figure 7.13 The SPECTRO iSORT metal analyser (Reproduced by permission of Spectro Analytical Instruments).

The MS-200 has a membrane inlet concentrator which allows a wide range of gases to be identified and measured from the low parts per billion range up to percentage levels. No operator maintenance or specialist knowledge is required for the vacuum system. The system also uses EI ionisation and a TOF mass analyser. The mass range is up to 1000 amu and sensitivity is <5 ppb (for benzene).

Applications for portable mass spectrometers include the measurement of toxic compounds in air[9].

Elemental Spectrometers

The SPECTRO iSORT from Spectro Analytical (Figure 7.13) claims to be the only commercially available battery-operated handheld metal analyser. Metal chemistry and alloy identification using fingerprinting software is possible in less than four seconds. Iron, titanium, cobalt, copper, nickel and aluminium can be detected without changing any spectrometer settings. The device can also now measure carbon in iron base. The measuring range extends from 275–554 nm and uses a CCD detector. The portable unit, which can be carried in a shoulder bag, weighs 4.5 kg and the handheld probe weighs 1.3 kg.

7.2 Separation Instruments

Gas Chromatography

Gas chromatography (GC), due to its wide applicability and ruggedness, has seen instruments reduced in size for many years now. Small GC instruments can be categorised as compact, portable/field, handheld or micro chromatographs[10]. Compact GC instruments weigh approximately 10–25 kg and have analytical capabilities on a par with benchtop GC. Portable/field GC instruments weigh approximately 5–15 kg, usually have their own gas supply and power and their main attribute is rapid analysis. Handheld GC weigh less than 3 kg, are usually dedicated systems capable of analysing only a limited number of compounds and are fully self-supported. Micro GC are specially designed for applications such as space investigation and have very stringent hardware requirements, such as the ability to withstand harsh environments and vibrations. The various attributes of and developments in portable GC instruments have been reported elsewhere[11, 12].

Figure 7.14 The 3000 micro-GC (Copyright 2006, Agilent Technologies Inc., reproduced with permission).

The Agilent 3000 micro GC line of instruments (Figure 7.14) enables field analysis of gas samples. This compact system has up to two channels and a rechargeable gas canister which lasts 30 hours. It is equipped with batteries and a charger for cordless operation. The detection limits using a TCD detector are typically in the low ppm range. The dimensions are $15.5 \times 36.4 \times 41.3$ cm and the maximum weight is 16.6 kg.

The CP-4900 from Varian Inc. (Figure 7.15) is a rugged, portable, laboratory-grade micro-GC for measuring the composition of gas mixtures. There are four GC channels, each with associated pneumatics, injector, narrow-bore column and detector. It has two gas cylinders. Low power consumption means that the GC can be taken to the sample where it works on rechargeable batteries for eight hours. It is 15 cm wide and weighs from 5.2 kg upwards. The 200 nL volume thermal conductivity detector (TCD) keeps peak broadening to a minimum and hence the CP-4900 is capable of peak area precision of 0.13 % RSD. One version of the Varian CP-4900 incorporates a differential mobility detector which employs a tunable ion filter to allow simultaneous detection of both positive and negative ions. This detector is connected in series with the TCD detector.

The Voyager GC from Photovac Inc. has a miniaturised electron capture /photoionisation dual detection system. It has three columns built in and an isothermal oven for fast analysis of up to 40 volatile organic compounds (VOCs). There are preconfigured assays for environmental, petrochemical and other applications. It has a refillable gas cylinder. The rechargeable batteries enable eight hours of use in the field but it can also operate from AC or DC power. It weighs 6.8 kg.

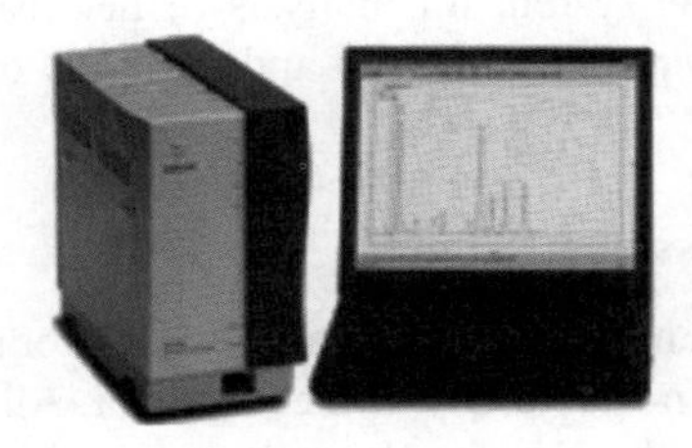

Figure 7.15 The CP-4900 micro-GC (Reproduced by permission of Varian Inc.).

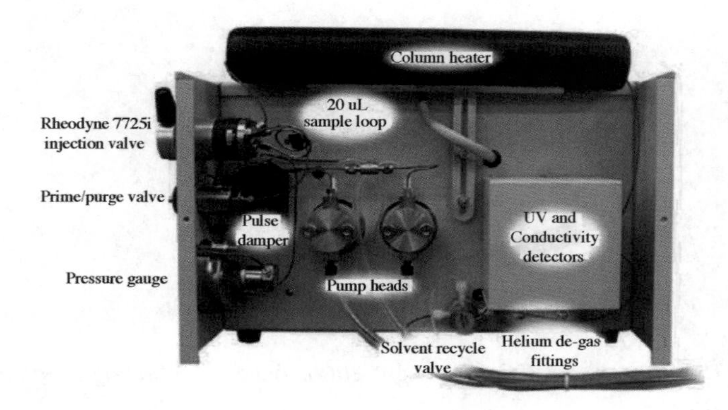

Figure 7.16 The 210D portable HPLC (Reproduced by permission of SRI Instruments).

High Performance Liquid Chromatography

High performance liquid chromatography (HPLC) equipment has been reduced in size, though the pump has been a challenge to miniaturise. An example of a compact, commercially available HPLC is the 210D from SRI Instruments (Figure 7.16). It comes with a fixed wavelength (254 nm) UV detector and a built-in, single channel data system. Extra channels can be made available for external detectors, which can be easily connected. The column heater will accept most HPLC columns up to 25 cm in length. The heater temperature is adjustable from ambient to 100 °C for a wide range of temperature-sensitive applications. The solvent recycling valve allows the solvent to be recycled when no peaks are eluting and can also be used as a single sample fraction collector. The storage compartment built into the front of the 210D holds the injection and priming syringes, spare parts and a few vials securely during shipping. The entire HPLC can run off 12-volt DC power such as from the cigarette lighter in a vehicle or storage batteries.

Capillary Electrophoresis

Because CE is a modular system that is easily put together with off-the-shelf components, it can be made small if required. The limiting component in terms of size is usually the power supply and the ruggedness of the system can be an issue. Also, there has not been the same level of demand for portable CE for field or point-of-care applications as there has for other techniques such as HPLC. Yuan *et al.* have described the use of a portable capillary electrophoresis system for analysis of neomycin compounds within four minutes[13]. The instrument was suitcase-sized and the LOD of neomycin B obtained was 7 ppm.

Gas Chromatography–Mass Spectrometry

Since both GC and MS instruments have been appreciably reduced in size, the natural next step was the development of portable GC–MS equipment. An example of a commercially available GC–MS system is the Hapsite, from Inficon, which is a compact quadrupole GC–MS designed specifically for the analysis of VOCs. The Hapsite weighs only 16.8 kg

Figure 7.17 *The Griffin 500 portable GC-MS (Photograph reproduced by permission of Griffin Analytical Technologies).*

and uses a sampling wand with an internal pump to collect the sample. The sample is pulled into a sample loop with variable injection capabilities. The column is a 30-meter OV-1 with a backflush column that allows the VOCs to get onto the column, then backflushes off the semivolatile compounds. This keeps the instrument free of contamination. The interface between the GC and MS is a methyl silicone membrane, which allows organics to migrate through to the MS while sweeping most non-organics out through the vent. The run time on the instrument is typically about 10 minutes. Since the column is isothermal and the heavier compounds never reach the analytical column, there is no cool-down time and the next run can be started immediately after the last for maximum throughput. The Hapsite GC–MS can also be used to identify and semiquantitatively measure unknown compounds using searchable spectra. As well as full scan mode, selected ion monitoring (SIM) can be used for an order of magnitude more sensitivity and the instrument can be used in MS-only mode. A headspace interface enables analysis of soil and water samples in the field.

The Griffin 500 from Griffin Analytical Technologies (Figure 7.17) is a compact system capable of GC–MS^2 with analytical performance equivalent or better than traditional laboratory ion traps and quadrupoles. It allows the analysis of both semivolatile and volatile organic compounds. The Griffin 500 has the capability to perform split/splitless liquid injections and has minimal consumable requirements. It can also handle air samples through the use of the X-Sorber handheld sampler which interfaces easily with the Griffin 500 and weighs only 1.5 kg. This instrument employs cylindrical ion trap (CIT) technology developed in the USA at Purdue University and commercialised by Griffin. It can operate off batteries or line power. The vacuum pump system and battery are completely contained in the instrument.

7.3 Imaging Instruments

As well as straightforward imaging instruments such as light microscopes, electron microscopes and scanning probe microscopes, combined imaging and spectroscopic instruments are becoming more and more popular in analytical laboratories. All of these are also being reduced in size for easier transport, space considerations and an associated drop in cost.

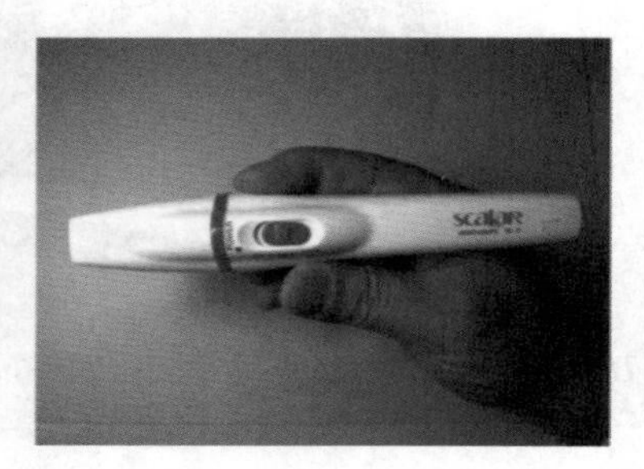

Figure 7.18 *The DG-3 portable digital microscope and VL-5 pen digital microscope (Photographs reproduced by permission of Scalar Corporation (www.scalar.co.jp and www.scalarscopes.com)).*

Optical Microscopes

In comparison to the traditional laboratory optical microscopes, new portable and handheld optical microscopes are tiny. For example, the new DG-3 portable digital microscope from Scalar (Figure 7.18) weighs only 500 g with battery and memory card. Among the features of this new device are a 9 cm LCD, a parfocal zoom lens that goes from 25× to 1000× and a 2.3 megapixel CCD sensor for sharp, clear images. The associated technology allows immediate data capture on site. Data is recorded on a memory card and may be transferred to a computer for graphic enhancement. The battery life is 1–1.5 hours depending on the settings. Scalar also produces the VL-5 'pen' digital microscope (Figure 7.18). It has a more limited magnification range and resolution but only requires two AA batteries and can be connected to a computer to download images.

Electron Microscopes

Electron microscopes have been difficult to reduce in size but progress has been made. Some sacrifices are expected in terms of resolution when compared to conventional electron microscopes but if this limitation does not adversely affect the application, then purchase of such an instrument is justified. The Hitachi TM-1000 Tabletop microscope gives magnification (up to 10 000×) an order of magnitude better than an ordinary optical microscope, requires no special sample preparation for hydrated, oily or nonconducting samples, and is easy to use with a three minute start-up time. It has a resolution of 30 nm and a 0.5 mm depth of field. Samples can be up to 70 mm in diameter and up to 20 mm deep. The dimensions of the instrument are 478 (W) × 564 (D) × 513 (H) mm and it weighs 81.5 kg excluding the pump (which is 4.5 kg). The detector used in the TM-1000 can show contrast due to different average atomic numbers of materials in the specimen.

Scanning Probe Microscopes

As with the electron microscope, there have been challenges in reducing the size of scanning probe microscopes but it has been achieved. In fact, due to their smaller size and

Figure 7.19 *The portable Caliber Mini SPM (Reproduced by permission Veeco Instruments Inc.).*

price, AFM instruments are now used in some teaching laboratories. The Caliber Mini SPM from Veeco (Figure 7.19) is portable and offers a number of imaging modes, e.g. contact mode, phase imaging and tapping mode. The Caliber is able to perform a variety of SPM applications on samples of various sizes for materials and surface science studies, polymer characterisation and biomaterials and inorganics. The small, handheld Caliber can image large samples without having to resize them. Also, the sample stage has a kinematic mount that allows users to scan a sample repeatedly. Caliber uses premounted tips that facilitate handling and laser alignment. The system also uses integrated optics with a colour camera that enables easy judgement of tip-sample separation. The instrument has a large field of view (1.5 × 1.5 mm) and an optical resolution of 10 μm.

A new portable, battery-operated AFM on the market is the Mobile S Cordless from Nanosurf (Figure 7.20). It weighs only 4.7 kg and the rechargeable battery can power it for up to five hours. It interfaces easily with a laptop. This allows the analyst to

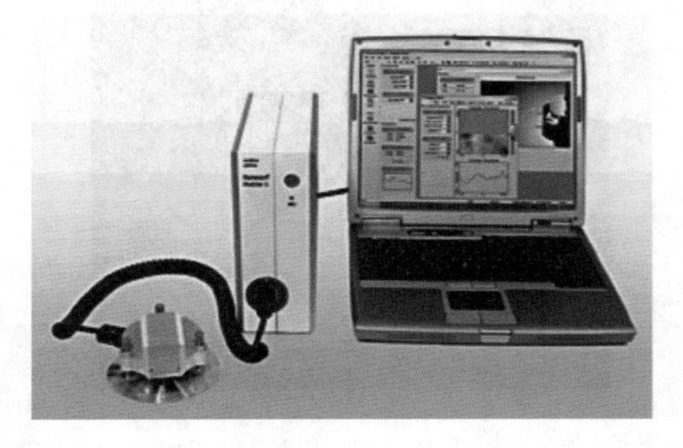

Figure 7.20 *The Mobile S Cordless AFM (Reproduced by permission of Nanosurf).*

perform measurements in the field, where simple transport and minimal set-up time are important. It has a scan range of over 100 × 100 μm laterally and 20 μm in height. The Mobile S Cordless can carry out a number of imaging and spectroscopy measurement modes at sub nm resolution.

Spectral Imaging Devices

Since both spectroscopic and imaging instruments have been reduced in size, development of portable spectral imaging equipment has been possible. For example, J&M produces the Tidas range of microscope spectrometers with UV to NIR applications. The measurement light is switched in and out of the microscope using fibre optic technology. Measurements can be carried out locally (in the micrometer range) as well as directly on the object. The Tidas MSP 400/800 spectrometer for microscopes is especially designed for forensic applications and can cover transmission, reflectance or fluorescence, depending on the chosen configuration. They can be interfaced with microscopes from different manufacturers. For fluorescence experiments, a xenon light source combined with a fast monochromator provides the fluorescence excitation. The emission spectra can be quickly scanned over the entire UV–Vis range.

The VS140 from HORIBA Jobin Yvon (Figure 7.4) can also be interfaced with a microscope to give an imaging spectrometer (Figure 7.21).

Portable spectral imaging systems for medical applications is a growing field. One example is the portable near infrared system for topographic imaging of the brains of babies[14].

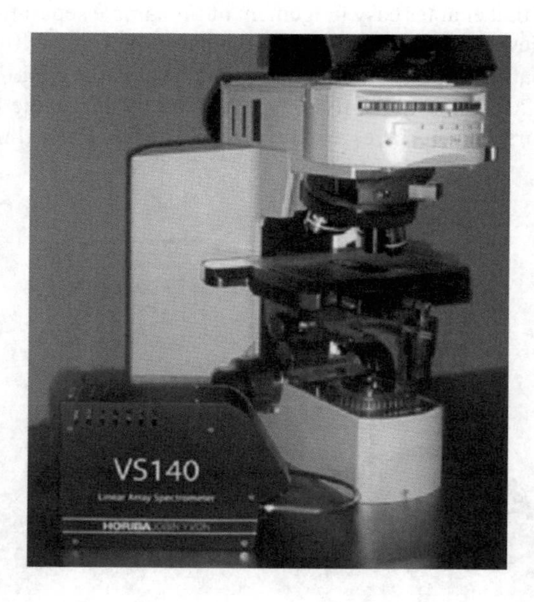

Figure 7.21 *The VS140 interfaced to a microscope to give an imaging spectrometer (Reproduced by permission of HORIBA Jobin Yvon).*

7.4 Electrochemical Instruments

pH Probes, ISEs and Multimeters

Both pH and ISE probes have always been small but have, historically, been attached to a pH meter that is often not so small. This has changed with the advent of handheld dataloggers and built-in LCD screens that allow the analyst to carry the pH or ISE probe to the sample, take the measurement and obtain a reading immediately. Calibration standards can be checked just before the measurement. pH and ISE meters that are commercially available range from handheld to pocket-size to devices that have a neckstrap for hands-free operation. Generally, the decrease in size equates to an decrease in accuracy and memory capability. Some meters allow the data to be downloaded on return to a central repository or PC.

The Lutron pH-201 is a low cost pocket pH meter ideal for use in the laboratory or in various industries, such as food processing, fishing and the paper industry. It is small in size and weighs only 153 g. It has a resolution of 0.01 pH unit and an accuracy of ±0.07 to ±0.2 pH units depending on where in the pH range the measurement falls. The even lighter ExStik PH100 from Extech (Figure 7.22) is a robust, waterproof pocket pH meter with a flat surface electrode that is useful for measuring pH on liquids, solids and semisolid surfaces, e.g. food, paper, cement and wet surfaces. It has a rugged, waterproof design

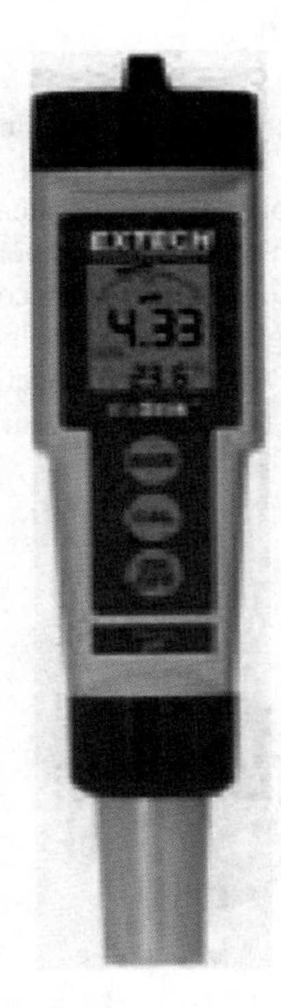

Figure 7.22 *The ExStik PH100 pH meter (Reproduced by permission of Extech).*

to withstand harsh environments and weighs only 110 g. It has automatic temperature compensation, can store up to 15 readings and alerts the analyst when the meter requires recalibration or the electrode needs to be replaced. Accuracy and resolution are both ±0.01 pH unit.

The Orion 4-Star portable is a combination pH/ISE meter from Thermo Electron Corp. that works with Orion ISE or ISFET electrodes for direct reading. It is waterproof and dustproof. Calibration can be manual or automatic. It delivers over 2000 hours of continuous operation. It can record and store up to 10 set-up parameters and calibrations and up to 200 data points. The pH resolution can be set to 0.1, 0.01 or 0.001. Relative accuracy is ±0.002 pH unit. For ISE measurements, relative accuracy is ±0.2 mV or 0.05 %, whichever is greater.

Hach has brought out a portable multiparameter meter (sensION156), which measures pH, oxidation reduction potential (ORP), dissolved oxygen, conductivity, total dissolved solids (TDS), salinity and temperature all-in-one. The sensION156 meter features quick calibrations with automatic temperature corrections. It can store 199 points of data and downloading is via a docking station, which converts the battery-powered sensION156 to line power for laboratory use. Batteries give 500 hours of continuous operation. Calibration for pH can be 1, 2 or 3 point-based and pH resolution can be set to 0.1, 0.01 or 0.001. Dimensions of the unit are $21.2 \times 8.7 \times 4.2$ cm.

Other multimeters which test for various combinations of analytical parameters, e.g. colour, turbidity and chlorine simultaneously are available.

Potentiostat

The portable PG580 potentiostat–galvanostat from Uniscan Instruments (Figure 7.23) is handheld in size and can be used in the laboratory connected to a PC or in the field working off battery power for complex electrochemical and corrosion experiments, such as very fast linear and cyclic voltammetry, chronoamperometry, square wave voltammetry, normal and differential pulse voltammetry. It has connections for up to five working electrodes when multiplexed. It weighs 300 g and the batteries last several hours. Generic sensors for easy connection to the PG580 are also available from Uniscan. The standard sensors comprise a standard three electrode configuration with a carbon working and

Figure 7.23 *The portable PG580 potentiostat–galvanostat (Reproduced by permission of Uniscan Instruments).*

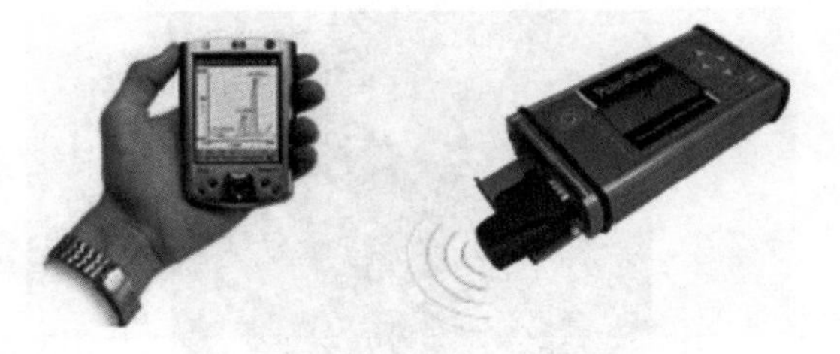

Figure 7.24 The wireless PalmSens potentiostat–galvanostat (right) and the pocket PC used to control it (left) (Reproduced by permission of Palm Instruments).

counter electrode and a silver/silver chloride reference electrode. These are useful for developing electrochemical sensors based around screen printed electrode technology.

A combination portable potentiostat–galvanostat from Palm Instruments is the PalmSens (Figure 7.24). The PalmSens can be connected to a palmtop PC and is used for sensors or cells with two or three electrodes, which can be micro- as well as macro-electrodes. It weighs 450 g with batteries which give eight hours of continuous operation. The response time of the instrument is short enough to apply fast techniques, such as square wave voltammetry, and it is suitable for a wide variety of experiments. There are eight current ranges from 1 nA to 10 mA with a resolution of 1 pA on the lowest current range. Its capacity can be extended with an optional multiplexer. PalmSens can be controlled by the palmtop PC by means of a wireless connection based on Bluetooth technology.

Conductivity Meter

Conductivity meters have been relatively easy to reduce in size and there are now a number of companies producing handheld meters. One example is the Enterprise 470 conductivity meter from Jenway (Figure 7.25). This meter is small and lightweight (370 g). It also measures total dissolved solids (TDS). Direct calibration is possible on standard solutions

Figure 7.25 The Enterprise 470 conductivity/TDS meter (Photograph courtesy of Barlworld Scientific–Jenway).

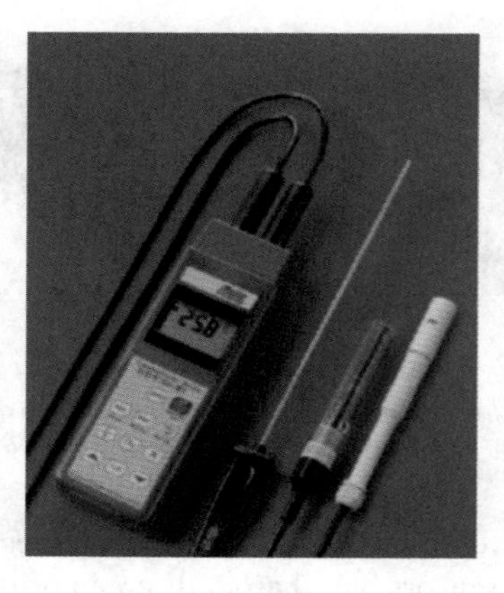

Figure 7.26 The HD 2106 conductivity–temperature meter (Reproduced by permission of Delta OHM).

or by direct cell constant entry. Both the conductivity and TDS accuracy is ±0.5 %. An electrochemical kit comprising the 370 pH meter, the 470 conductivity meter and the 970 dissolved oxygen meter supplied together is available from Jenway. The 470 conductivity meter displays results that are automatically temperature compensated and has 500 hours of battery life.

Extra features are often added to conductivity meters. The HD 2106 conductivity meter from Delta OHM (Figure 7.26), for example, measures the electric conductivity of liquids as well as resistivity, total dissolved solids and salinity. This instrument makes measurements using 4-ring or 2-ring conductivity probes. The measurement range of the cell is wide, from a few microSiemens (distilled water) to a fraction of Siemens (base or strong acid). Over the measurement range (0–199.9 μS), the resolution is 0.1 μS. Accuracy is ±0.5 %. Different probes are used for different ranges. The HD 2106 version 2 is a datalogger and can store 36 000 results. The device weighs approximately 470 g with batteries and the battery duration is about 200 hours.

References

1. Turza, S., Kurihara, M. and Kawano, S. (2006) Near infrared analysis of whole blood and plasma in blood-collecting tubes. *J Near Infrared Spec*, **14 (3)**, 147–153.
2. Agrawal, A., Cronin, J., Tonazzi, J. *et al.* (2006) Validation of a standardized portable fluorescence method for determining trace beryllium in workplace air and wipe samples. *J Environ Mon*, **8 (6)**, 619–624.
3. Zhang, J.X., Huang, W.D., Jin, S.Y. *et al.* (2005) Determination of carbofuran and paraoxon using a portable detector. *Fenxi Ceshi Xuebao*, **24 (4)**, 89–91.
4. Eidmann, G., Savelsberg, R., Blumler, P. and Blumich, B. (1996) The NMR MOUSE: a mobile universal surface explorer. *J Mag Reson A*, **122**, 102–109.
5. Perlo, J., Casanova, F. and Blumich, B. (2006) Single-sided sensor for high-resolution NMR spectroscopy. *J Mag Reson*, **180 (2)**, 274–279.

6. Perlo, J., Demas, V., Casanova, F. *et al.* (2005) High-resolution NMR spectroscopy with a portable single-sided sensor. *Science*, **308 (5726)**, 1279.
7. Muir, H. (2005) Hope for portable MRI scanners. *New Scientist*, **186 (2494)**, 9.
8. Savukov, I.M. and Romalis, M.V. (2005) NMR Detection with an atomic magnetometer. *Phys Rev Lett*, **94 (12)**, 123001.
9. Mulligan, C.C., Justes, D.R., Noll, R.J. *et al.* (2006) Direct monitoring of toxic compounds in air using a portable mass spectrometer. *Analyst*, **131 (4)**, 556–567.
10. Yashin, Y. and Yashin, A. (2001) Miniaturisation of gas-chromatographic instruments. *Anal Chem*, **56 (9)**, 794–805.
11. Jing, S.L., Zhang, Y. and Fan, Y.X. (2006) Characteristics of various portable gas chromatographers. *Yankuang Ceshi*, **25 (4)**, 348–354.
12. Eiceman, G.A., Gardea-Torresday, J., Dorman, F. *et al.* (2006) Gas chromatography. *Anal Chem*, **78 (12)**, 3985–3996.
13. Yuan, L.L., Wei, H.P., Feng, H.T. and Li, S.F.Y. (2006) Rapid analysis of native neomycin components on a portable capillary electrophoresis system with potential gradient detection. *Anal Bioanal Chem*, **385 (8)**, 1575–1579.
14. Vaithianthan, T., Tullis, I., Everdell, N. *et al.* (2004) Design of a portable near infrared system for topographic imaging of the brain in babies. *Rev Sci Instr*, **75 (10)**, 3276–3283.

8

Portable Instruments in Various Applications

8.1 Medical Applications

Medicine is a major area for portable analytical instruments. Many devices used in clinics, hospitals and surgeries must be at the very least mobile, e.g. on wheels, so that they can be brought to a patient quickly. However, the greatest growth area in medical analytical devices has been in diagnostics, home testing and long-term monitoring. Many of these devices are called point-of-care (POC) devices, as they are brought to where the test or care is required. The medical devices and diagnostics market is now estimated to be worth £16 billion per annum[1].

8.1.1 Point-of-Care Technology

Point-of-care (POC) technology is where small devices are brought to the patient as and when they are needed. They are sometimes called point-of-need or near-patient testing devices. The clinical need for faster turnaround time and the importance of minimising the length of stay for patients in hospital have provided the incentivies for instrumentation and diagnostics companies to develop point-of-care testing technologies. Some examples of POC situations are the bedside of a patient, an operating table in a theatre, a local clinic or general practitioner's office, in remote areas such as space or on ships at sea or even at home.

Benefits of POC technology include the lower costs of out-patient versus in-patient care, savings in both waiting time and consultation time in clinics, savings in laboratory analysis time and effort, the fact that it is often less invasive than hospital testing and that POC devices are more readily available to the public. Importantly, too, POC devices allow doctors to make decisions quickly based on the rapidly available results. Reimbursement from insurance providers is now possible for some POC devices, such as home glucose

Analytical Instrumentation: A Guide to Laboratory, Portable and Miniaturized Instruments G. McMahon

testing kits. The ease and speed of self-monitoring lets patients become more involved in looking after their own health.

The emphasis with POC is often on monitoring and prevention rather than cure. In the USA, use of POC testing has accelerated, with 60 % of hospitals in a recent survey performing POC coagulation testing[2]. The target analytes are some of the common chemical constituents of blood, plasma, serum or urine. Results can be qualitative or quantitative. Some of the main POC application areas are in blood glucose testing, blood coagulation monitoring and in diagnostic tests such as pregnancy, HIV and HCV testing.

8.1.2 Blood Glucose Testing

Diabetes is a chronic illness from which an estimated 200 million people worldwide suffer. It is also a growing problem. With diabetes, a person's blood sugar is elevated and this can lead to a number of complications. Blood glucose monitoring facilitates good diabetes control. Glucose levels are tested by diabetics as many as four or more times each day, depending on the patient and their precise condition. More blood glucose measurements are performed worldwide each day than any other single analytical test.

At least 25 different portable POC meters are commercially available in the USA alone. They differ from each other in several ways, including the size of the unit, the amount of blood needed for the test, the speed of the result, the size of the memory available in the instrument and the cost of both the meter itself and the test strips[3]. A recent paper compared four glucose meters in terms of their precision and accuracy[4]. Although the type of technology exploited affects the performance of a blood glucose meter, other factors such as the environment, other constituents in the sample, the operator and the test strips also affect instrument readings. Overall, a glucose meter gives a diabetic the freedom to self-test at home or when out and about. It also empowers them to take direct action immediately after the result if required, e.g. consumption of carbohydrate if the glucose level is low.

Because glucose is difficult to measure directly, it is normally first converted into another compound and then measured by indirect means. Testing is generally based on the reaction between glucose and a highly specific enzyme, which renders POC tests for glucose very accurate. Enzymes such as glucose oxidase, glucose dehydrogenase or hexokinase can be used. Of these, glucose oxidase is most commonly employed. Fortunately for glucose meter technology, glucose oxidase has proven itself to be a very robust enzyme that can be integrated onto test strips as a dry reagent and which regenerates itself on appropriate wetting by the sample. The reaction between glucose and glucose oxidase is shown in Figure 8.1.

Different manufacturers of analytical POC glucose meters use different constituents of these enzyme reactions for measurement but there are two principal technologies used: reflectance photometry and electrochemical technology. Reflectance photometry has replaced the older technique of colorimetry for glucose measurement. In reflectance photometry, glucose is catalysed by the embedded enzyme on the test strip. The oxidised glucose reacts with a dye to produce a blue coloured complex. The darker the complex, the higher the concentration of glucose. An LED shines light of a specific wavelength onto the complex on the strip. The amount of visible light reflected by the complex is then related

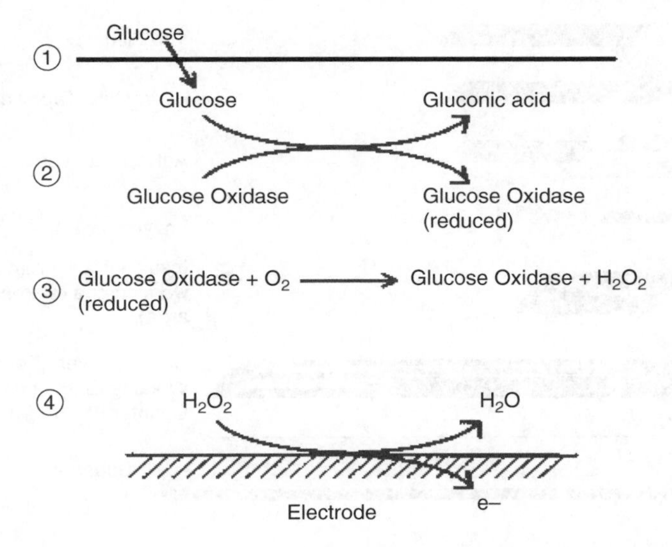

Figure 8.1 *Steps in the reaction between glucose and glucose oxidase.*

to the amount of glucose in the blood. The darker the complex, the more light it absorbs and the less light reflected.

Some issues with reflectance photometry include the fact that it requires larger sample volumes (1–3.5 μL of blood), longer processing times, frequent calibration and there is a slight risk of contamination of the blood by the strips or optics interface. The Accu-Chek Compact Plus from Roche Diagnostics (Figure 8.2) contains the glucose meter, teststrips and built-in lancet device all-in-one. Only a 1.5 μL blood sample is required and glucose is measured by reflectance photometry in eight seconds.

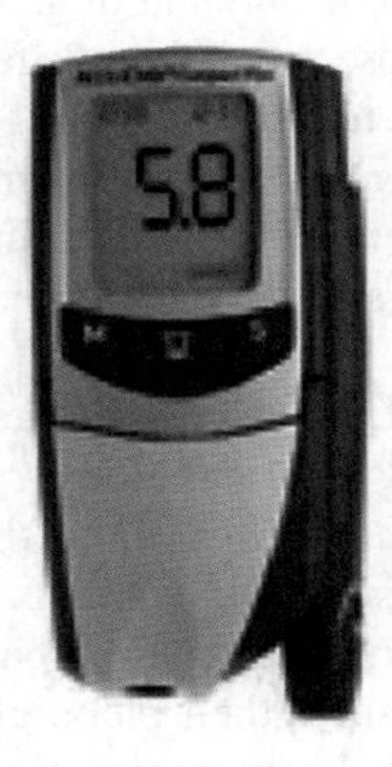

Figure 8.2 *The Accu-Chek Compact Plus (Reproduced by permission of Roche Diagnostics).*

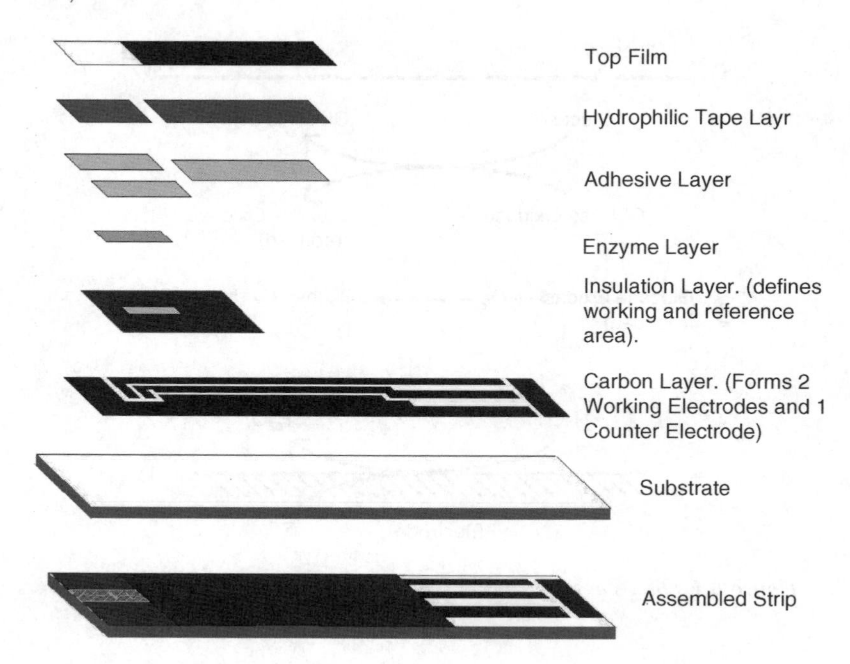

Figure 8.3 *An exploded view of the OneTouch® Ultra glucose test strip (Reproduced by permission of Lifescan).*

With the electrochemical technique, the blood glucose again undergoes oxidation, but this time the charge generated is measured by an electrode; this current is related to the glucose present in the sample. There are two different electrochemical approaches: amperometric and coulometric. In amperometry, the blood glucose reacts with a mediator in the presence of an enzyme to catalyse the reaction. A voltage is applied to encourage the electrons to flow between the working and reference electrodes. The size of the current generated is proportional to the amount of glucose present in the blood drop. The basis of the OneTouch® Ultra glucose meter from Lifescan is amperometric; it is used in conjunction with OneTouch® Ultra test strips. Only 1 μL of blood is needed and the test takes five seconds. The test strip contains the impregnated enzyme, which reacts with the glucose in the blood causing a current. Even though only a small percentage of the glucose is actually oxidised, the method is still very sensitive. An exploded view of the test strip is shown in Figure 8.3.

With coulometry, virtually all of the glucose in the sample is converted into electrical current, so it can be more efficient than amperometry and may require a smaller blood sample (<1 μL). Usually, glucose dehydrogenase is the enzyme involved. The CozMore system is based on coulometric testing and contains both a blood glucose monitor and an insulin pump that delivers any insulin required based on the result obtained. After a patient inserts a test strip and measures their blood glucose, the monitor instantly sends the results to the pump using infrared communication.

Point-of-care glucose testing normally works in the following way:

- Blood is taken by fingerstick or lancet (only μL of blood required).
- Blood is placed on or into a test strip and is sometimes drawn up by capillary action.
- The test strip is inserted into handheld meter.
- Blood interacts with dry reagents on test strip.
- The measurement is processed by instrument.
- The result is shown on a LCD panel.
- The device stores the result and may also have the capability to directly upload it to a centralised data repository.

Practically all portable glucose monitors can store at least 30 days of readings within the device. Most meters can, additionally, be connected via a USB to a computer or printer and the results downloaded or printed so that patterns or trends can be elucidated.

Recent developments in blood glucose monitoring instrumentation are focussing on less invasive testing for the patient and alternative technologies to photometry and electrochemistry. One such development is the Glucowatch Biographer from Animas Corp. (Figure 8.4). It is based on amperometric measurement and is the only noninvasive glucose monitor approved by the United States Food and Drug Administration (FDA). Instead of requiring blood to be withdrawn for testing, this device samples fluid from the skin by reverse iontophoresis, which involves applying a small electrical current (0.3 mA) to the skin for a few minutes through a replaceable pad (AutoSensor) that

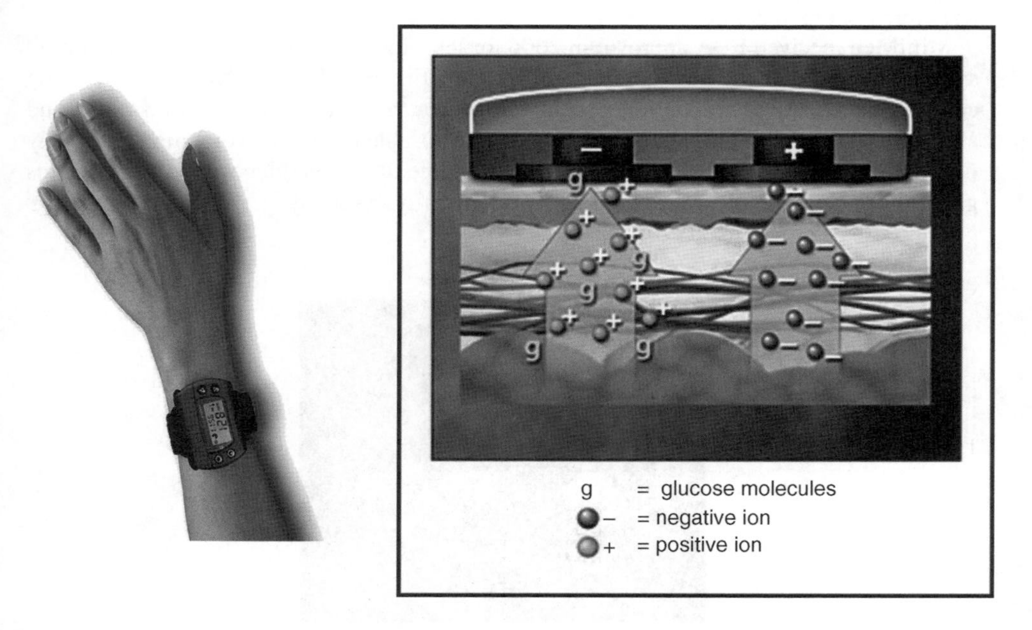

Figure 8.4 *The Glucowatch Biographer from Cygnus/Animas and a schematic illustrating how the glucose is collected and measured (Reproduced by permission of Animas).*

clips onto the back of the watch. The current pulls glucose to the skin surface. This glucose is collected at the iontophoretic cathode, where it reacts with glucose oxidase in the hydrogel producing hydrogen peroxide. At the end of the collection period the iontophoresis current is stopped and the current obtained from the detection of hydrogen peroxide is measured. The glucose concentration in this interstitial fluid is proportional to the concentration in blood. Iontophoretic extraction also eliminates the problem of interference from many species that might interfere with the measurements made directly in blood but there is a time lag of twenty minutes before a reading is available and iontophoresis can be irritating to the skin. Alerts and alarms can be set up for high and low blood sugar levels, for rapid changes in the blood sugar level and for any readings above or below target levels.

Continuous glucose monitoring systems (CGMS) are becoming more desirable since spot measurements may miss diurnal trends or certain peaks and troughs associated with meals or exercise. DexCom obtained FDA approval in 2007 for its second generation short-term CGMS and is seeking approval for its long-term CGMS. Both systems are based on the glucose oxidase enzyme reaction. The short-term system includes a small insertable or implantable sensor that continuously measures glucose levels in subcutaneous tissue just under the skin over three days or seven days. A small external receiver accepts the values from the sensor at specified intervals. Current blood glucose values, as well trend data can be displayed. The receiver also sounds an alert when glucose levels are too high or too low. A study by Garg *et al.* using the LTS sensor reported that when patients were provided with the information from the sensor over a period greater than 50 days, the patients spent 47 % less time hypoglycaemic and 25 % less time hyperglycaemic[5].

MiniMed received FDA approval in 2006 for its combination CGMS and insulin pump system called the Paradigm REAL-Time System (Figure 8.5). It includes the following components: a real-time continuous glucose sensor, a radio frequency transmitter and an insulin pump. The glucose sensor is a tiny disposable electrode worn subcutaneously (usually in the stomach) by the patient for up to three days. The glucose sensor measures glucose levels in the interstitial fluid amperometrically. As many as 288 glucose readings

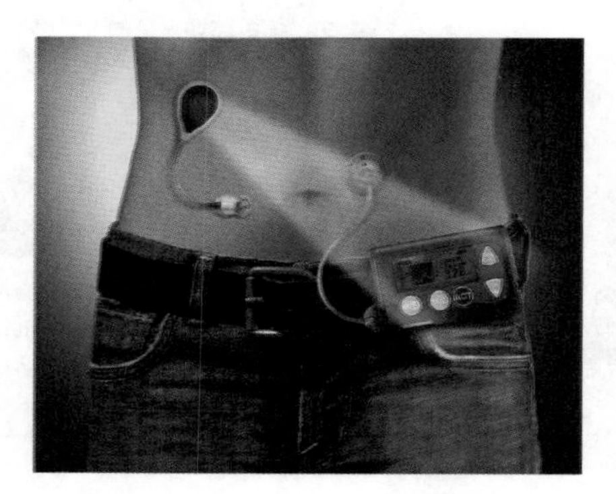

Figure 8.5 *The Paradigm REAL-Time System (Reproduced by permission of Medtronic MiniMed).*

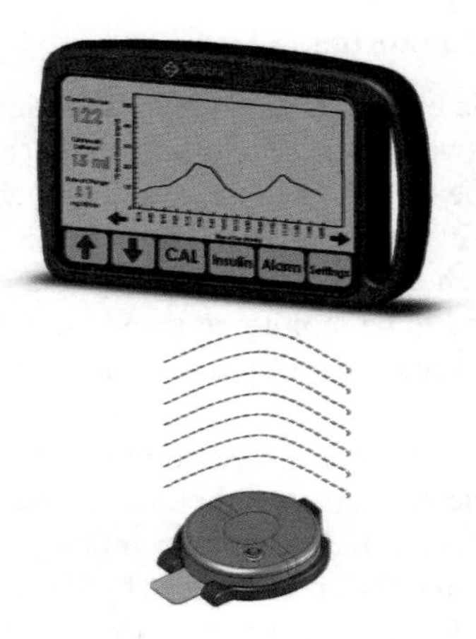

Figure 8.6 *The Symphony noninvasive gel-based sensor communicates wirelessly with the glucose meter (Reproduced by permission of Sontra).*

are recorded by the sensor each day, so the device can provide useful trend data. The pump displays glucose values every five minutes and delivers insulin when required.

Sontra has developed a noninvasive CGMS system in conjunction with Bayer Diagnostics, though it is not yet commercially available. The Symphony device (Figure 8.6) measures glucose diffusing through ultrasonically permeated skin continuously for up to twenty four hours. The product consists of a SonoPrep instrument, a glucose biosensor with RF transmitter and a glucose meter. The glucose meter provides blood glucose readings on demand in addition to trends and alarms. Symphony has a 90 % correlation to reference blood glucose and reliably detected hypoglycaemia. Product development is currently focused on improvements to the glucose flux biosensor and design of the wireless RF interface between the biosensor and glucose meter.

A. Menarini Diagnostics has developed the GlucoDay® S semi-invasive continuous glucose monitor, which is based on a microdialysis technique and is specifically for clinical use.

Other technologies are also being explored for glucose sensing. For example, Ciba Vision has developed a contact lens that can measure glucose fluorometrically in tears by holding a small detector up to the eye. The higher the fluorescence, the greater the glucose concentration. Raman spectroscopy is being investigated through the use of a laser to measure the amount of glucose in the finger. Another noninvasive glucose testing device is being developed by Smart Holograms. It uses sensor holograms to measure disease biomarkers in contact lenses, one of which is glucose for diabetes[6, 7, 1]. In Korea LG has released the KP8400 cellphone that doubles as a blood test for diabetics. Users can place a blood drop on a strip, place the strip in a special reader on the phone and get insulin and blood readings on the phone display. Readings can then be uploaded to an online database for retrieval later. Other noninvasive technologies are being explored for glucose monitoring in small, personal devices. These include near infrared, far infrared and radio wave impedance[8, 9].

8.1.3 Blood Coagulation Monitoring

The ability of blood to coagulate (clot) is normally important in the prevention of thromboembolic events. There are two important clotting tests, called prothrombin time (PT) and partial thromboplastin time (PTT), each measured in seconds. They measure the time taken for blood to clot under different circumstances. Millions of people worldwide take blood thinning medication such as warfarin every day. Patients on warfarin are usually on the drug long-term and need to be monitored using the PT test. This is normally done in out-patient clinics on a weekly to monthly basis but point-of-care testing is becoming more common.

In the hospital setting, blood thinning medication such as heparin is often given to patients during surgery and the patients would be monitored using the PTT test to make sure that the blood was not being under- or over-anticoagulated. Both of the tests are also employed to diagnose haemophilia and other clotting disorders. In the case of both warfarin and heparin, these drugs have a very narrow therapeutic range, so frequent monitoring is required and has clear benefits. POC coagulation monitoring allows rapid turnaround of results and timely dosage adjustments if required.

Blood coagulation monitoring is based on timing how long it takes for the blood to clot. The PT test is the time it takes plasma to clot after addition of tissue factor (usually obtained from animals) and is more commonly used than the PTT test. The reference range is approximately 7–16 seconds. The PT is often reported the as international normalised ratio (INR) with normal values being in the range 0.8–1.2. In one trial patients who used self-testing for a year achieved a similar level of control over their anticoagulation treatment and had less complications and haemorrhages than patients working solely through their anticoagulation clinic[10].

Point-of-care coagulation monitoring normally works in the following way:

- Blood is taken by fingerstick or lancet (only μL of blood required).
- Blood is placed on or into a test strip and is sometimes drawn up by capillary action.
- The test strip is inserted into handheld meter.
- Blood interacts with dry reagents on test strip.
- The measurement is processed by instrument.
- The result is shown on a LCD panel.
- The device stores the result and may also have the capability to directly upload it to a centralised data repository.

Various technologies are exploited to detect when the clot has formed, including electrochemical and photometric approaches. These are exemplified in the two different handheld monitors produced by Roche Diagnostics: the CoaguChek S and the CoaguChek XS. The CoaguChek S uses reflectance photometry to measure the PT from one drop of blood. The blood is applied to the test card and is drawn by capillary action into the reaction path where it mixes with thromboplastin and small paramagnetic iron oxide particles (PIOP). The blood sample keeps moving until the clot forms. The particle movement is monitored optically by detecting changes in reflected light from the surface of the test card. When the movement stops due to the fibrin network in the clot, the PIOP are immobilised, no further light changes are seen and the test is

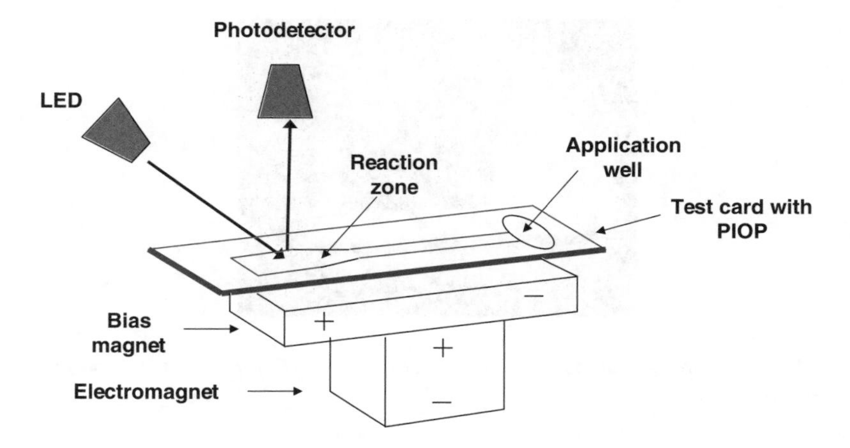

Figure 8.7 An image illustrating the reflectance photometric principle upon which a number of commercial coagulation meters are based.

complete (Figure 8.7). The dry reagents on the test card are different for the PT and PTT tests.

Roche's CoaguChek XS system is based on electrochemical technology. The test strip in this case contains thromboplastin reagent and a peptide substrate (Electrocyme TH). When blood is added to the strip, coagulation starts and thrombin is generated. The thrombin cleaves the peptide forming an electrochemically active compound that generates a signal. The time taken from application of sample to signal generation is used to calculate the INR value.

Other companies have in the past used alternative technologies, such as fluorescence. This is based on the fact that thrombin, which is formed during the clotting process, is a fluorescent molecule.

8.1.4 Other Point-of-Care Devices

There are other devices used in point-of-care situations that are portable but not necessarily small in size. For example, Spectromed and Spectrolab have collaborated to develop a new technology called Mediscreen, which is based on a portable FTIR and FTNIR spectrophotometer designed specifically for medical applications. The device, which is fully portable, fits into a briefcase and can be used to diagnose a wide range of medical conditions, including diabetes, specific cancers and skin disorders. In addition, it is possible to use the instrument for direct analysis of tissue in the operating theatre as a support technique for surgical procedures.

Horiba Jobin Yvon has an interesting portable instrument on the market called SkinScan (Figure 8.8). It is an *in-vivo*, fibre optic spectrofluorometer designed specifically to acquire fluorescence measurements directly from skin or hair. The fibre optic bundle transmits UV–NIR radiation to the sample and then collects the resulting fluorescence. Results are used to investigate the age and health of skin, and to evaluate sunscreens and cosmetics among other applications.

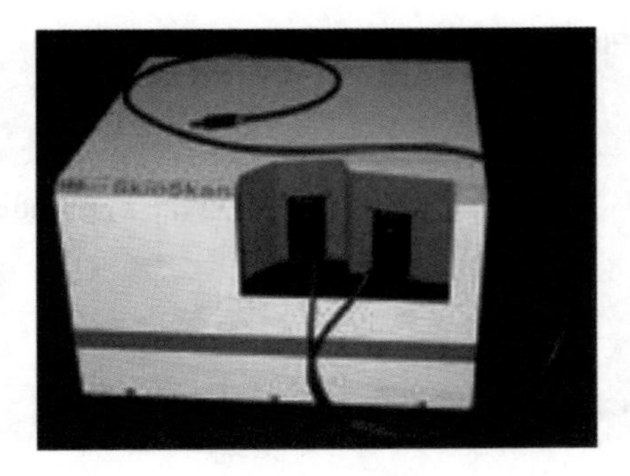

Figure 8.8 *The portable SkinScan spectrofluorometer (Reproduced by permission of HORIBA Jobin Yvon).*

There are many examples of the use of very small colorimetric devices for POC applications, such as urine dipsticks and test strips for blood sample screening. While not instruments as such, they are based on clever chemistries and reactions and are often single use, throwaway devices. The reagents required for the test are embedded and dried into a strip which is then dipped into the sample, e.g. urine or blood. The colour changes that result are compared to a colour chart, which is usually provided on the side of the container for the test strips.

One of the earliest such POC tests developed was the home pregnancy test. It is based on a qualitative immunological antibody assay. It relies on the determination of human chorionic gonadotrophin (hCG) levels in urine, which gives rise to a coloured band if positive. The device is very small and completely disposable. A quality control check is built in where the rest of the urine sample continues along the dipstick and interacts with a pH-sensitive indicator to show that the test has been performed correctly.

It has been reported that more than one-third of people tested for human immunodeficiency virus (HIV) in the USA do not return to receive their test results[11]. With the availability of rapid POC qualitative tests and immediate diagnosis, more infected patients will be given their results immediately. The OraQuick Advance test from Orasure Technologies can detect the antibodies to HIV from a saliva swab (or fingerstick blood) within 20 minutes. It is 99.8 % accurate for people who are not infected. It was the first POC HIV test to receive FDA approval in the USA for use in clinics and is now awaiting clearance for home testing.

Kits for home testing of HIV have been available in Europe from MiraTes since 2000[12]. The MiraTes HIV HomeTest uses fingerstick blood and has a sensitivity (percentage of HIV infected people that get a positive test) of more than 99.9 % and a specificity (percentage of people without HIV that get a negative test result) of more than 99.6 %.

Biochip-based HIV tests have been developed by researchers at the University of Texas and Harvard Medical School[13]. The chip is placed into a toaster-sized analyser that counts the number of CD4+ cells in the drop of blood. Results are available in ten minutes and correlate well with flow cytometry (the gold standard). The device, which is being marketed by LabNow, Inc. weighs less than 2.3 kg, requires no excess reagents and is battery-operated.

8.2 Environmental Applications

The environment is another major area for the use of portable analytical instruments. Many devices used in the field must be, at the very least, transportable so that they can be brought to a location quickly. Many of the instruments discussed in Section 8.1 can be used when analysing environmental samples. The greatest growth area in environmental field devices has been in monitoring applications.

8.2.1 Field Devices

There is a growing need for testing to take place outside the laboratory as environmental monitoring is becoming more important. These field-portable devices are analytical instruments that can be sent or brought to where they are needed in the 'field'. In terms of environmental applications, the field can be anywhere from ships at sea to remote locations, such as deserts or space or the tops of mountains, to dangerous environments, such as war zones or contaminated areas. Benefits of field testing include the fast turnaround of results (often in real-time), lower cost per test (usually), a saving of laboratory and personnel time, the use of smaller samples and improved safety for personnel with the advent of remote sensing.

The main field testing application areas are routine air, soil and water monitoring, pollution control and measurements taken directly at sources of contamination, at remote environmental locations and in hostile or dangerous territory. The emphasis with field devices is often on monitoring and control rather than reaction. A well-accepted definition for a field-portable device in terms of environmental applications is that[14, 15]:

- The combined weight is below 6.8 kg (15 lb) for the device and required supporting equipment.
- The size is less than 16 387 cm^3 (1000 cubic inches).
- There is enough battery life and consumables for eight hours of field operation.
- Data is retrieved and displayed without the need for a laptop in the field.
- The device is easy to use, weatherproof and can be manipulated even when wearing thick gloves.

8.2.2 Water Quality Monitoring

Water quality is an issue of international concern due to increasing contamination from a variety of sources. The most common types of water examined are surface waters, ground waters and waste waters. Common indicators of the quality of water are conductivity, pH, ion and nutrient measurements. The HI-991300 from Sheen Instruments Ltd (Figure 8.9) is a portable conductivity meter that measures conductivity as well as total dissolved solids (TDS), temperature and pH. For conductivity, it has a range from 0–3999 μS/cm with a resolution of 1 μS/cm. All readings are compensated automatically for any temperature variation. Calibration is carried out manually with provided solutions. The device can be operated with one hand and the rubber keypad is splash-proof. The HI-991300 is supplied complete with a conductivity probe with a one meter screened cable, a wrist strap for safety and the 9 V battery.

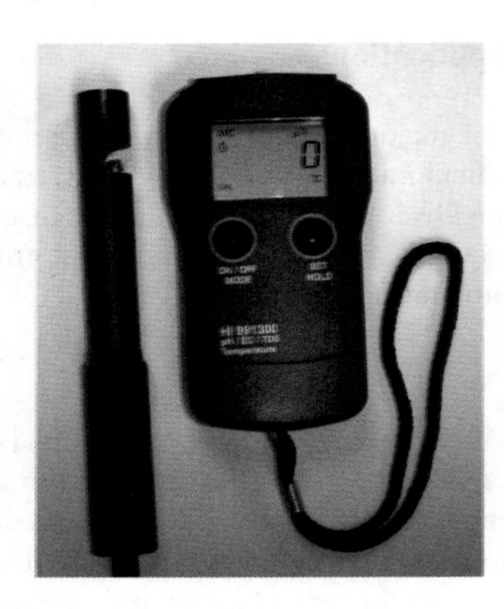

Figure 8.9 *The portable HI-991300 conductivity meter (Reproduced with permission of Sheen Instruments Ltd).*

Twin Cond is a conductivity meter, available from HORIBA Jobin Yvon, which features a waterproof flat sensor. Twin Cond can measure the conductivity of a solution from a single drop of sample and is therefore very suitable for both field measurements in rivers and lakes and for rain samples. It features a measurement range from 1 μS/cm to 19.9 mS/cm. It also has one-touch auto-calibration, automatic temperature conversion and a salinity conversion function (from 0–1.1 %). Multiparameter meters can test for pH, conductivity, total dissolved solids and temperature in the one device.

Specific ion colorimeters are very useful for the rapid analysis of ions such as ammonia, iron and manganese in surface waters. These lightweight, handheld devices are used extensively in environmental surveillance programmes. The Orion range of colorimeters from Thermo Electron Corp. features very small and lightweight instruments. A range of twelve models of the AQUAfast® II allows measurement of over 25 species in water samples based on colour chemistry. Sample concentration is easily read from the relatively large LCD. The microprocessor-controlled, battery-operated units weigh only 284 g and make these colorimeters useful for both laboratory and field use. Different units can measure different groups of chemicals, e.g. the AQUA 2005 measures aluminium, bromine, chloride, copper, sulfate and iron over two ranges. Another unit designed for drinking water analysis measures aluminium, chlorine (free and total), chlorine dioxide, fluoride and pH. The AQUAfast IV waterproof colorimeter can store up to ten user-defined methods and can save 100 data points in the field. It automatically detects the species to be measured, selects the method and the wavelength and initiates the measurement. Four wavelengths are possible: 420, 520, 580 and 610 nm with an accuracy of ±2 nm. The source lamp and detector are based on diodes.

A portable voltammetric water quality analyser for metal ions (cadmium, copper, zinc, nickel and lead) in sewage samples has been reported[16].

Figure 8.10 *The AgriSpec® portable NIR spectrometer with plant probe and leaf clip assembly (left) and in use in a soilpit (right) (Photos courtesy of Analytical Spectral Devices, Inc., Boulder, Colorado, USA. Copyright © Analytical Spectral Devices, Inc. All rights reserved.).*

8.2.3 Soil and Sediment Testing

Soil and sediment analysis is another important environmental assay. Levels of certain compounds in these samples can indicate contamination levels, their sources and whether remediation steps are working or not. Soil testing is also valuable in assessing the quality of soil for farming and for other agricultural applications. The AgriSpec® from Analytical Spectral Devices Inc. (Figure 8.10) is a rugged, field-portable NIR analyser designed specifically for analysis of a wide range of organic and inorganic materials, including the nutrient and moisture properties in soil. It is not necessary to dry or grind the soil samples since measurement of intact soil samples provides results similar to more processed samples for a wide range of soil properties. The AgriSpec® can also carry out chemical and mineralogical composition analysis of sediments. It has a spectral range of 350–2500 nm and resolution of 3–10 nm, depending on the wavelength. Its dimensions are $12.7 \times 36.83 \times 29.21$ cm (H, W, D) and it weighs 5.66 kg.

Copper has been measured in soil using a portable electroanalytical device based on stripping voltammetry[17]. In marine chemistry, sediments have been sampled and analysed in real-time by *in situ* solid-state voltammetric micro-electrodes[18]. The analyser was used to examine levels of various redox species and trace metals, e.g. iron and manganese, in salt marsh sediments and other matrices.

8.2.4 Air Monitoring

Determining the levels of gases, vapours and particulates in the air (both indoor and outdoor) is very important for a number of reasons, e.g. assessing industrial emissions, examining the quality of urban air and monitoring pollution. An instrument called the RAPID from Bruker is a long distance automated detection system, based on Fourier transform infrared (FTIR) technology, for remotely identifying most known chemical warfare agents and toxic industrial chemicals over distances of up to 5 km. Its dimensions are 50 × 33.1 × 38.6 cm (L, W, H) and its weight is 28.7 kg. This rugged detector may be mounted on several platforms and can even make measurements while moving. Another instrument from Bruker is the OPAG (open path gas analyser) 33 spectrometer, also based on FTIR, and which has been developed primarily for the remote detection of chemical pollutants in the atmosphere. It is composed of a sensor module, a controller module, power supply, telescopic sight and tripod. In the sensor module (which weighs 19 kg) are the spectrometer components, i.e. entrance window, auto-calibration unit, interferometer, parabolic mirror and IR detector. The spectral resolution of the OPAG 33 is 1 cm^{-1} or 0.5 cm^{-1} as required. Analysis is fast (<1 s). Positive identification of a chemical triggers an alarm and the display of the quantification results. The batteries allow four hours of field use.

A smaller instrument from Wilks Enterprise is the InfraRan Specific Vapour Analyser that is used to detect and measure single or multiple gaseous contaminants in ambient air. Based on mid-infrared technology, it can detect almost any compound in the region 2.5–14.5 μm. It is useful for detecting airborne solvent vapours, anaesthetic gases and refrigerants. Initial readings are available in 5–10 seconds and stable concentration results in under one minute. The device weighs 8.2 kg and batteries allow up to eight hours field use.

Even smaller in size is the AirPort MD8 from Sartorius; it is a battery-operated handheld device capable of quantitatively detecting airborne microorganisms and viruses. Volume flow is regulated by an integrated impeller wheel and sample volume can be from 25–1000 L. The instrument weighs 2.5 kg and its dimensions are 30 × 13.5 × 16.5 cm (L, W, H). The battery life is approximately 4.5 hours.

8.3 Security and Defence Applications

Analytical instrumentation now plays a big role in military applications. In fact, funding from the large defence agencies has contributed to the research into reducing the size of analytical equipment and making it more mobile and portable. The CBMS (chemical biological mass spectrometer) from Bruker is an example of an ion trap mass spectrometer that has been made rugged for field use. It is composed of a virtual impactor, pyrolyser, ion trap MS and a data system. It is still large, weighing 65 kg, but is mobile and robust. The CBMS is used to identify chemical warfare agents and to classify biological ones in three minutes.

Portable GC–MS is also used by the United States military in war zones. The instruments are often sent ahead of the troops, often on unmanned, robot-controlled vehicles to detect any airborne chemicals or toxins so that soldiers can avoid contaminated zones. The Viking 573 is a GC–MS for fast on-site detection of hazardous compounds from gases, solids, and liquids. This rugged GC–MS system weighs 39 kg and is designed for fast and simple on-site assessment of chemical catastrophes and environmental accidents involving organic compounds. It conducts definitive analysis of compounds and is capable of identifying more than 300 000 chemical compounds in minutes.

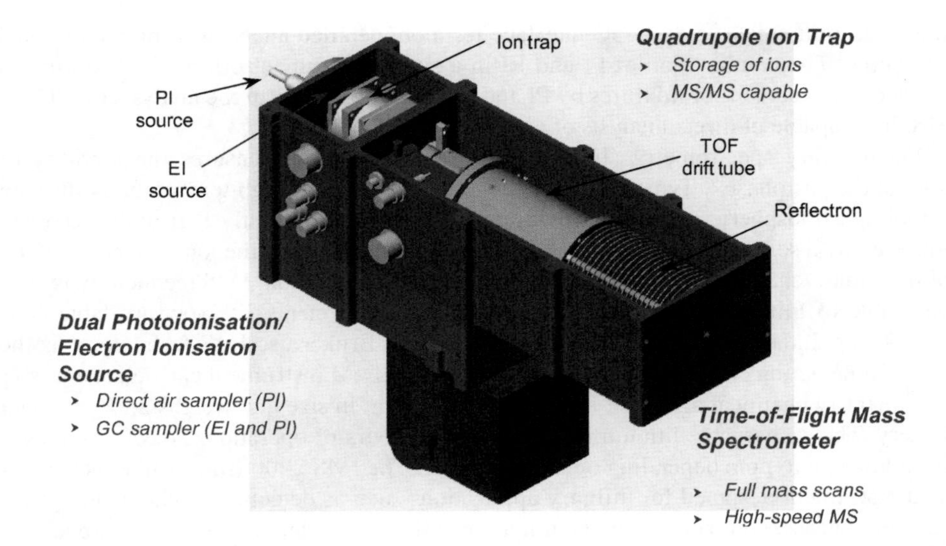

Figure 8.11 The FieldMate prototype combined EI-PI QitTof mass spectrometer (Image courtesy of Syagen).

Syagen has developed a prototype portable MS instrument called the FieldMate. It has both electron impact (EI) and photoionisation (PI) sources and a quadrupole ion trap, time-of-flight (QitTof) mass analyser. It has MS–MS capability and can be interfaced with a GC (Figures 8.11 and 8.12). It weighs just under 32 kg and its dimensions are $45.7 \times 40.6 \times 35.6$ cm. FieldMate permits direct air and liquid sampling using the PI source.

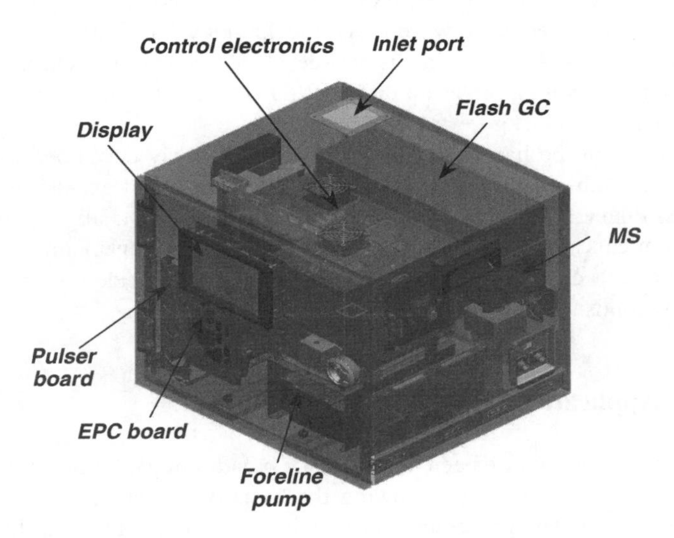

Figure 8.12 The FieldMate prototype system (Image courtesy of Syagen).

It also accommodates a flash GC interface for more detailed analyses using the optional EI source. The combination of PI and EI in a single instrument offers the flexibility to conduct rapid screens of mixtures by PI and subsequently confirm the analysis by GC–EI MS. It is capable of direct analysis of air, liquid and soil samples.

Ion mobility spectrometry (IMS) is a technology which measures the mobility of ions in the gas phase[19]. Typically the drift time of ions under high temperatures through a tube at atmospheric pressure is measured. The sample is firstly thermally desorbed followed by a soft chemical ionisation process. The drift time of the ions is characteristic of their mass/charge ratio and the cross-sectional area of the ion. IMS technology is very amenable to miniaturisation and very small IMS instruments are now available. The RAID (rapid alarm and identification devices) from Bruker use IMS technology as the basis for detecting chemical warfare agents. The RAID-M instrument can detect, classify and quantify chemical agents. It is $40 \times 11.5 \times 16.5$ cm in size and weighs 2.9 kg without battery. The rechargable lithium battery gives six hours of operation. Sensitivity ranges from low ppb to ppm depending on the substance. The IMS 2000 from Bruker is a hand-held instrument designed for military applications such as detection of chemical agents, e.g. nerve gases, confirming decontamination and ensuring an area is clear of toxic substances. It weighs less than 2.5 kg and has an anti-shock system built-in to facilitate its mounting on a vehicle.

Similar IMS-based devices are also being used at airport security checkpoints. A passenger's clothing or luggage is swabbed and the swab inserted into a small heated chamber. Inside the chamber, any traces of organic compounds evaporate, mix with a carrier gas and are ionised. The ionised gas moves through an electrical field in the drift tube colliding with other ions at atmospheric pressure. Smaller ions reach the detector first. If the spectral signature matches certain known substances, further analysis will be carried out for definitive identification and quantitative measurement.

The Model 4300 zNose is a portable GC that can detect and measure levels of many types of vapours, toxins, explosives, narcotics and other compounds. It is the first GC to be based on surface acoustic wave (SAW) technology. The battery pack allows field operation for six hours. Analysis time is between 1–60 seconds and sensitivity is low ppb for most compounds with a precision of 5 % RSD. The combined weight of the sampler, chassis and charger is 14.6 kg.

Even smaller than the handheld devices above is a newly developed pen-like device which can detect sub-milligramme amounts of peroxide-based explosives[20]. The prototype costs less than €23 per unit. A suspect sample is placed on a silicone rubber test pad. Three test chemicals are sequentially injected into the transparent chamber in the pen and a blue–green colour change occurs on reaction with any peroxide present in the sample within three seconds.

8.4 Other Applications

A variety of technologies have been used to test a breath sample for alcohol, one of which is infrared. A schematic diagram showing the basis on which such a device works is shown in Figure 8.13. The procedure starts with a person blowing into the tube and the absorption band at 3.44 μm (3400 nm) measures alcohol. A correction is made for acetone

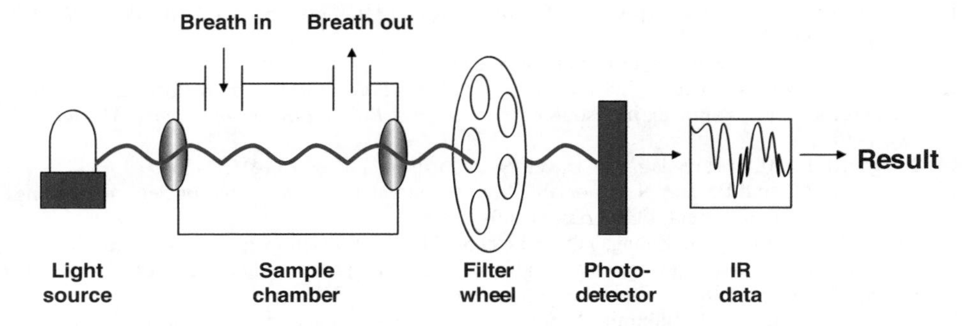

Figure 8.13 *Schematic diagram showing the operation of a breath alcohol analyser.*

at 3.37 μm and the percentage blood alcohol is displayed on a readout. The light source provides IR radiation which enters the breath sample chamber via a lens. The radiation passes through the breath sample and the more alcohol present in the sample, the greater the IR absorption. The modified radiation leaves the chamber via a lens and passes through a narrowband IR filter wheel. From there, it reaches the detector where the IR signal is generated and a final result is displayed.

One such portable instrument that is based on IR for measuring breath alcohol is the Intoxilyzer-8000 from CMI. It uses dual wavelengths for measurement and takes less than one minute to deliver a result. When the performance of the protable Intoxilyzer-8000 (7.7 kg) was compared with that of the approved stationary Intoxilyzer-5000 instrument (13.6 kg), it was shown that the portable instrument performed well enough to be used for evidential breath alcohol testing[21].

References

1. Davies, E. A smart move for holograms. *Chemistry World*, May 2006, 54.
2. The top 100 point-of-care manufacturing companies worldwide, Research & Markets Report, Oct 2003.
3. www.diabetesnet.com/diabetes_technology/blood_glucose_meters.php
4. Lippi, G., Salvagno, G.L., Guidi, G.C. *et al.* (2006) Evaluation of four portable self-monitoring blood glucose meters. *Ann Clin Biochem*, **43 (5)**, 408–413.
5. Garg, S.K., Schwartz, S. and Edelman, S.V. (2004) Improved glucose excursions using an implantable real-time continuous glucose sensor in adults with Type 1 diabetes. *Diabetes Care*, **27**, 734–738.
6. Pioneering a picture of health. UKWatch, 13 Mar 2006, 15.
7. Dean, K.E.S., Horgan, A.M., Marshall, A.J. *et al.* (2006) Selective holographic detection of glucose using tertiary amines. *Chem. Comm*, 3507–3509.
8. www.medihealthdme.com/education/diabetes_non.htm
9. www.diabetesnet.com/diabetes_technology/new_monitoring.php
10. Mendenez-Jandula, B., Souto, J.C., Oliver, A. *et al.* (2005) Comparing self-management of oral anticoagulant therapy with clinic management: a randomized trial. *Ann Intern Med*, **142 (1)**, 1–10.

11. FDA Gives Preliminary Approval to OraSure's Rapid HIV Test. Kaiser Network HIV/AIDS Daily Reports, May 16 2002.
12. www.mirates.com/read/hiv_test?submenu=260
13. Rodriguez, W.R., Christodoulides, N., Floriano, P.N. *et al.* (2005) A microchip CD4 counting method for HIV monitoring in resource-poor settings. *Public Library of Science-Medicine*, **2 (7)**, 1–10.
14. Norgaard, J. (2005) Have lab, will travel. *Environ Protect Mag*, **16 (4)**, 32–35.
15. Ebersold, P.J., and Barker, N. A portable high speed gas chromatograph for field monitoring, article 1291, Environmental Expert.com, July 2003.
16. Zhang, S.Z., Yang, X.Q., Zhang, J.Q. and Chen, Y.L. (2006) Determination of cadmium, copper, zinc, nickel, lead in sewages by portable HK-1 automatic water quality analyzer. *Yankuang Ceshi*, **25 (2)**, 185–188.
17. Beni, V., Ogurtsov, V.I., Bakunin, N.V. *et al.* (2005) Development of a portable electroanalytical system for the stripping voltammetry of metals: Determination of copper in acetic acid soil extracts. *Anal Chim* Acta, **552 (1–2)**, 190–200.
18. Luther III, G.W., Glazer, B.T., Ma, S. *et al.* (2007) Use of voltammetric solid-state (micro)electrodes for studying biogeochemical processes: laboratory measurements to real time measurements with an *in situ* electrochemical analyzer (ISEA). *Marine Chemistry*, In Press.
19. Li, F., Xie, Z., Schmidt, H. *et al.* (2002) Ion mobility spectrometer for online monitoring of trace compounds. *Spectrochim Acta Part B*, **57 (10)**, 1563–1574.
20. Milgrom, L. 'Pen' detects peroxide-based explosives. *Chemistry World*, July 2005, 7.
21. Razatos, G., Luthi, R. and Kerrigan, S. (2005) Evaluation of a portable evidential breath alcohol analyzer. *Forens Sci Int*, **153 (1)**, 17–21.

Section III

Process Analytical Instrumentation

The Drive for Process Analysis

Process analysis means analysing samples from a process in 'real-time' as distinct from waiting until the end of the process to take samples for subsequent analysis in a laboratory. In recent years, this has meant the migration of certain analytical testing from the laboratory into production and direct measurement at sampling points during manufacturing. The data obtained are used to control the process itself, and hence the quality of the products made. Process analysis is now a substantial area for use of analytical instruments.

Across many industries, there has been a push to carry out analysis during the process (on-line) rather than taking samples away for investigation by conventional methods (off-line), for the reasons of saving time and improving quality. There is also the added advantage of the real-time visibility into the process. As well as the actual analytical measurements, additional steps such as sampling, pretreatment, manipulation and use of the results are usually required and can also be integrated. Because of the often harsh environments encountered, process analytical systems must meet higher standards of ruggedness and simplicity than laboratory-based instruments.

Traditionally, physical measurements of pressure, viscosity, flow rate and temperature were the only indicators that a process was running smoothly. However, of late there has been a surge of interest in process instruments that can give more specific information, such as kinetics, reaction stage, concentration and presence of impurities. Certain process instruments, such as gas chromatographs and mass spectrometers, have been used and proven in the petrochemical industry for some years. The continuous nature of the petrochemical processes and the profit margins involved made investment in process analytics in this industry a worthwhile endeavour. The knowledge and insights gained by the petrochemical industry have made it easier for other industries to embrace the technology.

The main driver for the more recent development of process analytics has come from the pharmaceutical industry, in response to the Process Analytical Technology (PAT) initiative issued by the United States Food and Drug Administration (FDA). This has come about to enable manufacturers to reduce costs while providing customers with the assurance that they are receiving high-quality, pure drugs. Process Analytical Technology is defined by the FDA as: 'A system for designing, analysing and controlling

Analytical Instrumentation: A Guide to Laboratory, Portable and Miniaturized Instruments G. McMahon

manufacturing through timely measurements (i.e. during processing) of critical quality and performance attributes of raw and in-process materials and processes with the goal of ensuring final product quality'[1]. The PAT initiative is consistent with the current FDA belief that quality cannot be tested into products, but should be built-in or by design. Other regulatory initiatives contributing to the process analytics revolution include the clean fuels initiative, which limits the sulfur levels in gasoline, diesel and other fuels[2, 3]. As a result of this, refineries have started to use more process analysers to allow them to ensure the concentrations of sulfur in their products are kept at low levels.

Process Analytical Technology involves using process analytical chemistry tools that are analytical instruments or devices that can carry out process analysis on-line or in-line to achieve the aim of obtaining real-time results. PAT also uses feedback process control strategies, information management tools and process optimisation strategies. The advantages to the pharmaceutical (and other) industries in embracing PAT include:

- Increasing quality and purity of product.
- Increasing productivity and efficiency of process.
- Increasing the possibility of real-time release of product.
- Reducing reworks, rejected batches and waste materials.
- Meeting regulatory requirements, e.g. good manufacturing practice (GMP) and good laboratory practice (GLP).

Real-time analysis means that generally the process being analysed is better understood, samples can be tracked, events do not go unnoticed, control is maintained at all times and the quality of product is consistently high.

However, this means that the expectations of the analytical instruments needed to achieve this aim are stringent. Aside from actually being able to accurately detect the analyte(s) of interest, the equipment must be robust, reliable, operationally simple, have a foolproof readout, be easy to maintain and be flexible. Reliability is most vital as downtime or operation without control are very costly. This PAT drive, in some cases, has resulted in analytical chemists putting on hard hats and working with engineers to incorporate analytical instruments into the sides of manufacturing vessels, into lines and tanks, into fermentation vats and into carrier pipes[4]. Special enclosures are sometimes needed to house analytical equipment to protect it from the process environment, e.g. the broth or liquor. Cooling systems may even be required for the analytical systems if reaction temperatures are very high. Some instruments can be adapted for the job extremely easily while others are very dependent on the new environment in which they find themselves.

On-line analytical instruments should ideally be:

- Reliable.
- Smaller and more robust than laboratory instruments.
- Capable of fast analysis.
- Self-calibrating or capable of remaining calibrated for long periods.
- Simple in design, easy to install and maintain.
- Able to withstand harsh environments (rugged and robust).
- Able to tolerate different kinds of samples.
- Nondestructive so that the sample can be returned to the process as it was.
- Automated and capable of unattended operation.

On-line analytical instruments may be:

- Less versatile.
- More dedicated (specific).
- Specially housed.

When converting laboratory-based assays into process assays, there are a number of considerations to be made. With laboratory analyses, qualified analytical scientists carry out the testing, extra pretreatment steps to improve sensitivity or selectivity are always possible and the instruments, which are fairly versatile in their use, are housed in an environment where conditions are gentle and controlled. With process testing, the equipment must work with little or no human intervention and must be rugged with few moving parts, only very simple pretreatment steps can normally be performed on-line (such as dilution or filtration) and the instruments, which are usually dedicated to one kind of assay, are often housed in harsh, unstable conditions. Dealing with the large amount of data is also an issue, as is the associated increasing use of chemometrics[4]. Value for money must also be considered, as sometimes the cost of installation, maintenance and operation is just too high to justify the change-over. While some instruments lend themselves well to process analysis and may even work well or better when adapted in this way, others require a paradigm shift in thinking to render them suitable for the job[5].

References

1. www.fda.gov/cder/OPS/PAT.htm
2. www.epa.gov/diesel/
3. www.fe.doe.gov/programs/fuels/index.html
4. Marasco, C.A. (2005) Pharma's process analytical technology. *Chem Eng News*, **83 (8)**, 201–206.
5. Kamal, S.Z. (2005) Process analytical technology – productivity gains and risk minimisation via process intimacy. *Gases and Technology*, July/August, **4 (4)**.

9

Process Analytical Instrumentation in Industry

9.1 In-Process Sampling

Obtaining a representative sample from a process for on-line analysis is a huge challenge for chemists. The interface between the process stream and the analytical instrument is key, and no matter how sophisticated and sensitive the analyser, if the sampling section is not well thought out, the system will not be successful. Because processes vary and can have very extreme temperatures, pressures and other conditions, samples should be taken carefully. The sampling system should take into consideration that:

- The sample must be representative of the process, i.e. homogeneous.
- There must be a sufficient amount of the sample for the analyser.
- The sample must be presented to the instrument in as short a time as possible in a suitable form for analysis. This may mean pretreatment steps, e.g. filtering, cooling, dilution etc.
- The sample must be eliminated from the analyser to waste or returned to the process stream unchanged.

Traditionally, a sample was taken from a tank, vat or pipe and carried to a laboratory remote from the process for analysis on conventional benchtop instruments. This takes time and when results are finally available the process may have been halted unnecessarily or, worse, have continued for a number of hours when results may have indicated that it needed to be stopped and remedial action taken. To avoid such costly delays, there has been a drive to bring the instruments to the process rather than the other way around. There are a number of ways of taking samples from a process for assay (Figure 9.1):

Analytical Instrumentation: A Guide to Laboratory, Portable and Miniaturized Instruments G. McMahon

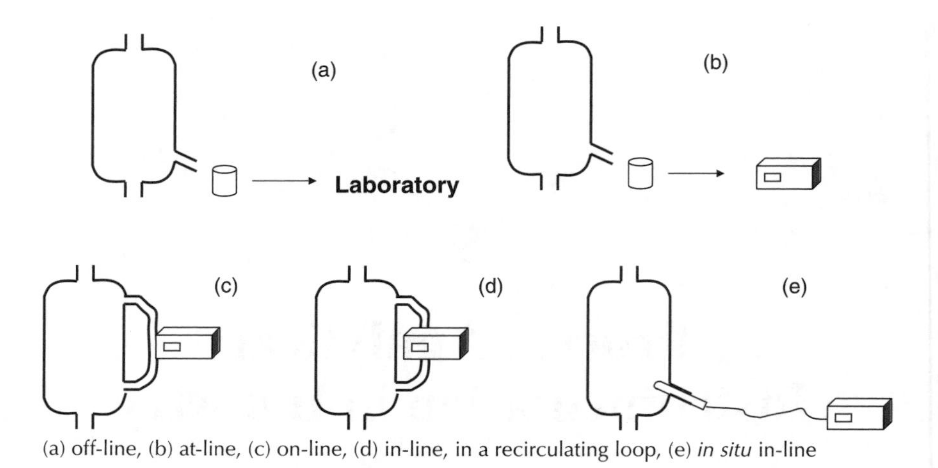

(a) off-line, (b) at-line, (c) on-line, (d) in-line, in a recirculating loop, (e) *in situ* in-line

Figure 9.1 *Schematic representations of sampling.*

- off-line samples
- at-line samples
- on-line samples
- in-line samples.

The most common method would be off-line or 'grab' samples, which are taken intermittently and transported to the laboratory for analysis (Figure 9.1(a)). The procedure can be slow and the process may be finished before the results are available. This is more suited to quality control samples taken at the end of the process. Increasingly, samples can be taken at-line (Figure 9.1(b)). These types of samples are also taken intermittently but analysed in an instrument that is very close to the process, i.e. in the plant itself. The next type of sample (on-line samples) are taken from the process (usually automatically) and transferred directly into the analytical instrument for analysis without human intervention (Figure 9.1(c)). Pretreatment of the sample may also be carried out automatically as part of the assay.

In-line analysis is an extension of on-line analysis where the contents of the process are continually sampled and analysed by instrumentation, e.g. a sensor, *within* the process equipment (Figures 9.1(d) and 9.1(e)). In-line sampling and analysis can be invasive or noninvasive. These sensors or probes are often installed in a bypass test stream as opposed to being inserted into the main process stream (Figure 9.1(d)). This is called extractive sampling. Extractive sampling enables representative sampling of the process while allowing for the sample preparation chamber or flow-through sensor to be serviced, cleaned or calibrated as required without affecting the main process. This technology is usually set up as a 'fast loop' system, so that a sufficient flow of sample past the entrance of the of the analyser is representative of the actual flow in the reaction stream or vessel. The sample is returned to the main process stream after measurement. Alternatively, there is *in situ* sampling, which in its purest form is a probe immersed directly into a batch vessel, line or recirculating loop (Figure 9.1(e))[1]. Miniaturisation is making more in-line analysis possible.

In the past, analytical measurements during a process were always performed off-line or, at best, at-line. In recent years, the trend is towards on-line analysis. This has a significant impact on the way scientists use analytical instruments and how they can adapt existing equipment to work in the often harsh environments of processes in the pharmaceutical, food and other industries.

It is important to consider the sampling environment because chemical process streams may be hot, corrosive, full of salts, particulates etc., so samples may require cooling or filtration on a continuous basis. If the process stream is under pressure or corrosive, a special sampling cell may be needed, e.g. ATEX rated (for explosion control) and/or one made of resistant materials. Another important consideration is where to sample in a process. It may be that multiple sampling points can be covered by one analyser or it may be that each sampling point requires its own analyser. This would impact on the sampling location(s). Also, the best position for a sampling probe must be carefully considered if the most representative sample is to be taken. For example, if a stainless steel tank containing a liquid broth is to be sampled, a sampling probe can be immersed into the top of the vessel, attached to a recirculating feed pipe on the side of the vessel or inserted into a feed or drain pipe at the bottom of the vessel (Figure 9.2). Alternatively, a noncontact approach may be taken where the probe is not in direct contact with the sample at all. Sometimes special bypass pumping devices are needed to keep fresh sample supplied to the analyser.

Naturally, the sampling of solids, liquids and gases requires different approaches. Often, noncontact spectroscopic probes are used for powders, tablets and thick slurries. Gases can often be sampled in the same way as liquids but may require some preconcentration prior to analysis, especially if levels of the analyte being examined are low. An example of a commercially available system for sampling gases is available from ABB. It is called the DRS2170 Dynamic Reflux Sampler and is used to couple with GC instruments, photometers and other process analysers. It conditions samples *in situ*. The condensable components are removed and are used to support particulate removal. The unit

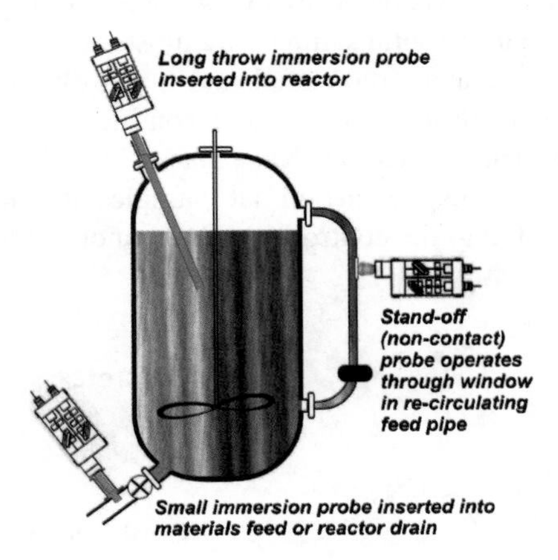

Figure 9.2 *Some of the possible sampling points for a liquid in a reaction vessel (Reproduced by permission of HORIBA Jobin Yvon).*

is a self-contained sampling device suitable for mounting at or near the process pipeline tap. It allows for high sample throughput and is self-cleaning.

9.2 In-Process Analysis

Once the sampling protocol has been decided, the next step is analysis of the sample(s). Of course, sampling and analysis may also be integrated. In the past 50 years, the type of assays that were done in-process exploited the nonselective properties of the process stream, such as density, viscosity and conductivity. This monitoring has been achieved by both automatic and automated instruments. Selective properties of the process stream, such as chemical composition, were usually measured by taking grab samples and examining them in a laboratory by off-line techniques as they are more difficult to adapt for process stream analysis.

Automatic and automated instruments can be differentiated as follows: automatic instruments tend to perform specific operations at given points in a process or analysis to save time or effort, e.g. robotics, while automated instruments tend to control some part of a process without human intervention and do this by means of a feedback mechanism from sensors. For example, an automatic conductivity detector might continuously monitor the conductivity of a process stream, generating some alarm if the conductivity goes outside a preset limit. An automated detection system could transmit the measured conductivity values to a control unit that, by utilising a feedback mechanism, adjusts relevant process parameters, e.g. temperature or cycle time, to maintain the conductivity of the stream within the preset limits.

True process analytics are based on automated systems. Automated instruments must be smaller, more rapid and robust than laboratory instruments and designed for unattended operation. Those most commonly used are based on spectroscopic, separation and electrochemical analytical techniques. Many of these are incorporated into or combined with flow injection analysis (FIA) systems in order to work well. Not all analytical methods lend themselves to automation. Analyses involving gases and liquids are most successfully automated while those using solid samples are most difficult to automate. And there will always be certain assays that are too complex or too costly to automate.

There are two types of analyser: discrete and continuous (Figure 9.3). In discrete analysers, samples are taken at selected intervals, placed in individual containers and any required sample preparation steps are carried out. Samples are then analysed in batches, with the information being fed to the controller and operator in the usual way. Obviously

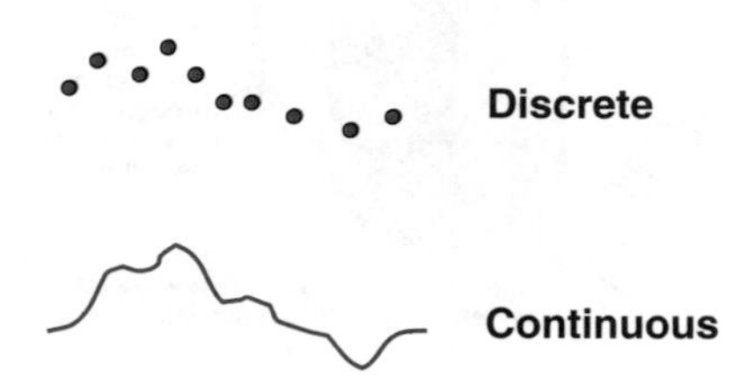

Figure 9.3 *The data points from discrete and continuous analysis.*

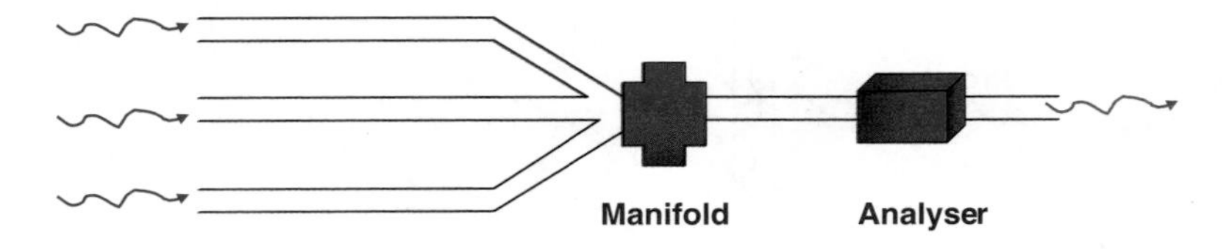

Figure 9.4 Directing a number of streams with similar composition into a common manifold which leads to the analyser.

there will be periods where results are unknown and, therefore, transient errors may be missed and not corrected for. Instruments that work in this way include HPLC as it sometimes needs individual (discrete) samples. Discrete analysers are most often used when *selectivity* is key to the process.

Continuous analysers make continuous measurements directly in the flowing process stream, in a bypass test stream or inside a reactor vessel. This works best when no pretreatment of the sample is required although simple manipulations, such as reagent addition, sample filtration or dilution, can be carried out prior to the analyser by a sample preparation chamber. Actual determinations are carried out in a flow-through sample cell.

Often, direct sensing instruments are used, e.g. probes or electrodes. If there are a number of streams with similar composition, they can all be directed into a common manifold that leads to the flow-through analyser (Figure 9.4). Each stream can be sampled in sequence. This provides rapid analyses for several streams. Continuous measurement means that the instrument constantly measures some physical or chemical property of the sample and yields dynamic information that is a continuous function of time. Continuous analysers are used when *speed and sensitivity to changes* are key to the process.

Continuous analysers are automated and operate on-line or in-line. They are usually incorporated into a control loop which operates by means of a feedback mechanism (Figure 9.5). The system is composed of three main parts:

- The analyser that measures/monitors the variable being controlled.
- The controller that compares the measured variable against a set point (reference value) and feeds the information to an operator.
- The operator that activates some device such as a valve to bring the variable back to the set point.

A continuous analyser is attached to a sampling line and thereafter continuously and automatically obtains a signal proportional to the instantaneous concentration of a selected component in the flowing stream. The information acquired is automatically used to set the process environment controllers and to take any corrective action needed to control the process. These actions might be to close a valve, cool the stream, allow more diluent to be added, speed up mixing etc. Thus continuous analysers carry out the function of the control laboratory but in real-time and more efficiently. Continuous analysers are employed in many situations such as routine analysis, monitoring and on-line process control.

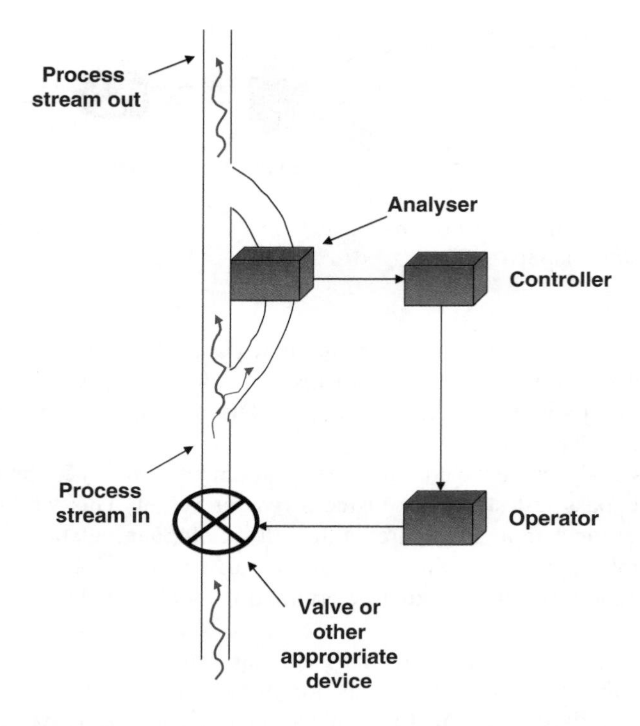

Figure 9.5 *Schematic diagram of a continuous analyser connected to a control loop that operates by means of a feedback mechanism.*

In recent years, it has become possible to measure more selective properties on-line by incorporating instruments such as potentiometry (Figure 9.6), IR and GC into the process stream. These provide a *dynamic* rather than historic measure of what is going on with the process. The data are fed back to controllers; this permits better regulation of the process

Figure 9.6 *On-line low-level measurements of sodium using the Orion ionplus 2111LL on-line ion-selective sodium electrode (Reproduced by kind permission of Thermo Electron Corp).*

stream composition than using non-selective methods such as pH, conductivity, turbidity, viscosity etc.

Developments in the portability and miniaturisation of laboratory instruments along with progress in the electronics industry and computer technology have all contributed to the development of this field. Hence, automated continuous on-line process control has become a major factor in the efficient operation of many large-scale chemical processes. Having these systems in place reduces the margin of operational errors due to the feedback nature of the assay and increases safety when personnel do not have to come in contact with potentially harmful substances. Preset control specifications require that the systems are highly flexible, yet easy to operate and maintain. Some process analysers can even check the performance of the analysers themselves and carry out calibration checks. However, there are also hurdles to this technology, such as the expense of replacing existing equipment and apprehension about reliability. But the advantages outweigh the disadvantages and so, where possible, there is a concerted effort to convert laboratory-based assays to on-line assays with automated continuous analysers.

When converting laboratory-based assays to on-line assays, it is important to use the simplest technology that fulfils the requirements, to use analysers that can automatically perform calibration and routine checks where possible and to use those that may even be able to monitor other analyser operations. The number of sampling points in the process to be measured must be decided, as must the exact monitoring locations. It is important that only relevant information should ever get to the operator and, as far as possible, changes made to the process based on analytical results should be taken care of in the background.

What this drive towards increased use of process analytics has meant for the pharmaceutical business is a complete rethink of the way assays are done. In many cases the analytical systems need to be added on to existing plant and equipment, and this retrofitting can be technically difficult and expensive to install. Following development of a process assay, there is a changeover phase, during which implementation and optimisation of the assay takes place. However, laboratory analysis will still be done until the new assay has been fully validated. When the assay is fully converted to a process one, it may then be handed over from the analytical group to engineering, or even to maintenance when it becomes fully routine.

In a pharmaceutical plant that has embraced PAT, the testing protocol for the process looks very different to that of a plant that still carries out off-line assays (Figure 9.7). Spectroscopic analysis is the most popular method especially for dry products such as raw materials and drug powders[2]. Near infrared (NIR) and Raman, due to the noncontact nature of these techniques, are prevalent. All of the various types of process analysers are discussed in the following sections.

9.2.1 Flow Injection Analysis

Principle

Flow injection analysis (FIA) is based on the injection of a defined volume of liquid sample into a moving stream of a suitable liquid. It was developed to overcome the disadvantages of batch assay and continuous-flow analysis forms of chemical analysis and as a means of automating wet chemical reactions. In batch assays, the sample is mixed

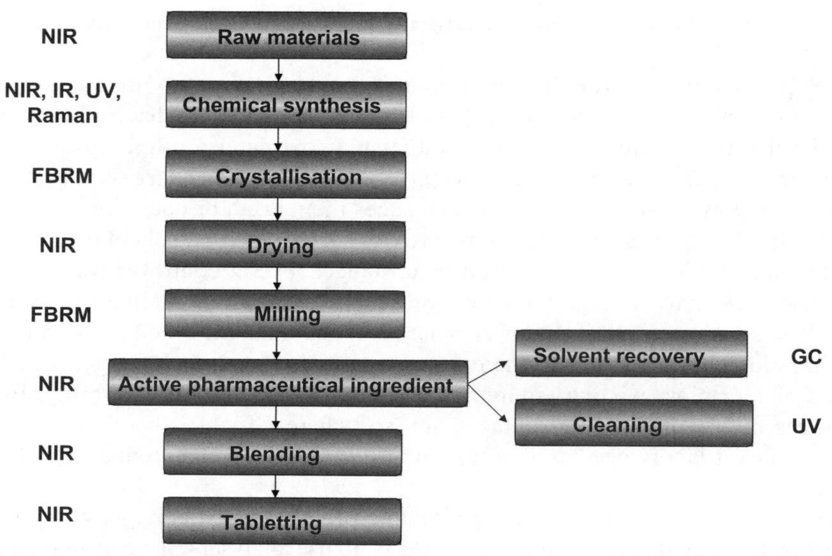

NIR = near infrared spectroscopy, IR = infrared spectroscopy, UV = ultraviolet spectroscopy, Raman = Raman spectroscopy, FBRM = focused beam reflectance measurement, GC = gas chromatography.

Figure 9.7 *Typical steps involved in a chemical synthesis and possible associated process analytical testing that can accompany each step.*

with appropriate reagents in a container, equilibrium is reached and a measurement is then taken. In continuous-flow analysis, the system is stationary while the sample flows through it, mixing with reagents along the way. Samples are separated from each other using air bubbles. In both cases, steady-state conditions (physical and chemical mixing) must be reached before measurement.

In FIA, the sample is mixed with reagent by conduction and diffusion in the flowing carrier stream and a chemical reaction occurs. The detector continuously monitors the flow and records a transient signal (peak) as the sample flows through the flow cell. Physical and chemical equilibrium are often not reached by the time the analyte passes through the detector cell but as long as the measurement is reproducible, the signal is reliable. As steady-state conditions are not obtained, it is timing of all events along the flow path that is most important in this technique. To this end, small-bore tubing should be used and mixing volume (dispersion) and flow rates must be rigidly controlled.

There are many advantages to using FIA in on-line and in-line analysis. It has a very high sample throughput (hundreds of samples per hour) and results are obtained as quickly as seconds after injection. Starting up and shutting down times are fast. The technique is useful in automated systems as the instrumentation is simple, cheap and robust. It enables many reagent (wet chemistry) based methods as well as sample processing steps, such as dissolution/digestion, dilution, phase extraction, reagent addition and on-line reaction, to be integrated and automated into process systems. FIA is often the means by which electrochemical, spectroscopic and separation techniques are accommodated in process

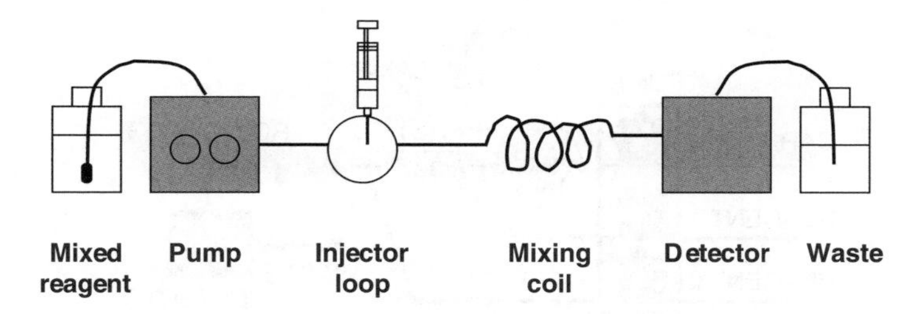

Figure 9.8 *Schematic diagram of the instrumentation for flow injection analysis.*

analysis. There is minimal contact between the operator and reagents, which are contained in a semi-enclosed system, thus there is less of an issue with contamination and exposure to toxic reagents. There is also the ability to combine FIA with sensitive detectors to increase the sensitivity of the process analysis. Particularly when FIA is used in on-line monitoring, the baseline can be monitored by the system as an indication of how the assay itself is working and on-line calibration can be carried out by simple injection of an occasional standard. And, finally, FIA enables chemists to make some analytical measurements in flowing systems that might not be done in any other way.

Instrumentation

FIA today usually consists of a high quality pump to move the liquids around the system, an injection valve, an autosampler, a reaction coil and a flow-through detector (Figure 9.8). The peristaltic pump allows for pulse-free flow and multi-line operation. Flowrates tend to be from 0.5–4.0 mL/min. For injecting the sample, loop injector valves, similar to those employed in HPLC, are common in FIA but dedicated FIA-injector valves can also be employed. Autosamplers are often used. As small sample volumes are normally injected (1–500 μL), only small volumes of reagent are required, which in turn means that only small amounts of waste are generated. The reaction coil may be heated. Filters should be used to keep particulates out of the system. Tubing is usually plastic with internal diameters less than 1 mm and the tubing lengths are kept short to reduce unwanted dispersion. However, in the case of wet chemical reactions, a balance must be achieved between the need to allow sufficient dispersion to enable thorough mixing and the need to prevent too much dispersion, as unnecessary dilution of the sample adversely affects the detection limit. The tubing must be well maintained and cleaning of sample lines is important.

A schematic diagram of a working FIA system is shown in Figure 9.9. FIA can be used with many types of detector including electrochemical, e.g. pH probes, ISEs and conductivity, and spectroscopic, e.g. UV–Vis, infrared and fluorescence. Diode array detectors allow the monitoring of many components simultaneously, and coupled with chemometrics can be a very fast and information-rich technique. The detector itself should have a small cell volume to avoid undue dispersion.

Information Obtained

The information obtained depends on the information required by the analyst and the analyte under investigation. It can be qualitative or quantitative. The reagents employed

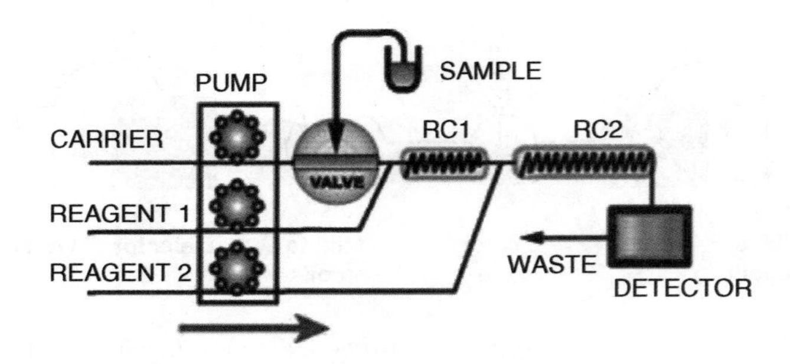

Figure 9.9 *Schematic diagram of an FIA system with an injection valve, peristaltic pump, two reaction coils (RC1 and RC2) and a UV–Vis detector; there are two reagents (reagent 1 and reagent 2) and a carrier buffer (Image provided by kind permission of FIAlab Instruments).*

can be specific for one analyte or applicable to a number of compounds depending on the reaction. The versatility of FIA allows the analyst to choose the important parameters for the assay, e.g. is sensitivity more important than selectivity, is speed of result the most important criterion?

One common and simple spectrophotometric example is the determination of chloride. Samples containing chloride (Cl^-) are injected via a valve into the carrier solution containing a single mixed reagent (mercury (II) thiocyanate and iron (III) ions ($Hg(SCN)_2$ and Fe^{3+})) that is being pumped through the system. The chloride reacts with the mercury (II) thiocyanate liberating thiocyanate ions (SCN^-) which then react with the iron (III) ions in a single mixing coil to form the red iron thiocyanate complex ($Fe(SCN)^{2+}$). The absorbance of the carrier stream is continuously monitored at 480 nm in a micro flow cell at the detector and the peak height is proportional to the concentration of the chloride. However, as can be seen from Figure 9.9, FIA can easily run an experiment with two reagents and two mixing coils and, in fact, many other more complicated variations on the same theme.

Developments/Specialist Techniques

Kinetic discrimination and kinetic enhancement are two procedures that can be used in conjunction with wet chemical FIA. Kinetic discrimination works well when it is known that there are differences in the rate of reaction between the reagent and the analyte of interest and the rates of reaction between the reagent and the interferents in the sample. If, for example, the analyte reacts more quickly with the reagent than the interferents, then the timing for measuring of the signal by the detector can be set so as to exploit this. Kinetic enhancement is when the direction of the reaction is manipulated to favour measurement of the analyte of interest with improved signal-to-noise.

Until recently, samples for FIA were already extracted, filtered, centrifuged or pretreated in some way prior to assay. However, some sample preparation and preconcentration steps can now be accommodated in FIA. Some examples are on-line liquid–liquid extraction, solid phase extraction and ion-exchange procedures. In this way, FIA is managing to convert some traditionally labour-intensive steps into automated operations that have higher precision and faster throughput. FIA can also tolerate other sample types, such as fermentation broth samples and even gases through the use of silicon membrane separators and gas diffusion systems, respectively.

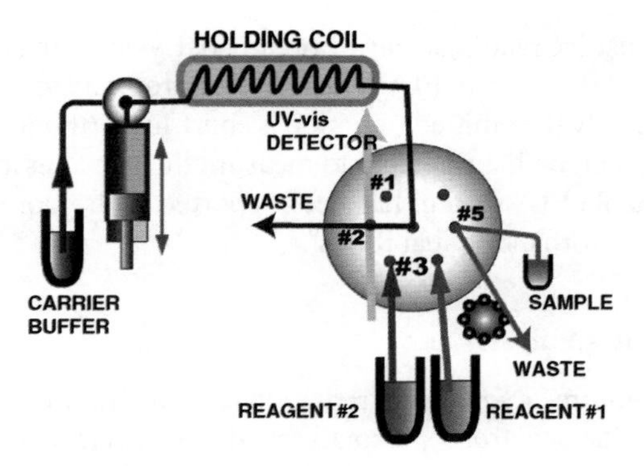

Figure 9.10 *Schematic diagram of an SIA system with a 6-port valve, syringe pump, holding coil and a UV–Vis detector; there are two reagents (reagent 1 and reagent 2) and a carrier buffer (Image provided by kind permission of FIAlab Instruments).*

Sequential injection analysis (SIA) is version of FIA based around a multiport valve, which facilitates the moving around of liquids through different channels at different times and in different directions, and a syringe pump, which permits reversal of flow. It differs in operational aspects from FIA (Figure 9.10). Coordinated by a computer, the ports are programmed to deliver sample and reagent volumes into a holding coil where they are stacked as required. Next, the ports switch to send the volume plugs in the reverse direction for mixing and detection. One great advantage of SIA is that very small volumes can be precisely controlled using the computer. There is less reagent required and less waste generated.

Another technique that can be used in FIA is the stopped-flow method, where the sample is stopped at a certain point in the flowing system and held for a period. The reason for this can be to allow extra reaction time without excessive dilution or to allow real-time monitoring of the reaction product formation in the detector flow cell.

Lab-on-valve is another development of FIA and SIA; this is discussed further in the section on miniaturisation of analytical instruments.

Applications

There are many applications of the use of FIA in industry as it is a technique that is very amenable to flowing systems and the detectors used can be very robust, e.g. ISEs. FIA is very important in the food and beverage industry, for example in the analysis of ethanol using photo-oxidation and spectrophotometric determination[3] and in the analysis of peracetic acid and hydrogen peroxide during the cleaning process[4]. FIA also has many uses in agricultural[5], environmental[6] and biotechnology processes[7]. FIA is often used with electrochemical detectors, e.g. ISEs and amperometric sensors, and in fact is particularly suited to this type of detector as their response is diffusion-controlled and, often, steady-state conditions take a while to be reached. An example of the use of flow injection amperometry is the determination of paracetamol in pharmaceutical drugs using gold electrodes[8]. The detection limit was 0.2 μg/mL and 180 samples could be analysed per hour.

By delivering the sample using FIA and therefore by manipulating the sample exposure time, more reproducible and often more selective (due to discriminating against the interferences) and sensitive readings may be obtained. By changing the flow rate, the linear range can be expanded if required. FIA is also very appropriate to the use of luminescence detection, since generally the emitted radiation is short-lived (though very intense). Again, with FIA, the timing can be manipulated to measure the signal at maximum emission for each sample. A portable FIA system has been reported with fluorescence, luminescence and long pathlength absorbance capabilities[9].

9.2.2 Spectroscopic Analysis

All spectrometers require a source of radiation, a wavelength selector and a detector. For most current on-line spectroscopic analyses, the instruments tend to be derived from laboratory-based ones. They can be specially enclosed for protection in the harsh plant environment and use fibre optic cables to transport radiation to and from the sample, which is usually quite far away from the spectrometer (Figure 9.11). Alternatively, they can be mounted directly onto the process point with fibres linking the sampling probe and the specially rated spectrometer housing. It is estimated that the cost of installing a spectroscopic analyser is three times the cost of the instrument itself. Hence, there has been a move to redesign more rugged analysers from scratch instead of adapting existing instrumentation for the job[10]. Many process analytics are custom-built for a particular application.

Microspectrometers which focus on the bottom-up approach rather than the top-down approach are being developed and hence are being made small and rugged enough for the harsh environments of process analytics. The most commonly employed spectroscopic process analysers, in order of popularity, are NIR, mid IR and Raman. Each has its advantages and disadvantages for various applications. There are a number of designs for the optical probes used with process spectrometers depending on whether the measurement involves collecting transmitted (absorbed) light, reflected light or scattered light. The transmission probe can be configured such that the optical fibres are facing each other across the sample or are side by side in the same housing. One waveguide transports

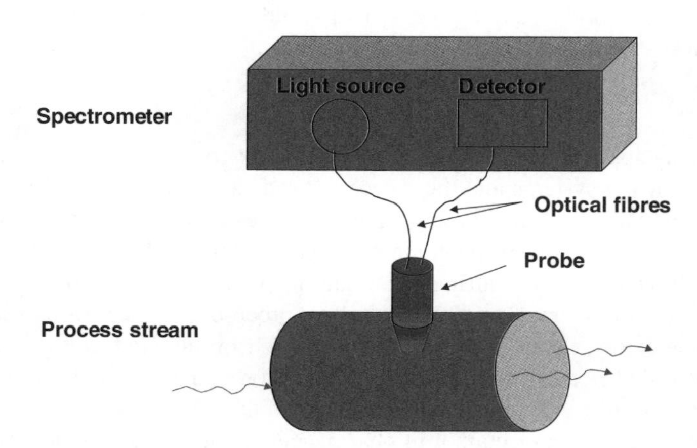

Figure 9.11 *Schematic diagram of a spectrometer coupled via a fibre optic probe to a process stream for on-line analysis.*

the radiation to the sample and the second returns the sample-modified radiation to the detector. The attenuated total reflectance (ATR) probe allows incident radiation to be reflected by the sample back into the probe and to be picked up by the returning optical fibre. The scattering probe is composed of a fibre bundle where the central optical fibre delivers the light to the sample and the surrounding fibres pick up the scattered light and carry it back to the detector. These designs are illustrated in Figure 9.12.

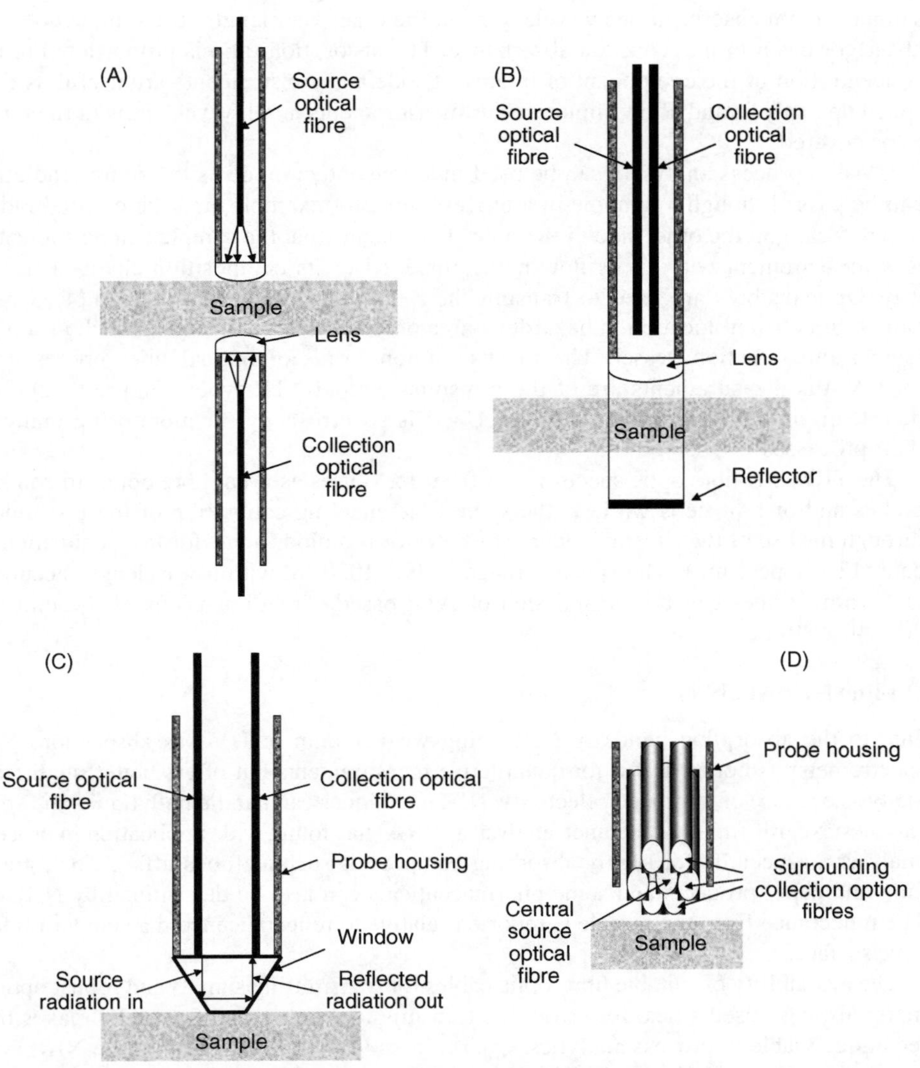

(A) Transmission probe with facing configuration, (B) Transmission probe with side-by-side configuration, (C) Attenuated total reflectance (ATR) probe, (D) Scattering probe.

Figure 9.12 *Examples of fibre optic-based probes for on-line spectroscopic analysis.*

On-line UV–Vis Analyser

On-line UV–Vis analysers can be used to monitor liquid or gas streams. They are reliable, rugged and support continuous flow. They are sometimes called photometric analysers when a simple filter-based spectrometer is used instead of a monochromator. This type of analyser measures the UV or Vis radiation transmitted through the sample at two different but fixed wavelengths to provide a continuous measurement of the absorption ratio. The two wavelengths are chosen so that the component of interest in the process sample stream absorbs at one wavelength. At the other wavelength, the sample does not absorb or has minimal constant absorbance. The absorption ratio is proportional to the concentration of the component of interest. Diode array instruments are useful as they speed up analysis and allow simultaneous measurements at all wavelengths in the range to be covered.

UV–Vis process analysers can be used in a noncontact mode as the source radiation can be passed though a window in a process pipe, for example, and the emitted radiation detected on the other side of the pipe. This means that the sample can be subjected to a measurement without its flow being impeded or its composition changed in any way. Optical fibres are used to transmit the radiation to and from the sample, which can be in a distant location, a hazardous area or in a harsh environment such as a high temperature reaction vessel. The most common forms of optical fibre probes used for UV–Vis measurements are of the transmission and ATR types (Figure 9.12). The petroleum and chemical industries use UV–Vis spectroscopy for monitoring many of their processes[11].

The TIDAS II fibre optic spectrometer from J&M uses external fibre optics to connect probes and/or flow cells while at the same time enabling calibration of the instrument through the use of the cuvette holder. The detector is a diode array for fast acquisition of data (12 ms/spectrum). The spectral range is 190–1020 nm with a wavelength accuracy of 0.3 nm. Probes can be transmission or ATR-based. The dimensions of the unit are $20 \times 48 \times 49$ cm.

On-line NIR Analyser

Due to the absorption bands in NIR being weaker than in UV–Vis absorption, NIR spectrometry is not as useful for quantitative measurements but offers better qualitative analysis because of improved selectivity. NIR techniques can handle both liquid and solid samples. Near infrared reflectance analysis (NIRA) has found wide application in process analysis[12], especially for highly absorbing compounds such as foodstuffs[13]. Coal, grain, pulp and paper products and some pharmaceuticals can also be determined by NIRA[14]. The reflectance from the sample is reported relative to reflectance from a standard reference surface.

The availability of suitable fibre-optic cables for transmitting signals and robust optical materials, e.g. fused silica, for use in the transmission probes and flow cells makes this technique viable in process analytics. Quartz loses very little radiation in the NIR, even over long distances. Using optical fibres in an ATR probe configuration, the sample can be remote from the analyser. The absorption of NIR can be related to concentration also and so transmission probes can be used. Most in-process NIR instruments employ a tungsten–halogen source, a scanning grating and a single detector with a resolution varying from

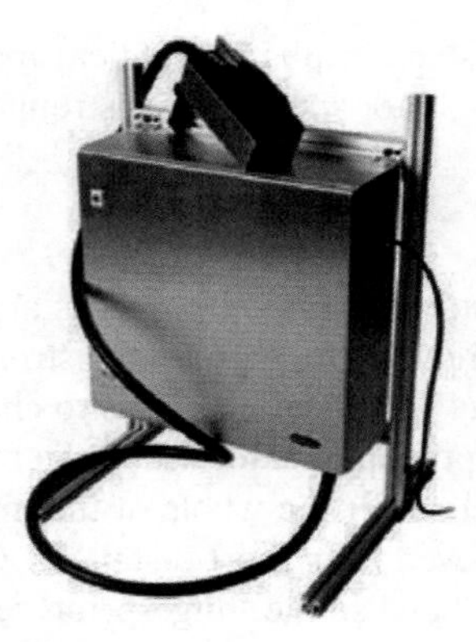

Figure 9.13 *The QualitySpec® iP NIR spectrometer with handheld probe (Photo courtesy of Analytical Spectral Devices, Inc., Boulder, Colorado, USA. Copyright© Analytical Spectral Devices, Inc. All rights reserved).*

2 nm to 10 nm. Some are based on filters instead of monochromators or interferometers because of their inherent simplicity. Detector arrays, e.g. InGaAs, can be used but are more expensive. Chemometrics, multivariate analysis and complex algorithms are often used to aid quantitative measurement[15].

The QualitySpec® iP from Analytical Spectral Devices Inc. is an example of a commercially available NIR for in-process analysis (Figure 9.13). It has a spectral range of 350–2500 nm (across six modular configurations). It was developed for on-line process and quality control and can be integrated into existing plant IT systems. It is housed in an industrially-rated box to withstand harsh environments. It can come with a heater or cooler for temperature control and has a variety of monitoring head/probe options. The instrument is capable of simultaneous multi-component analysis and automatic calibration. Data collection is fast with 10 spectra per second over the full spectral range. The instrument weighs 18.1 kg and the probe heads vary, e.g. the handheld probe weighs 4.5 kg.

Another example of a NIRA instrument is the the IntegraSpec from Axsun Technologies, a pre-dispersive NIR spectrometer platform which is small and easy to deploy in process streams. The Axsun battery-operated spectrometer can be used for NIRA in reflectance and transmission modes for an array of process applications. The small optical bench is only 1.4 cm long and is created using miniaturised technology and lithography techniques[16]. It houses a source, which is a superluminescent light emitting diode (SLED), a tuneable Fabry–Perot filter, which can be used before or after the sample (pre- or post-dispersive) and a gallium–arsenic (GaAs) detector. Spectral sensitivity is available (through the range of models) up to 1970 nm, the wavelength accuracy is ±0.025 nm and the spectral resolution is 3 cm^{-1}. The dimensions of the fully-packaged IntegraSpec platform vary depending on application but are of the order of 17.8 × 11.4 × 6.2 cm.

There are many published applications of the use of NIRA in processes such as qualitative and quantitative analysis of liquid phase systems. Organic solvents have very specific overtone bands in the first carbon–hydrogen stretching region from 1600–1800 nm which can be exploited for their assay[17]. There are also many applications for the monitoring of blending, drying and other steps in pharmaceutical manufacture[18]. In these types of applications, temperature and agitator speed have a strong effect on the NIR data.

On-line IR Analyser

Infrared spectrometry is currently exploited in process analysis but less so than near IR and Raman spectrometry. The reasons for this are the strong absorbances of most mid IR bands and the sensitivity of mid IR optical materials to chemical erosion. There is also a relative lack of practical fibre optic options for use in the mid IR range since silver halide and chalcogenide glasses, which cover the whole of the mid IR region, can attenuate the radiation by as much as 95 %, even over short distances. Other fibres such as zirconium fluoride cut off below $2500\,cm^{-1}$ and so the 'fingerprint' region information is lost.

Rugged hardware and reproducible results are very important criteria for in-process analytics so another issue with FTIR can be the interferometer design. The original Michelson interferometer can be subject to misalignment. Some designs have flat mirrors, dynamic alignment, rotary scanning, refractive scanning, birefringent scanning and other features in order to overcome alignment issues. These many types of interferometer each have their own advantages and disadvantages. However, if these hurdles can be overcome, when IR works for a particular process, it is extremely useful. Compared to UV–Vis measurements, infrared is more selective but less sensitive if used quantitatively. As a technique, it is most useful for samples in the gas phase. On-line and in-line FTIR is very popular in the oil refining, food[19], environmental[20] and chemical industries[21, 22, 23].

Infrared analysers fall into two categories: dispersive and non-dispersive. The dispersive IR analyser works by using two specific wavelengths which pass through the flow cell. One of the wavelengths is specific for the component of interest and at the other the sample does not absorb. The ratio of these readings can be related to the concentration of the component being measured. This type of instrument handles liquids as well as gases. Tunable laser diodes provide the monochromatic IR sources and a small thermocouple may be the detector. The non-dispersive IR analyser is used for the selective analysis of gas process streams. The sample molecules attenuate the wavelengths absorbed by the detector and the difference in power between the sample and reference beams is related to concentration. It is rugged and reliable and does not require monochromatic radiation. It has better sensitivity than the dispersive IR analyses since all the radiation from the IR source passes through the sample. Filter photometers are technologically very simple. A filter is usually used for each compound of interest and they can be on a filter wheel or equivalent.

The PCM (process control monitor) 5000 is an FTIR instrument from Analect/Applied Instrument Technologies that measures the physical and chemical properties of liquid, solid and gaseous in-process samples. The heart of the system is a rugged optical head coupled with a patented fibre optic system. The system is configured for continuous, unattended operation in harsh and hazardous environments. The PCM 5000 FTIR uses close-coupled sample cells ($7000–450\,cm^{-1}$). The instrument is configured with a self-diagnostic module to monitor the critical parameters of its own operation.

The MonARC FTIR system from Mettler Toledo identifies and quantifies chemical species based on changes in the fingerprint region of their IR spectrum. It also has a built-in ultrasonic cleaner for the diamond sensor which increases reproducibility and reduces downtime. Various types of fibre optic cable and probe heads are available depending on the IR range required. The instrument weighs 45 kg and can be transported on wheels. It can tolerate extreme temperatures (up to 200 °C) and pressures. ReactIR is another product from the same manufacturer for smaller scale processes.

ABB produces a number of process analytical technologies. In the IR range it has a multiwave photometer and also the FTPA2000–300 industrial FTIR. The ABB Process Multiwave IR Photometer provides continuous on-line measurement of vapour or liquid components in simple or complex process streams. The filter wheel of the photometer accepts up to eight wavelengths; hence the photometer can compensate for many types of interference and handle multiple component applications. The unit has a solid state detector, a spectral range of 2.5–14.5 μm and weighs 36.3 kg. The FTPA2000-300 is designed for use with remote, real-time monitoring of continuous and batch processes, extractive sampling systems and is a multi-point, multi-stream FTIR analyser.

On-line Raman Analyser

The Raman technique has been readily adapted for on-line process analysis, especially in the pharmaceutical industry[24]. It has the benefits of mid IR, e.g. the ability to identify compounds from the vibrational fundamentals, without the constraints of mid IR, e.g. the limitations of the optical materials that can be used. Its popularity is also due in part to the excellent throughput of optical fibres for the radiation required for Raman, i.e. in the Vis and NIR regions. This use of optical fibre probes (Figure 9.14) facilitates easy in-line analysis because the sample can be remote from the instrumentation, even to hundreds of metres in distance. Fibre optic multiplexers are also available, allowing many samples to be analysed sequentially. Small laser diode sources and CCD detectors can be attached to the optical fibres and changed as required, rendering the overall device small and flexible. Radiation from the laser diode light source is transmitted to the sample by optical fibre

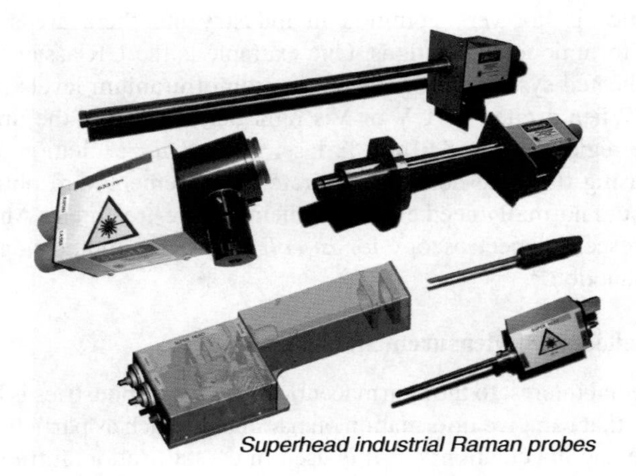

Figure 9.14 *Selection of Superhead industrial Raman probes from HORIBA Jobin Yvon; all use a single fibre to deliver the excitation radiation and a second fibre for transporting the signal back to the detector (Reproduced by permission of HORIBA Jobin Yvon).*

and focused at the radiation/sample interface. Scattering probes based on a fibre bundle can be used for the Raman measurements. After interaction with the sample, scattered radiation is carried back to the detector by separate optical fibres. The technique can be noncontact and is nondestructive. A notch or edge filter is incorporated to remove the source laser line.

HORIBA Jobin Yvon produces two Raman process analysers, the RPA-HE and the RPA-AX. They provide coverage from 150–3200 cm^{-1}. They both have no moving parts and use an air-cooled CCD detector. They can be configured with lasers at 532, 633, 785 and 830 nm (only the RPA-AX). The main difference between the two systems is that the RPA-HE system can obtain multiplexed information from up to four fibre probes simultaneously while RPA-AX can be multiplexed to up to 64 separate channels of information. They are used in the semiconductor industry and for polymer characterisation and pharmaceutical manufacturing.

Two other commercially available in-process Raman systems have been developed by Kaiser for on-line or at-line analysis of powders, slurries and other types of process samples. The first is the RamanRxn3 PAT Analyser, which can be wall-mounted or put on wheels for use in a laboratory or plant situation. It can be coupled with multiple probes, which themselves can be a noncontact or immersion type. It is self-calibrating. The second system is the PhAT System Raman Analyser, which is smaller in size, much more portable and developed mainly for analysing solids in pilot scale manufacturing.

Raman is very useful for the discrimination of polymorphs[25] and is also suitable for monitoring many steps in chemical processes[26]. In terms of sampling, Raman can be carried out using an immersion probe or by non-contact techniques. These have been compared and contrasted for blending and wet granulation steps in a pharmaceutical application, with the large spot noncontact device yielding better results due to its greater sample volume and increased robustness[27].

On-line Fluorescence Analyser

On-line fluorescence is not very common in industry but there are some specially-designed systems for unique applications. One example is the UR assay from HORIBA Jobin Yvon, a dedicated system for the determination of uranium levels (0.1–40 ppb) in process samples. When excited by UV or Vis light below 450 nm, the uranyl ions emit green light in the region 490 to 540 nm. A laser is used for excitation, a PMT as the detector and by using the time-domain, accurate measurements are obtained even on samples which would normally need extensive chemical pre-treatment. Another example is the use of fluorescence spectroscopy for *E. coli* bioprocess monitoring[28] and recombinant protein production[29].

Focused Beam Reflectance Measurement (FBRM)

A technique of special interest to the pharmaceutical and other industries is FBRM. It is an on-line instrument that can give information on parameters such as particle size[30], surface area, filtration rate and flocculation[31, 32]. It is used in crystallisation, filtration and drying applications. It works by scanning a laser beam in a circular motion over a surface at high speed. Light is back scattered by the particles and picked up by a detector. The duration of time that it takes for the backscattered light to return to the detector is called chord length and this can be related to characteristics of the particles being studied. Crystal diameter and other data can be obtained. The Lasentec FBRM from Mettler Toledo can be combined with particle video microscopy (PVM) on the same probe for insertion into a vessel or reactor. The image resolution from the PVM can confirm dimension and shape for particles, cells, droplets, and bubbles ranging from 2 μm to 1000 μm, thus adding to the characterisation of the particles being studied[33].

On-line MS Analyser

There are a few challenges when trying to set up on-line MS. The pressure in a mass spectrometer is much less than atmospheric pressure so flow restrictors must be used when introducing a sample. Typically, samples are introduced into a mass spectrometer in the gas phase, though it is possible to introduce a liquid sample and convert it to the gas phase (see section on MS). The process stream can be introduced into the MS in a number of ways:

- capillary inlet
- molecular leak
- porous frit
- membrane interface.

The most common ionisation chamber is the electron impact (EI). The most common mass analyser employed is a quadrupole or magnetic sector. In terms of detectors used, they are based on the Faraday cup or electron multipliers. Process analyser mass spectrometers are mostly used for light gases such as hydrogen, oxygen, nitrogen, carbon dioxide and water; this renders them very suitable for monitoring fermentation processes but they also have useful applications in the petroleum, pharmaceutical, environmental and heavy manufacturing industries.

For analysis of low molecular weight gases, the MGA (Multiple Gas Analyser) iSCAN from Applied Instrument Technologies provides real-time compositional analysis of process gas streams. The MGA iSCAN is a magnetically scanned double focusing magnetic sector mass spectrometer. The sample gas is subjected to EI ionisation. Then the positively charged ions are accelerated out of the ion source and into the analyser. The first part of the analyser, the electric sector, sorts the ions by their energy. The second part, the scanning magnetic sector, sorts the ions by their momentum. There are two detectors, the choice of which is based on the concentration of the gas being detected.

Also designed for the analysis of low molecular weight gases, the VG Prima δB process MS from Thermo Electron Corp. is based on scanning magnetic sector technology and has a mass range of 1–200 amu (Figure 9.15). It is $1.5 \times 0.7 \times 0.65$ m in size and weighs about 300 kg so it is large but when set up takes less than one second to measure each gas component and it can handle up to 250 sample streams.

AMETEK's ProMaxion process mass spectrometer can monitor up to 32 components. Automated sample switching at the multiport allows unattended analysis of process and calibration gases. Different calibration and analysis methods can be assigned to each sample port. A membrane inlet system is incorporated for ambient gas sampling. The available mass range is 1–200 amu. The detection range is 10 ppm to 100 % with a Faraday cup detector; lower LODs are possible with an electron multiplier.

Figure 9.15 The VG Prima δB process mass spectrometer (Reproduced by Kind permission of Thermo Electron Corp.).

On-line mass spectrometry has been implemented in pharmaceutical processes for monitoring raw materials and products[34]. In this particular application, dilution of the samples is carried out by a membrane interface coupled directly to the atmospheric pressure chemical ionisation source of a quadrupole mass spectrometer for real-time analysis. Continuous on-line MS has also been used for monitoring fermentation processes in the brewing industry[35].

On-line NMR

Nuclear magnetic resonance has historically not been associated with process analytics due to its size, expense and the expertise required to both operate the instrument and interpret the data. However, NMR has a number of advantages that make it worth considering for certain applications. For example, NMR is nondestructive and fairly independent of the sample's physical state or environmental condition. NMR does not require the measurement probe to be inserted into the process liquors, which avoids fouling. And recently, small, dedicated low-field (<60 MHz) NMR systems based on permanent magnet technology, which are easier to operate and maintain than the high-field laboratory-based systems, have been developed. With design improvements in the magnet, electronic and data processing components of an NMR spectrometer, smaller, more portable NMRs are now being exploited in a number of at-line and on-line applications[36].

Noninvasive NMR analyses does not have the optical path and contamination restrictions experienced with on-line infrared systems and eliminates the need for solvents, columns, carrier gases and/or the separations required with on-line chromatographic systems. NMR analysis is also real-time, with typical detection of any protonated analyte at the 0.1–100 % level. Magnetic resonance imaging (MRI), which is more typically associated with medical applications, can also be used to monitor more physical aspects of chemical processes[37].

On-line Atomic Spectroscopy

Atomic emission spectrometry can be used for on-line monitoring of incineration[38] and other processes where elemental analysis is vital. Developments in the area of on-line atomic absorption spectrometry (AAS) have been reported[39]. SPECTRO produces on-line systems for elemental analysis from aluminum to uranium over a wide concentration range (ppm to %). These are based on optical emission spectrometry (OES) or X-ray fluorescence. Measurement is continuous and can be nondestructive. There are a number of clever sampling designs that can accommodate a variety of process samples from solids to slurries to liquids.

9.2.3 Separation Analysis

On-line GC Analyser

The first of the separation techniques to be used in process measurement was gas chromatography (GC) in 1954. The GC has always been a robust instrument and this aided its transfer to the process environment. The differences between laboratory GC and process GC instruments are important. With process GC, the sample is transferred directly from the process stream to the instrument. Instead of an inlet septum, process GC has a valve, which is critical for repetitively and reproducibly transferring a precise volume of sample into the volatiliser and thence into the carrier gas. This valve is also used to intermittently introduce a reference sample for calibration purposes. Instead of one column and a temperature ramp, the set up involves many columns under isothermal conditions. The more usual column types are open tubular, as these are efficient and analysis is more rapid than with packed columns. A pre-column is often used to trap unwanted contaminants, e.g. water, and it is backflushed while the rest of the sample is sent on to the analysis column. The universal detector – thermal conductivity detector (TCD)–is most often used in process GC but also popular are the FID, PID, ECD, FPD and of course MS. Process GC is used extensively in the petroleum industry, in environmental analysis of air and water samples[40] and in the chemical industry with the incorporation of sample extraction or preparation on-line. It is also applied for on-line monitoring of volatile products during fermentation processes[41].

There are many vendors of process GC instruments, each with their own advantages and disadvantages, as well as special application areas. One process GC on the market is the FXi Series5 process GC from Hamilton Sundstrand Applied Instrument Technologies (Figure 9.16). This instrument continuously analyses, measures and reports the concentrations of components in process streams of gases and liquids. It is employed in a variety of applications in the natural gas, refining, petrochemical, chemical and pharmaceutical industries. The analysers also yield the data for their own maintenance and self-diagnostics information. The system consists of an analyser section and a microprocessor-based controller. The FXi Series5 can be configured with a number of detectors such as a TCD, an FID or an FPD for measuring sulfur compounds. The unit weighs 118 kg.

The PGC2000 from ABB includes an isothermal oven which contains the analytical columns, the detectors and the sample valves. Any two of the following detectors can be installed: TCD, FID and FPD. The dimensions of the system are $49.6 \times 34 \times 117.5$ cm and it weighs 73 kg.

Peak Laboratories LLC has introduced a new process GC–the PP1–which is based on reducing compound photometer (RCP) technology. RCP technology reduces analytes in a heated mercuric oxide (HgO) chamber, which releases mercury vapour into the photometer cell for direct measurement using UV light absorption. This is sometimes referred to as

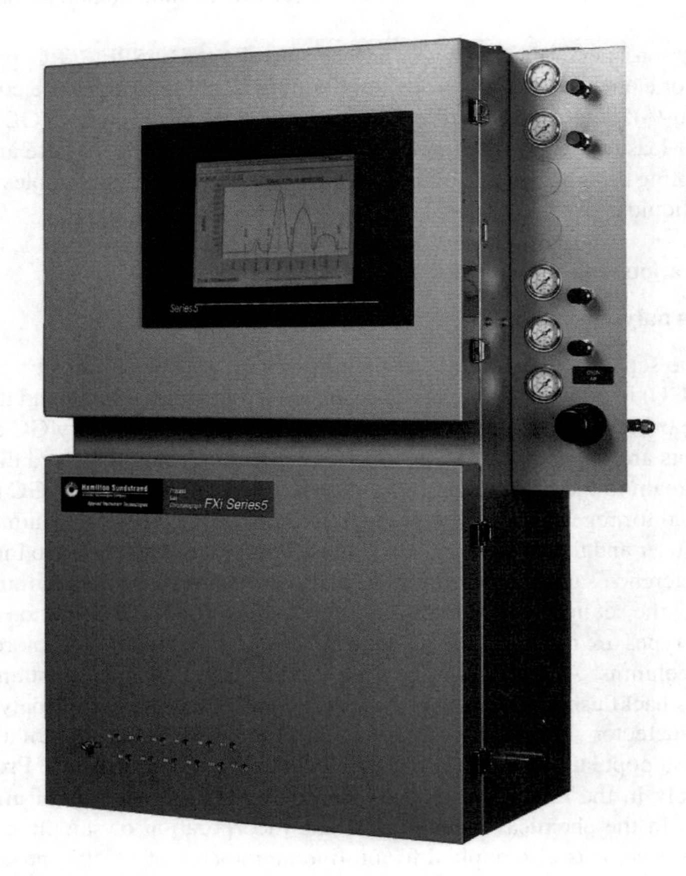

Figure 9.16 The FXi Series5 process GC (Reproduced by permission of Hamilton Sundstrand Applied Instrument Technologies).

GC–ML (mercury liberation). The RCP design can measure very low levels (sub-ppb) of hydrogen and carbon monoxide in air with a cycle time of only two minutes. Using a 1 mL sample loop, the measurement range is 1 ppb to 5 ppm. The PP 1 weighs only 11.3 kg and has very compact dimensions (68.6 × 43.2 × 17.8 cm).

On-line HPLC Analyser

HPLC has been adapted for use in process analytics but it has been more challenging to develop the methods due to issues around sampling, sample preparation, maintenance of both the column and mobile phase, cost and hazards associated with solvents as well as pressure difficulties. The two companies that marketed on-line HPLC in the 1980s, Applied Automation and Waters, did not sell enough instruments to continue those products. Recently there has been renewed interest in the technique and solutions to the previous problems are being found. Recycling the mobile phase and slowing the flow rates increases assay precision, reduces maintenance and helps the mobile phase to last longer. Employing shorter, narrower columns requires less solvent without compromising efficiency. Using shorter or monolithic columns at faster flow rates – high speed liquid chromatography – works well for certain applications[42]. On-line HPLC is often used in conjunction with ultrafiltration membranes to eliminate high molecular weight components and in applications requiring high selectivity. The criteria for setting up process HPLC are:

- Simple, industrial design.
- Rugged and reproducible results.
- Easy to calibrate and maintain.
- Unattended, continuous operation.

Though there are not as many examples of in-process HPLC as there are of in-process GC, as a technique it is very useful for on-line analysis in industry. One example describes an on-line continuous HPLC with diode array and mass spectral detection for process monitoring of sulfonated azo dyes and their intermediates in bioreactors[43]. Waste water samples can be analysed on-line for chloride, fluoride, nitrate, nitrite and sulfate using ion chromatography (IC)[44]. A Dionex system has been used for on-line HPLC and/or IC of industrial processes such as copper plating, sodium during a polishing excursion and a fermentation broth[45]. The system was fitted with conductivity, absorbance and electrochemical detectors and the mobile phase could be delivered either isocratically or by gradient. Furthermore, there are reports in the literature of continuous on-line assays using HPLC which can be adapted for process monitoring and control. Some of these include the monitoring of haloacetic acids via membrane extraction[46] and the analysis of rifabutin[47] and sulbactam[48].

9.2.4 Imaging Analysis

Spectral Dimensions has produced the Blend Monitor, which claims to be the first system based on chemical imaging that operates on a process scale. The prototype, based on NIR spectroscopy, takes spectral images of the powder 'at-line' through a quartz window integrated into the blender itself. The data obtained allow decisions to be made as to whether the blending is complete or more time is required. The process does need to be stopped for the measurements to take place but samples do not need to be removed and, of course, the technique is nondestructive. It is hoped the next prototype will allow measurements to be taken continuously while the blender is turning so that the process does not need to be interrupted.

9.2.5 Electrochemical Analysis

On-line Potentiometric Analyser

The first analytical instruments adapted for use in in-process measurements were electrochemical pH meters used as immersion probes. On-line potentiometric analysers can give continuous, real-time results for various analytes in a process[1]. They are rugged ISEs that are not affected by the colour or turbidity of the process stream. Arrays of potentiometric sensors can even be used in fermentation broths[49]. The sample can be taken into a loop, passed through a filter to protect the ISE surface and measured. A feedback mechanism allows control of other parameters in order to keep the process in check. Applications

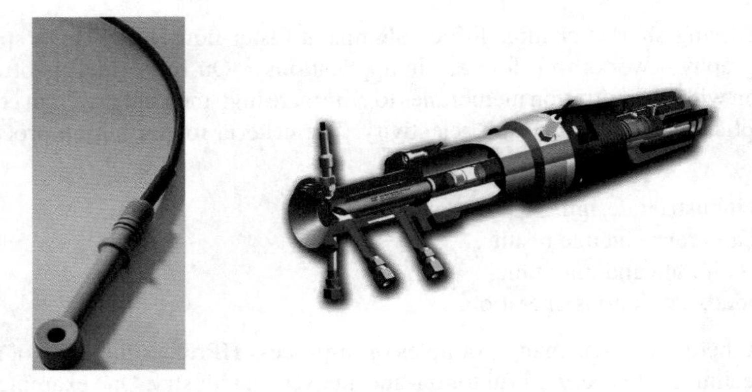

Figure 9.17 The InPro 7200 conductivity sensor and the InTrac 798 flushing chamber housing the 120 mm electrode/sensor. The flushing chamber allows in situ steam sterilisation of the electrode (Reproduced by permission of Mettler Toledo GmbH).

include monitoring of the sodium ion in the water–steam system of power plants[50]. Both ammonia and creatinine can be determined in a flowing solution by an ammonia gas-sensing electrode[51].

On-line Conductimetric Analyser

Conductivity is an important on-line and in-line measurement in a number of industries. For example, Mettler Toledo GmbH produces the InPro 7200 series of conductivity sensors (Figure 9.17). The wide bore doughnut design means that the sensor is very resistant to contamination and fouling. It weighs only 150 g and the cables extend 6 m. The temperature range can be up to 200 °C and the conductivity range up to 2000 mS/cm with a resolution of 0.01 mS/cm.

Scrubbers are important in a number of processes for treating harmful emission gases, vapours, odours and particulates. One type of scrubber, called a Cloud Chamber Scrubber, uses clouds of electrically charged water droplets to attract sub-micron particulates and fumes. On-line conductivity is an important measure of the efficiency of such a scrubber and is carried out on the recirculating water. Hence, conductivity probes are widespread in a number of manufacturing processes. However, the use of conductivity sensors is also prevalent in environmental applications where they are used for continuous monitoring of air[52]. A conductometric sensor for on-line testing of six haloacetic acids in aqueous samples in two minutes has been developed[53]. This modified electrode on a flow-through chip has been used for haloacetic acid determination in domestic and commercial drinking water samples.

9.3 Laboratory Integrated Management Systems

Analytical laboratories produce vast amounts of data which need to be accessed by many different people, e.g. quality personnel, clients, managers and inspectors.

Electronic databases are slowly replacing paper-based systems. Regulated laboratories must follow strict procedures with respect to sample logging, archiving and testing reports. Laboratory information management systems (LIMS) have been developed to both control and monitor the analytical work being carried out, as well as for storage and retrieval of that data. There are dedicated LIMS for certain scientific areas and companies[54, 55]. Everything from sample receipt to final reports can be carefully tracked and there is a full audit trail, which it is extremely important to be able to access during audits. LIMS can be as small or as big as required by the company using it. Some LIMS can do everything from reading barcoded samples to controlling user access levels, automatically backing up data at predefined intervals, sounding an alarm if a certain process goes out of control, allowing searches of data and monitoring trends in results over time. Hence LIMS impact on production processes and on-line monitoring also[56].

Direct interfacing of analytical instruments is another aspect of LIMS, where the system controls the equipment such as balances, pH meters and even more complex instruments such as chromatographic system. The use of LIMS means that standard operating procedures must be followed, any changes made by the user are tracked and require explanation or even approval and the methods and data are uploaded electronically. Once properly tested and validated, LIMS can bring about a huge gain in productivity and efficiency, especially in laboratories carrying out regulated work.

References

1. Vojinović, V., Cabral, J.M.S. and Fonseca, L.P. (2006) Real-time bioprocess monitoring: Part I: In situ sensors. *Sens Actuat B*, **114 (2)**, 1083–1091.
2. Pan, L., LoBrutto, R. and Zhou, G. (2006) Characterization of moisture-sensitive raw materials with simple spectroscopic techniques. *Talanta*, **70 (3)**, 661–667.
3. Perez-Ruiz, T., Martinez-Lozano, C., Tomas, V. and Iniesta, M.T.(2005) Flow injection determination of ethanol using online photo-oxidation and spectrophotometric detection. *Microchim Acta*, **149 (1–2)**, 67–72.
4. Pettas, I.A. and Karayannis, M.I. (2004) Simultaneous spectra–kinetic determination of peracetic acid and hydrogen peroxide in a brewery cleaning-in-place disinfection process. *Anal Chim Acta*, **522 (2)**, 275–2.
5. Adcock, J.L., Barnett, N.W., Gerardi, R.D. *et al.* (2004) Determination of glyphosate mono-isopropylamine salt in process samples using flow injection analysis with tris(2,2'-bipyridyl)ruthenium(II) chemiluminescence detection. *Talanta*, **64 (2)**, 534–537.
6. Sakai, T., Piao, S., Teshima, N. *et al.* (2004) Flow injection system with in-line Winkler's procedure using 16-way valve and spectrofluorimetric determination of dissolved oxygen in environmental waters. *Talanta*, **63 (4)**, 893–898.
7. Sohn, O.J., Han, K.A. and Rhee, J.I., (2005) Flow injection analysis system for monitoring of succinic acid in biotechnological processes. *Talanta*, **65 (1)**, 185–191.
8. Pedrosa, V.A., Lowinsohn, D. and Bertotti, M. (2006) FIA determination of paracetamol in pharmaceutical drugs by using gold electrodes modified with a 3-mercaptopropionic acid monolayer. *Electroanalysis*, **18 (9)**, 931–934.
9. Li, Q.Y., Morris, K.J., Dasgupta, P.K. *et al.* (2003) Portable flow-injection analyzer with liquid-core waveguide based fluorescence, luminescence, and long path length absorbance detector. *Anal Chim Acta*, **479 (2)**, 151–165.
10. Crocombe, R.A. and Flanders, D.C., Atia, W. (2004) Micro-optical instrumentation for process spectroscopy. *Proc. SPIE*, **5591**, 11–25.

11. Zhao, Y.X. and Dong, Y.H. (2003) Study on kinetic spectrophotometry determination of trace lead. *Fenxi Kexue Xuebao*, **19 (6)**, 573–575.
12. Oelichmann, J. (2002) What makes near-infra-red spectroscopy so attractive? *GIT Labor-Fachzeitschrift*, **46 (8)**, 906–907.
13. Sahni, N.S., Isaksson, T. and Naes, T. (2004) In-line near infrared spectroscopy for use in product and process monitoring in the food industry. *J Near Infrared Spec*, **12 (2)**, 77–83.
14. Luypaert, J., Massart, D.L. and Vander, Heyden, Y. (2007) Near-infrared spectroscopy applications in pharmaceutical analysis. *Talanta*, **72 (3)**, 865–883.
15. Blanco, M., Castillo, M. and Beneyto, R. (2007) Study of reaction processes by in-line near-infrared spectroscopy in combination with multivariate curve resolution: Esterification of myristic acid with isopropanol. *Talanta*, **72 (2)**, 519–525.
16. Crocombe, R.A. (2004) MEMS technology moves process spectroscopy into a new dimension. *Spectroscopy Europe*, **16 (3)**, 16–19.
17. Qin, F.L., Min, S.G. and Li, N. (2003) Determination of *n*-hexane, cyclohexane and toluene in carbon tetrachloride solvent by near-infrared spectroscopy. *Guangpuxue Yu Guangpu Fenxi*, **23 (6)**, 1090–1092.
18. Parris, J., Airiau, C., Escott, R. *et al.* (2005) Monitoring API drying operations with NIR. *Spectroscopy*, **20 (2)**, 34–41.
19. Wilks, P. (2004) Process monitoring: moving from the laboratory to the line. *Spectroscopy*, **19 (9)**, 24–34.
20. Coates, J. (2005) A new approach to near- and mid-infrared process analysis. *Spectroscopy*, **20 (1)**, 32–42.
21. Pollanen, K., Hakkinen, A., Reinikainen, S.-P. *et al.* (2005) IR spectroscopy together with multivariate data analysis as a process analytical tool for in-line monitoring of crystallization process and solid-state analysis of crystalline product. *J Pharm Biomed Anal*, **38 (2)**, 275–284.
22. Martin, P. and Holdsworth, R. (2004) High-resolution infra-red spectroscopy for *in situ* industrial process monitoring [overview]. *Spectroscopy Europe*, **16 (5)**, 8–15.
23. Druy, M.A. (2004) Applications for mid-IR spectroscopy in the pharmaceutical process environment. *Spectroscopy*, **19 (2)**, 60–63.
24. Chalus, P., Roggo, Y. and Ulmschneider, M. (2006) Raman spectroscopy: a tool for process analytical technology (PAT). *Spectra Analyse*, **35 (252)**, 17–22.
25. O'Brien, L.E., Timmins, P., Williams, A.C. and York. P. (2004) Use of *in situ* FT-Raman spectroscopy to study the kinetics of the transformation of carbamazepine polymorphs. *J Pharm Biomed Anal*, **36 (2)**, 335–340.
26. Weyer, L.G., Yee, D., Pearson, G.I. and Wade, J. (2004) Installation of a multiplexed Raman system in a chemical process. *Appl Spec*, **58 (12)**, 1500–1505.
27. Wikstrom, H., Lewis, I.R. and Taylor, L.S. (2005) Comparison of sampling techniques for in-line monitoring using Raman spectroscopy. *Appl Spec*, **59 (7)**, 934–941.
28. Franz, C., Kern, J., Pötschacher, F. and Bayer, K. (2005) Sensor combination and chemometric modelling for improved process monitoring in recombinant *E. coli* fed-batch cultivations. *J Biotechnol*, **120 (2)**, 183–196.
29. Hisiger, S. and Jolicoeur, M. (2005) A multiwavelength fluorescence probe: Is one probe capable for on-line monitoring of recombinant protein production and biomass activity? *J Biotechnol*, **117 (4)**, 325–336.
30. Kougoulos, E., Jones, A.G. and Wood-Kaczmar, M.W. (2005) Modelling particle disruption of an organic fine chemical compound using Lasentec focussed beam reflectance monitoring (FBRM) in agitated suspensions. *Powder Technology*, **155**, 153–158.
31. Blanco, A., de la Fuente, E., Negro, C. *et al.* (2002) Focused beam reflectance measurement as a tool to measure flocculation. *TAPPI Journal*, **1 (10)**, 14–20.
32. Ge, X.M. and Bai, F.W. (2006) Intrinsic kinetics of continuous growth and ethanol production of a flocculating fusant yeast strain SPSC01. *J Biotechnol*, **124 (2)**, 363–372.
33. Kougoulos, E., Jones, A.G., Jennings, K.H. and Wood-Kaczmar, M.W. (2005) Use of focused beam reflectance measurement (FBRM) and process video imaging (PVI) in a modified mixed suspension mixed product removal (MSMPR) cooling crystallizer. *J Crystal Growth*, **273 (3–4)**, 529–534.

34. Clinton, R. and Creaser, C.S. (2005) Real-time monitoring of a pharmaceutical process reaction using a membrane interface combined with atmospheric pressure chemical ionisation mass spectrometry. *Anal Chim Acta*, **539 (1–2)**, 133–140.
35. Tarkiainen, V., Kotiaho, T., Mattila, I. *et al.* (2005) On-line monitoring of continuous beer fermentation process using automatic membrane inlet mass spectrometric system. *Talanta*, **65 (5)**, 1254–1263.
36. Maiwald, M., Fischer, H.H., Kim, Y.-K. *et al.* (2004) Quantitative high-resolution on-line NMR spectroscopy in reaction and process monitoring. *J Mag Reson*, **166 (2)**, 135–146.
37. Mantle, M.D. and Sederman, A.J. (2003) Dynamic MRI in chemical process and reaction engineering. *Prog Nucl Mag Reson Spec*, **43 (1–2)**, 3–60.
38. Timmermans, E.A.H., de Groote, F.P.J., Jonkers, J. *et al.* (2003) Atomic emission spectroscopy for the on-line monitoring of incineration processes. *Spectrochim Acta Part B*, **58 (5)**, 823–836.
39. Burguera, M. and Burguera, J.L. (2007) On-line electrothermal atomic absorption spectrometry configurations: recent developments and trends. *Spectrochim Acta Part B*, In Press.
40. Wortberg, M., Ziemer, W., Kugel, M. *et al.* (2006) Monitoring industrial wastewater by online GC–MS with direct aqueous injection. *Anal Bioanal Chem*, **384 (5)**, 1113–1122.
41. Diamantis, V., Melidis, P. and Aivasidis, A. (2006) Continuous determination of volatile products in anaerobic fermenters by on-line capillary gas chromatography. *Anal Chim Acta*, 573–574, 189–194.
42. Gomis, D.B., Nunez, N.S. and Alvarez, M.D.G. (2006) High speed liquid chromatography for in-process control. *J Liq Chrom Rel Technol*, **29 (5–8)**, 931–948.
43. Plum, A. and Rehorek, A. (2005) Strategies for continuous on-line high performance liquid chromatography coupled with diode array detection and electrospray tandem mass spectrometry for process monitoring of sulfonated azo dyes and their intermediates in anaerobic–aerobic bioreactors. *J Chrom A*, **1084 (1–2)**, 119–133.
44. Hodge, E.M. (2000) Online monitoring of wastewater using ion chromatography. *J Chromatogr Sci*, **38 (8)**, 353–356.
45. Doyle, M.J. and Newton, B.J. (2000) Chromatography with online HPLC and ion chromatography for process control. CAST, **22**, 9–12.
46. Wang, X.Y., Kou, D.W. and Mitra, S. (2005) Continuous, on-line monitoring of haloacetic acids via membrane extraction. *J Chrom A*, **1089 (1–2)**, 39–44.
47. Blanco Gomis, D., Sanchez Nunez, N., Andres Garcia, E. *et al.* (2005) High speed liquid chromatography for in-process control of rifabutin. *Anal Chim Acta*, **531 (1)**, 105–110.
48. Blanco Gomis, D., Sanchez Nunez, N., Andres Garcia, E. *et al.* (2003) Fast high-performance liquid chromatography method for in-process control of sulbactam. *Anal Chim Acta*, **498 (1–2)**, 1–8.
49. Legin, A., Kirsanov, D., Rudnitskaya, A. *et al.* (2004) Multicomponent analysis of fermentation growth media using the electronic tongue (ET). *Talanta*, **64 (3)**, 766–772.
50. Li, Y.S., Xing, C.X. and Yang, L.L. (2005) Determination of trace sodium in the water–steam system of power plants using an FIA/ISE method with an automatic penetration and alkalization apparatus. *Anal Sci*, **21 (3)**, 273–279.
51. Suzuki, H. and Matsugi, Y. (2004) Microfabricated flow system for ammonia and creatinine with an air-gap structure. *Sens Actuat B*, **98 (1)**, 101–111.
52. Vianello, F., Boscolo-Chio, R., Signorini, S. and Rigo, A. (2007) On-line detection of atmospheric formaldehyde by a conductometric biosensor. *Biosens Bioelectron*, **22 (6)**, 920–925.
53. Suedee, R., Intakong, W. and Dickert, F.L. (2006) Molecularly imprinted polymer-modified electrode for on-line conductometric monitoring of haloacetic acids in chlorinated water. *Anal Chim Acta*, **569 (1–2)**, 66–75.
54. Karlsson, C. and Holgersson, S. (2003) ForumDNA, a custom-designed Laboratory Information Management System. *International Congress Series*, **1239**, 783–786.
55. Resin, A. (2002) A quality laboratory information management system implementation at a worldwide level. *Spectra Analyse*, **31 (229)**, 31–33.
56. Kershner, V. (2002) LIMS and the new business imperative in manufacturing: production-critical information management. *Am Lab News*, **34 (17)**, 22–22.

Section IV

Miniaturised Analytical Instrumentation

The Drive for Miniaturised Analytical Instrumentation

There is a big drive for miniaturisation because of the many advantages it offers, such as shorter analysis times, better control, high throughput, reduction in chemicals and sample required, lower cost and remote monitoring applications. Some devices can be made so small that they can be integrated into larger instruments or incorporated into on-line or in-line systems or arrays. They can be transported easily and entirely new techniques and concepts are becoming available to analysts as the spatial dimensions are reduced.

Often these small systems are capable of taking a sample, preparing the sample, isolating or separating the sample and measuring the sample all in one platform. Hence, they are called total analysis systems (TAS), a term that emerged in the early 1980s. Later in the same decade, it was thought that very small TAS could offer even better and faster performance. This concept was termed μTAS (micro total analysis systems)[1]. Technologies coming to the fore aiding such miniaturisation included optical fibres, LEDs, photodiodes, mechanical engineering, microfabrication technologies and wireless communications. With μTAS, sample handling, separation and detection are carried out by a single instrument of reduced weight and they range in size from a few millimetres to a few microns. Integration of all the required steps into one device via miniaturisation is the goal. Often, as an added advantage, analysis times of minutes are reduced to seconds and selectivity and sensitivity may improve.

Lab-on-a-chip (LOC) and lab-on-valve (LOV) devices are versions of μTAS where fluids are manipulated to give a complete assay on a microfabricated chip. Their main goal is the scaling down of laboratory processes onto a chip-based platform.

There are many application areas for μTAS including, to name but a few, environmental monitoring and control[2], high-throughput biomedical screening, process control in various industries and defence and security situations. Developments in this exciting area are constantly being reported[3, 4].

References

1. Manz, A., Graber, N. and Widmer, H.M. Miniaturized total chemical analysis systems: A novel concept for chemical sensing. *Sens Actuat B*, **1 (1–6)**, 244–248 (1990).
2. Marle, L. and Greenway, G.M. Microfluidic devices for environmental monitoring. *TRAC*, **24 (9)**, 795–802 (2005).
3. Vilkner, T., Janasek, D. and Manz, A. Micro total analysis systems. Recent developments. *Anal Chem.*, **76 (12)**, 3373–3386 (2004).
4. Fiorini, G.S. and Chiu, D.T. Disposable microfluidic devices:fabrication, function and application. *Biotechniques*, **38**, 429–446 (2005).

10

Chip-based Instrumentation

10.1 The Development of Chip-based Analytical Devices

Manz *et al.* invented the term µTAS in 1990[1]. µTAS was a natural follow-on from micro electro mechanical systems (MEMS), which were developed earlier by the microelectronics industry and involved the integration of mechanical elements, sensors, actuators, and electronics on a common silicon substrate through microfabrication technology. MEMS are also micron-sized devices like µTAS but tend to be based more on sensors rather than on total analysis systems. MEMS often contain moving parts, are used to sense or manipulate the physical environment and are prevalent in physics and engineering fields. MEMS are also called micro system technology (MST) in some parts of the world. They have been used as physical sensors of pressure, temperature and vibration. MEMS are also employed as blood pressure monitors, accelerometers (for airbags, pacemakers and games), microactuators for data storage and ink-jet printers and as electronic noses.

The first µTAS devices were based on GC[2]. Soon after, HPLC-based µTAS platforms began to appear and then CE followed. In chip CE, an injection was performed and an electrophoretic separation of a sample mixture (different fluorescent dyes) and all liquid handling was achieved using electro-osmotic flow. Since then, electrophoresis and electrokinetic fluid handling have been the cornerstones of many miniaturised analytical devices.

The term µTAS is sometimes interchanged with lab-on-a-chip (LOC), more often when the manipulation of fluids is involved. µTAS and LOC range in size from a few microns to a few millimetres. The technique of µTAS is interdisciplinary; it combines the advantages of chemical sensors and the resolving power of modern benchtop analytical systems and is constantly evolving. The main advantage of µTAS is integration of the entire separation process onto one analytical microdevice, so early efforts focused on micropumps and valves to manipulate fluids inside a microfabricated structure. In such a fluid-based µTAS,

Analytical Instrumentation: A Guide to Laboratory, Portable and Miniaturized Instruments G. McMahon

some separation may occur or liquid reagents may be mixed with the sample during analysis. The detector is then triggered to gather information by measuring a selective signal from the analyte(s) of interest. The μTAS may be able to carry out some processing of the information given by the sensors. Another miniaturised flowing system, but without separation, is lab-on-valve (LOV), which is flow injection analysis (FIA) based on a chip and a small multiport valve. Like LOC, all the steps of the assay are carried out on the small platform.

Many analytical assays that used to be very labour intensive have been reduced in size and work off a small chip. Examples include polymerase chain reaction (PCR)[3], DNA analysis and sequencing[4, 5, 6], protein separations[7] and cell manipulations[8, 9, 10].

10.2 Challenges for Chip-based Analytical Devices

There are many formidable challenges for chip-based analytical devices that, if not taken into account, can cause difficulties in the design and development of the new generation of tiny instruments. Four of the main issues are discussed below including microfluidics on chip, the problems associated with installing small components onto chips, getting samples onto the chip and detected there and, finally, understanding that processes on a microscale are not simply a matter of scaling-down normal macroscale processes. Lab-on-a-chip may also lack the flexibility of full-scale instruments but this may be solved by using chip arrays.

10.2.1 Moving and Mixing Fluids on a Chip

Moving Fluids on a Chip

Dealing with real fluids on the microscale is difficult. Bubbles, dust and surface tension can cause real problems due to the small size the fluid encounters. In fact, a whole science called microfluidics has built up and books have been written on the subject[11, 12]. Challenges to all microfluidic systems include:

- hydrodynamics of fluids in small channels;
- converting macro volumes to micro volumes;
- chemical resistivity of the silicon/polymer wafer to a variety of chemical compounds;
- technology hurdles of immobilising the substrate materials to form the final μTAS structure.

Liquid flow can be laminar or turbulent in nature. In laminar flow, fluids move in straight lines; in effect the adjacent layers 'slide' over one another. Hence, different liquids can flow side by side and the only way for them to mix is by diffusion. With turbulent flow, there are unstable stream-lines and it is unpredictable but mixing is more easily achieved. The type of flow that will predominate can be predicted mathematically according to the Reynolds number. In microchannels, such as in μTAS, the Reynolds number is low and so laminar flow dominates. Hence, diffusion in these narrow channels is practically the only way for mixing of the fluids to occur. The smaller the system and the narrower the channels, the faster the diffusion. Under typical conditions, flow rates in the microfluidic channels are about 1 μL/s. At these rates, and in contrast to macro conditions, the dispersion of fluids is limited and highly predictable. As a result, it is possible to get very exact, reproducible data.

A difficulty arises from the fact that liquids are typically transferred in microlitre or millilitre volumes, whereas microfluidic devices work with nanolitre or picolitre volumes because of the size of reaction chambers and channels, which typically have microscale dimensions.

The two main ways of actually moving fluids around a chip are by pressure-driven flow or electro-osmotic flow and each has its own advantages and disadvantages. Pressure-driven (hydrodynamic) flow has a parabolic flow velocity profile (Figure 10.1) where flow is fastest in the centre of the plug and almost zero at the wall of the channel. This profile can lead to dispersion, dilution and band broadening which can hinder detection and throughput. Another disadvantage is that the high pressures that are sometimes required can actually damage the chip itself. However, an advantage to pressure-driven flow is that both charged and uncharged molecules, even cells, can be moved and that both aqueous and non-aqueous solutions can be pumped. One way of overcoming the issue of interfacing pressure-driven flow with a chip is to split the flow into a number of channels or columns such that each has a small enough volume to be compatible with the chip geometry.

Electro-osmotic flow (EOF) is another possible flow mechanism and is used in capillary electrophoretic separations. The flow velocity profile is more plug-like, in contrast to pressure-driven flow (Figure 10.1). This type of flow is driven by ions in an electric field against a capillary so no external pump is required. EOF is generated when electrodes attached to a voltage power supply are placed in the reservoirs at each end of a channel and activated, so generating electrical current through the channel. Under these conditions, charged fluids will move by a process known as 'electro-osmosis'. EOF is dependent on buffer composition, e.g. ionic strength, pH and requires high electric fields (the voltage).

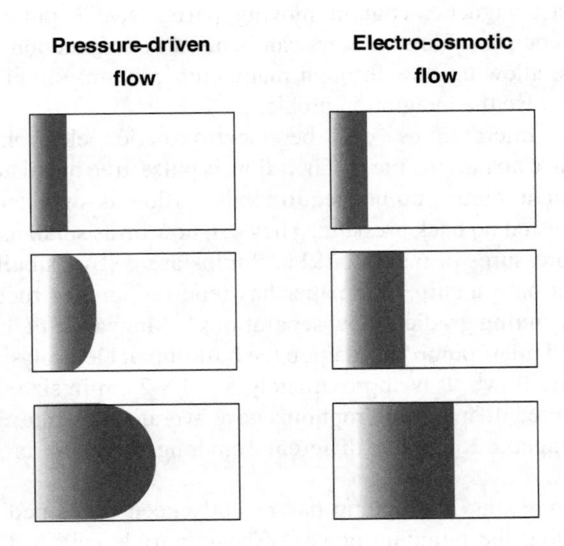

Figure 10.1 *Illustration of the parabolic flow velocity profile in pressure-driven flow and the plug-like flow velocity profile in electro-osmotic flow.*

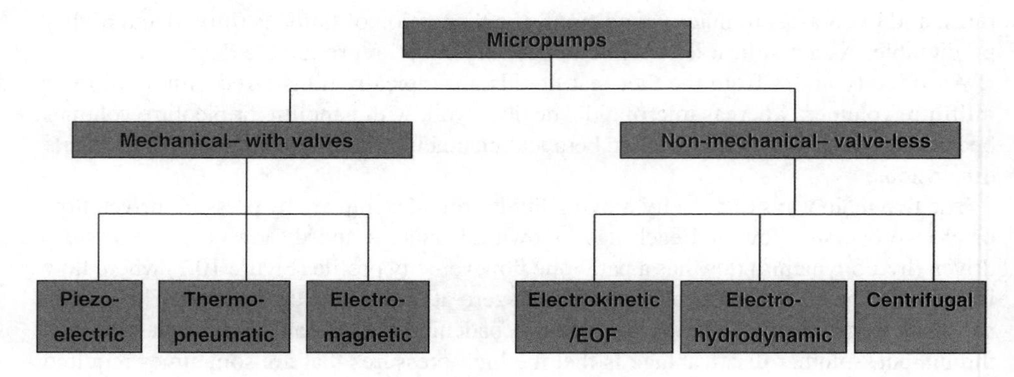

Figure 10.2 *The different types of micropumps.*

This flow works best for moving aqueous fluids. It works in glass and in several types of plastics, e.g. PDMS and PMMA, but moves more slowly in plastic. Another electrokinetic phenomenon known as 'electrophoresis' also occurs in these channels. While the blunt flow profile eliminates some of the band broadening and other issues associated with pressure-driven flow, there are some disadvantages, such as the requirement for high voltage and the changes that can occur on the surface of the channel, e.g. proteins adhering to the wall.

Other possible means of fluid flow include capillary (surface tension)[13], centrifugal[14, 15], ultrasonic, electromagnetic, electro-hydrodynamic and pneumatic force.

The main obstacle to HPLC on a chip has been the lack of integratable high-pressure pumps and valves. Mechanical micropumps (Figure 10.2) can be piezoelectric, thermo-pneumatic or electromagnetic, contain moving parts, have a pulsed flow and their fabrication can be complex. Flow splitters can be used in conjunction with conventional mechanical pumps, allowing flow through many chip columns simultaneously but this obviously does not solve the integration problem.

Non-mechanical micropumps can be electrokinetic, electrohydrodynamic or centrifugal; they have no moving parts. Their flow is pulse-free but historically pressures are low. Electrokinetic pumps do not require valves (flow is switched using voltages), have pulse-less flow and no back pressure. They can now be as small as 3–4 cm in length and can generate pressures of 6.9×10^5 kPa. Pumps are getting smaller with time and some are able to fit onto a chip. One paper has reported an electrochemical pumping system capable of running gradient flow separations[16]. Munyan *et al.* have developed an electrically actuated micropump based upon the build-up of electrolysis gases to achieve pressure-driven fluid flow[17]. It is approximately $5 \times 3 \times 2$ cm in size. Under an applied potential of 10 V, three different micropumps gave average flow rates of 8–13 μL min^{-1} for water being pumped through five different 2 cm long, 5500 μm^2 cross-sectional-area channels in PMMA.

Another very interesting micropump has recently been developed by Tanaka *et al.* and is based on using the pumping power of heart muscle cells[18]. The device, which involves coating a hollow silicone polymer sphere with the beating muscle cells, is only 5 mm in diameter. The heart cells require only glucose and oxygen to function and the prototype worked continuously for five days. Another novel chip-sized micropump with no moving parts uses a wire to send an electrical voltage to two immiscible liquids in a tiny narrow column (the width of a hair). Opposite charges are applied to each side of the column, which makes the fluids oscillate and cause the pumping action[19]. Researchers at Massachusetts Institute of Technology in the USA have developed a micropump which requires only battery power (a few volts)[20]. In their new system, known as a three-dimensional AC electro-osmotic pump, tiny electrodes with raised steps generate opposing slip velocities at different heights, which combine to push the fluid in one direction, like a conveyor belt. It is predicted that fast flows (mm/s), comparable to pressure-driven systems, can be attained.

Mixing Fluids on a Chip

A continuing problem for researchers is how to mix different fluids efficiently on the microscale. With flow in microchannels being laminar in nature, mixing is typically achieved by diffusion, although this can be poor. Alternatives to diffusion are being proposed. Sudarsan and Ugaz have reported some interesting means of mixing liquids[21,22]. They found that two liquids, if travelling fast enough, would swap positions as they travelled around a 180° bend in a channel. By repeatedly splitting and recombining the fluids, the authors achieved 90 % mixing between the two liquids. Whitesides *et al.* used herringbone design ridges etched into the channel floor to rotate the flow of two liquids in different directions, thereby mixing them[23,24].

10.2.2 Fitting Components onto a Chip

Due to the compactness of LOC systems, extensive integration of functionality and the small volumes involved, many parts must now be made available in a miniaturised form. Hence there are many components and tools to aid the fitting together of all the components onto such a small platform. These include:

- optical fibres;
- light emitting diodes;
- photodiodes;
- miniature components, valves and interconnects;
- microfabrication techniques;
- wireless communications.

Optical Fibres

Fibre optic cables (Figure 10.3) transmit light along waveguides. They were developed for the communications industry but are very useful for transmitting light to and from spectrometers and other devices. These cables also allow the sample to be far removed from the measuring instrument. Fibre optics can lose radiation as they get longer, especially in the UV region. Efficient fibre optic cables are available for Raman analysis as the light involved is usually in the visible and NIR regions where they have excellent throughput. Hence Raman techniques have been more easily miniaturised.

Optical fibres with selective reagents immobilised onto the tips can also be used as optical probes (optodes) that are physically dipped into a liquid sample or brought into

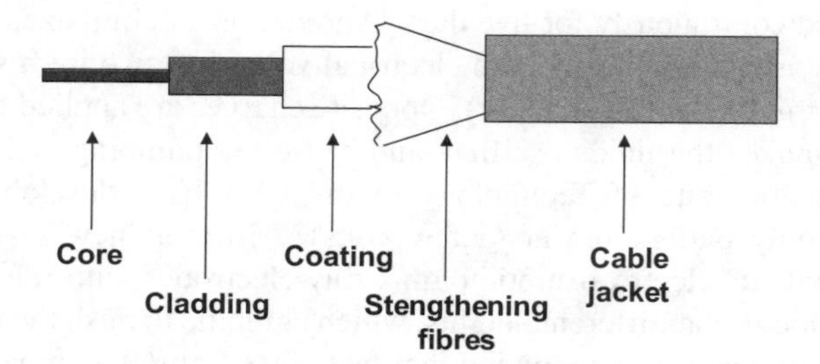

Figure 10.3 Schematic diagram of an optical fibre.

contact with a solid sample. In this set-up, they perform the function of electrochemical sensors without the need of a reference electrode. A good example is the immobilisation of an indicator dye onto the end of a fibre optic cable for which the spectrum will change according to pH.

Light Emitting Diodes

Light emitting diodes (LEDs) are tiny light bulbs that fit easily into an electrical circuit (Figure 10.4). They are semiconductor devices that emit UV, visible or IR light when an electric current passes through them. The light is not particularly powerful, but in most LEDs it is monochromatic and hence quite pure. The output from an LED can range

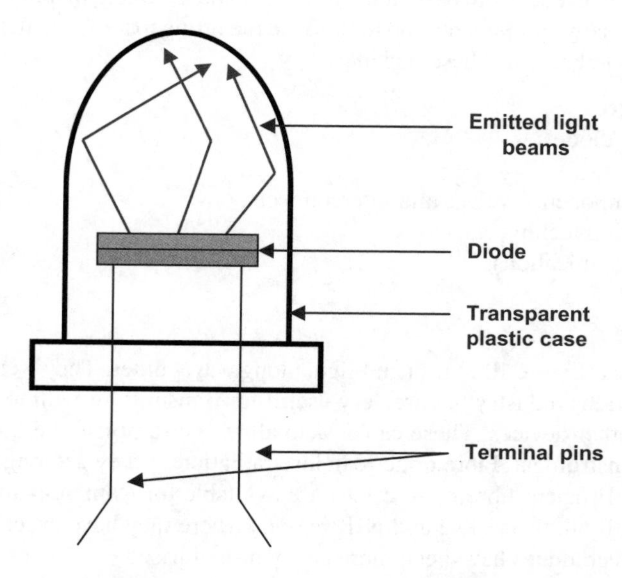

Figure 10.4 Schematic diagram of a light emitting diode.

from red (700 nm) to blue-violet (about 400 nm). Some LEDs emit IR radiation and more recently UV. The advantages of LEDs include:

- low power – most types can operate with battery power;
- high efficiency – most of the power supplied is converted into radiation;
- minimal heat production;
- long life since they do not have a filament that will burn out.

Superluminescent LEDs (SLEDs) are LEDs pumped at high current so that stimulated emission occurs. These miniature light sources are approximately 1000 times brighter than a tungsten–halogen light bulb.

Developments in the field of LEDs are occurring all the time. There are new white organic LEDs[25], new green polymer LEDs[26] and new blue polymer LEDs[27]. Ashwell *et al.* are developing molecular diodes from a bilayer of electron-rich and electron-poor molecules[28].

Photodiodes

Photodiodes are similar to LEDs but with reversed bias. They are semiconductor devices that, instead of emitting radiation, absorb UV, visible or IR light and register this as a change in electrical signal. These tiny detectors fit easily into an electrical circuit. An avalanche photodiode allows the normally small signal to be multiplied enabling better detection limits.

Miniature Components, Valves and Interconnects

There is now a selection of off-the-shelf miniaturised mechanical components available, such as actuators, micropumps, valves, filters and connectors. Integration and miniaturisation have been a huge success in the electronics/computing sector. Silicon chips, transistors etc. have been getting smaller since the late 1970s. Pumping and injecting devices, fluid connections, valves, mixing coils, capillary columns, filters, heaters, coolers and detectors, e.g. temperature, pressure, flow, pH, optical and electrochemical, are all now available in a much reduced size for μTAS development. Care must be taken to use bonding solutions and adhesives that do not allow any leaks to occur, that can withstand pressure if required and do not contaminate the device or the samples in any way.

Upchurch Scientific produce Nanoport™, which is a line of assemblies that bond to chip surfaces and provide a fluid connection *via* PEEK (polyetheretherketone) tubing. Other interconnects such as Luer fittings and connectors for HPLC may also be used. Sandia National Laboratories has also developed a line of interconnects and ferrules specifically for microfluidic devices.

Microfabrication Techniques

Many microfabrication techniques make use of integrated circuit (IC) technology developed originally for the electronics and communications industries. But recently, a whole new area of research has sprung up around making the tiny μTAS devices using microfabrication. Many of the techniques can and have been adapted for glass, silicon and plastic chips:

- photolithography
- wet etching
- reactive ion etching
- plasma etching
- conventional machining
- soft lithography
- micro milling
- casting
- injection moulding
- hot embossing
- laser ablation.

Some of the above, like photolithography and wet and dry etching, require clean rooms. They were adapted from the IC industry and are precise but somewhat expensive and inflexible. Using these techniques, the substrate, e.g. silicon, is taken and first coated with a metal film and a photoresist layer using deposition processes. A photomask can then be produced by lithography and placed onto the substrate. Photolithography works by transferring a pattern to a photosensitive material by selective exposure to a radiation source. The substrate is exposed and only the photoresist areas not covered by the photomask are removed. The substrate is then etched, whereby the exposed metal coated areas are also removed. A further etching step removes some of the substrate itself making the channels deeper. This step can be controlled in order to shape the channels to render their profiles curved or angular as required. Etching either occurs when a liquid is added that will dissolve the material (wet etching) or by dry etching, when the substrate is put into a reactor to break the gas molecules into ions, which react with the material being etched. Finally a lid is bonded onto the channels if needed.

Newer techniques such as soft lithography allow the scientist to construct a three-dimensional device with networks of channels and do not require clean rooms. The 'dry film resists' technique enables production of hybrid biochips with active elements in both the top and bottom substrates[29]. The ultimate choice of technique depends on a number of factors, such as the availability and capability of the technologies, cost and speed of manufacture and the substrate being used.

It is worth remembering that microfluidic channels do not have to be straight but can be curved or serpentine in shape, especially when accommodation of longer channels/columns is required. Even square cornered ones exist[30]. A new fabrication technique which enables the manufacture of raised canal-like channels on chips has been reported by Sandia National Laboratories[31].

Wireless Communications

These allow remote sensing, signal processing and data handling and get rid of the need to have cables connecting the platform to a computer. Bluetooth started as cable replacement technology and enables the exchange of data wirelessly between telephones, PCs and portable instruments. Other examples include Radiowave, IR, GPS (Global Positioning System) and GSM (Global System for Mobile communications) technologies.

10.2.3 Sampling and Detection Strategies

Sampling Strategies

Sampling strategies play a big part in the design and development of any chip-based analytical device[32]. However, they are often overlooked as research effort focuses on downsizing the analytical technique. Sampling on-chip can be particularly challenging where the samples are biological in nature but clever ways of coping with these more complex matrices are constantly being reported. The first step is getting the sample onto the chip and, in many cases, this is done by the user. For medical and biological applications, a sample is often taken and a small aliquot placed into a sample well or onto a sample pad on the chip. Some chip-based devices can be immersed into a sample but most require the procedure above.

Once on the chip, the next step is usually some sort of treatment of the sample such that it is made suitable for analysis and detection. Even aqueous samples should be filtered prior to assay. Typically, sample pretreatment takes the form of various fluid handling steps such as concentration[33], filtration[34], extraction[35, 36], ion exchange[37] and desalting[38]. Combinations and permutations of these steps are also reported[34]. Even degassing can be achieved on chip using ultrasonic-induced cavitation[39].

There are many reports in the literature about interfaces and devices for preparation of blood samples on chip. For example, Jandik *et al.* used a laminar fluid diffusion interface to replace centrifugation and consequently reduced the preparation time from between 30–60 minutes to only five minutes[40]. The H-filter has also been used to clean up biological samples[41]. A sample, e.g. blood, is put into a reservoir at one end of one post of the H, and a diluent such as water or saline is placed in the reservoir at the other end. The two parallel laminar streams will flow along the crossbar of the H as a result of hydrostatic pressure. Smaller, more mobile analyte molecules will cross the interface between streams quickly, whilst heavier particles remain in the carrier stream. Consequently, by controlling the fluid velocity and the length of the channel, the process can be optimised. There are also devices for separating plasma from blood on a chip[42, 43].

Detection Strategies

Detection can take place without separation (specific detection) or after separation (nonspecific detection). There are very few specific chip-based sensors that detect only one component to the exclusion of all others. It is more common to have some pretreatment or separation of the sample prior to detection.

Spectroscopic techniques are popular as a means of detection on chips. Examples include the determination of flavins[44] and DNA[45] by fluorescence. Spectrophotometric techniques are often used for biological samples[46]. Mass spectrometry has also been used. Benetton *et al.* coupled electrospray ionisation MS to a chip[47] while Sillon *et al.* developed a low cost mass spectrometer which incorporated the ionisation chamber, filter and detector on the chip[48]. A fibre optic coupler has been developed as a detector. The dual optical fibre configuration (one transmitting, one receiving, (Figure 10.5)) in the chip forms the microchannel as well as the detector itself and measures refractive index changes but can also be used to measure absorbance[49].

Electrochemical detection is common in microchips and can be potentiometric, voltammetric or conductimetric. An ultrathin ISE has been employed on a chip for the

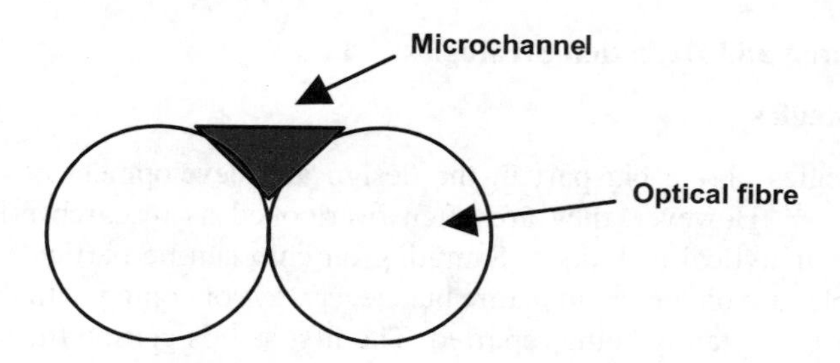

Figure 10.5 *A dual optical fibre configuration can serve as both microchannel and detector.*

detection of copper ions[50]. Both amperometry and voltammetry have also been realised. Examples include the analysis of carbohydrates, amino acids and antibiotics using pulsed amperometric detection[51], dopamine using a three-electrode amperometric detector[52] and neurotransmitters by sinusoidal voltammetric detection[53]. Using conductivity, Galloway *et al.* have obtained detection limits for various anionic compounds that were on a par with those obtained using indirect laser fluorescence detection[54]. Work has been done on new materials for electrochemical sensing for microfluidic platforms[55].

10.2.4 Understanding Processes on the Microscale

Many physical processes do not scale uniformly in the microscale domain i.e.:

- gravity
- friction
- combustion
- electrostatic forces
- van der Waals forces
- Brownian motion
- quantum mechanics.

At the millimetre scale, gravity, friction, combustion and electrostatic forces predominate. At the micrometer scale, it is mainly electrostatic forces, van der Waals forces and Brownian motion that are at work. Forces related to volume, like weight and inertia, tend to decrease in significance while forces related to surface area, such as friction and electrostatics, tend to increase in significance. Surface tension, which depends upon an edge, and viscosity both become very significant and impact hugely on the movement of liquids[56] as discussed in the previous section. Hence, sometimes these processes can aid the development of chip-based devices while in other cases they present limitations.

10.3 Chip-based Analytical Instruments

10.3.1 Lab-on-valve Flow Injection Analysis

Lab-on-valve (LOV) is a development of FIA whereby a microchannel chip (microconduit) is combined with a switching valve. It is a type of μTAS system. The function of the

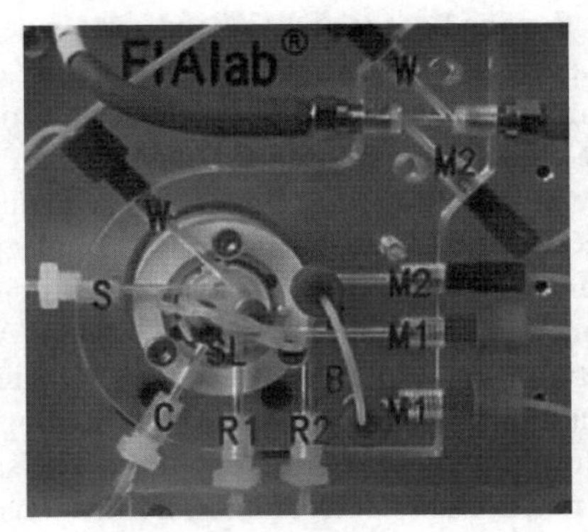

W = waste, S = sample, SL = sample loop,
B = bridge, C = carrier buffer/solvent,
R = reagent, M = mixing coil.

Figure 10.6 *The FIAlab Lab-On-Valve® integrates all connections, sample loop and flow cell into one simple manifold (Image provided by kind permission of FIAlab Instruments).*

microconduit is to carry out all the necessary analytical operations, e.g. mixing at T-junctions, separation in microcolumns or channels, preconcentration and detection at miniaturised detectors. An example of a LOV configuration is shown in Figure 10.6. Making the manifolds smaller drastically reduces the amounts of sample and reagents required and, hence, waste produced. Miniaturising such instruments challenges the detectors for they are now working with sample volumes in the microlitre–nanolitre range. However, for miniaturised optical detection techniques, optical fibres can be used. Normally, one optical fibre brings radiation to the sample cell and the other carries transmitted radiation back to the detection device. Other types of detectors may be on or off the chip, depending on size.

While LOV can have similar applications to LOC, it has found a particular niche in on-line sample pretreatment methods where the valve allows automation and miniaturisation of these traditionally labour intensive steps. Examples include solid phase extraction, ion exchange and solvent extraction. Compared to macrosample pretreatment, LOV offers the following advantages:

- ease of automation;
- improved speed and throughput;
- reduced use of reagents and sample and reduced waste output;
- lower risk of contamination of sample due to closed system and removal of manual steps;
- better precision due to automation;
- better selectivity for some reactions.

Disadvantages of LOV can include clogging, which because of the small channel dimensions can be a big problem if even fine particles are introduced. Hence, filtering of sample prior to analysis on chip is a requirement. Another issue can be the incompatibility between the continuously flowing liquids on LOV and certain detectors which prefer gases or discrete/stopped flow analyses.

Some of the interesting applications of LOV include the manipulation of beads and microcolumns for sample clean-up or separation. In particular, one version of the system has received considerable attention and is called microsequential injection-lab-on-valve (μSI–LOV). The μSI–LOV device can be used for sample pretreatment alone or in sequence with analysis and detection (usually absorbance or fluorescence). It has also been employed as the front end of a miniaturised CE system. An interesting use of μSI–LOV has been in conjunction with suspensions of specially coated beads. These beads can be manipulated by the multiport valve to flow into the holding coil and from there into the detector flow cell to form a temporary microcolumn. The sample is then mixed with the bead column and any reactions are detected, e.g. adsorption over time. After measurements have been made, the beads are sent to waste. The formation and removal of these temporary microcolumns has advantages over the use of fixed columns such as:

- no risk of sample carryover;
- no build-up of back pressure due to repeated injections;
- no problems with contamination or column bleed over time;
- improved precision as surface properties of column same every time.

The bead suspension is held between two outlets containing small PEEK tubing fittings, which allow liquids to flow freely but entrap the beads. A syringe pump is used to pump the beads into the temporary microcolumn. These bead columns can be used in many ways in the LOV system – from solid phase extraction to reaction columns.

10.3.2 Spectroscopic Devices

There are many examples of on-chip spectrometry in the literature. Some of these are described below.

On-chip MS

While mass spectrometers have been reduced in size, they have not been made sufficiently small to be integrated completely onto a microfluidic chip as yet. However, advances have been made in the interface between chip and spectrometer such that the analytes can be ionised as they elute from the chip, generally through a specially designed nozzle or tip. Currently two types of chip systems are used in conjunction with atmospheric pressure ionisation (API) MS – out-of-plane and in-plane devices. The first type has hundreds of nozzles integrated onto a single substrate, from which ionisation is established perpendicular to the chip. The second type builds in a microchannel at the end of which ionisation is generated on the edge of the microchip. Both types of chip have been used in a wide variety of applications from the direct analysis of drugs in plasma[57] to proteomics[58, 59] to carbohydrates[60].

Most examples in the literature exploit electrospray ionisation but some research groups have reported the use of atmospheric pressure chemical ionisation (APCI) on-chip. In one such example, the chip was composed of two wafers, one made of silicon wafer one made of pyrex glass[61]. The silicon wafer contained the nebuliser gas inlet, vapouriser channel and a nozzle. The sample inlet from the LC column was directly connected to the

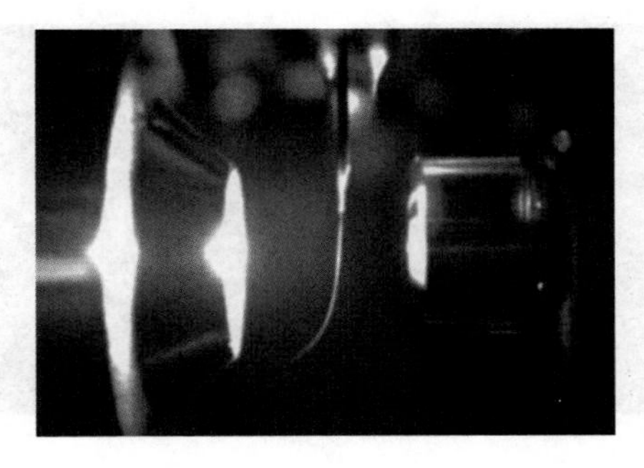

Figure 10.7 *A chip being moved into spray position in mass spectrometer (Copyright 2006, Agilent Technologies. Reproduced with permission.).*

vapouriser channel of the APCI chip. The nozzle then formed a narrow sample plume, which was ionised by the external corona needle, and the resultant ions were analysed by the mass spectrometer. When tested with neurosteroids, the chip performed with good repeatability and linearity. Research effort has also been invested in the design of the tips on the chip substrates[62, 63, 64].

Microfluidic chips and Compact disks have also been successfully coupled with MALDI–TOF mass spectrometry[65, 66].

Commercially, Agilent Technologies produces chips for both direct infusion into a mass spectrometer and for HPLC–MS applications. The chips accommodate nanoflow rates with an electrospray ionisation source, are about the size of a credit card and are reusable. The infusion chip is for collecting direct MS or tandem MS data. The protein HPLC chip has both a sample enrichment and C18 separation column on the chip, as well as the connections and spray nozzles for electrospray. There are also a small molecule chip and a glycan chip. The chip being used is placed in a Chip Cube MS interface which positions the sprayer tip perpendicular to the MS inlet (Figure 10.7).

Another commercial system for proteomics research is the TriVersa NanoMate from Advion BioSciences Inc. It integrates LC–MS, fraction collection and chip-based infusion in one platform. Use of such chip systems is widespread in the literature[67].

In terms of actually bringing a mass spectrometer onto the microscale, Sandia National Laboratories has made great strides forward. It has designed and built an ion trap mass analyser consisting of 1×10^6 micron-sized cylindrical ion traps (CITs) and has micro-fabricated parallel ion trap arrays (Figure 10.8). The array of micro ion traps is a freely suspended air gap structure fabricated in tungsten using silicon-based semiconductor and MEMS microfabrication methods. Simulations of their CIT indicate useful trapping efficiencies at low RF voltages (from a few volts to a few tens of volts). The influence of initial ion temperatures and RF heating due to nonlinear fields was found to be of added importance on this size scale.

On-chip NMR

At the University of Twente in the Netherlands, a multichannel chip has been designed with an integrated NMR detection coil and has yielded promising results at low field

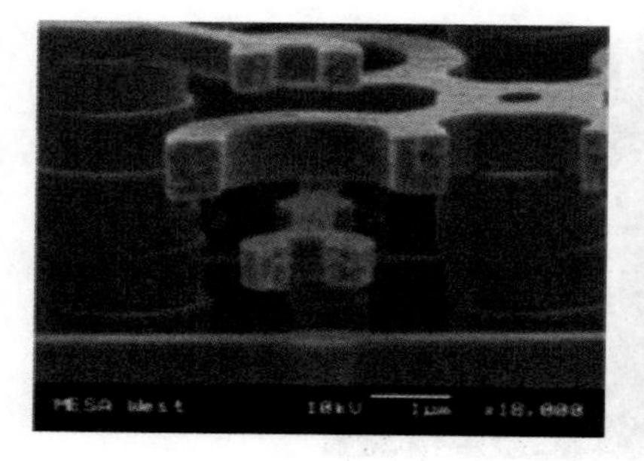

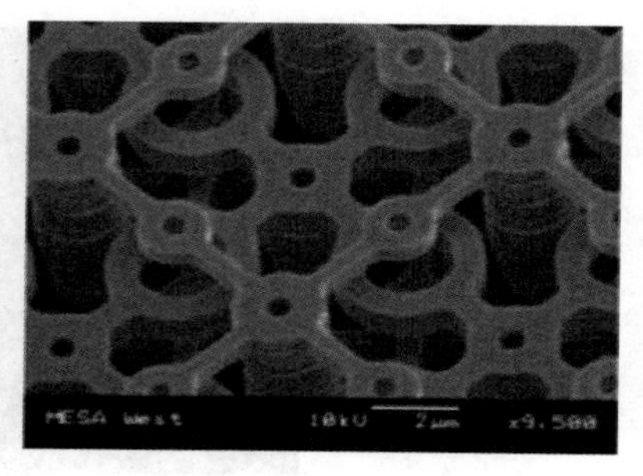

Figure 10.8 *Cross-section of a micron-sized cylindrical ion trap and an SEM image of an array of micro ion traps (Reproduced by permission of Sandia National Laboratories).*

strengths[68]. The potential use of these chips at higher magnetic fields is currently under investigation as is their optimisation for analysis and monitoring of specific reactions. A French research group has reported the design, fabrication and preliminary testing of a planar microcoils (500 × 500 μm) tuned and matched at 85.13 MHz (proton frequency at 2 T) associated with a micromachined channel in silicon. The group used these microcoils as NMR radio frequency detection coils and was able to analyse small sample quantities introduced into the microchannel[69].

Hilty *et al.* have used remote NMR with microfluidic chips to obtain profiles of gas flow in these devices[70]. Remote detection of the NMR signal both overcomes the sensitivity limitation of NMR and enables noninvasive measurement of microfluidic flow. Although used for gases, it can be applied to liquids also.

10.3.3 Separation Devices

Initially the main reason for miniaturising separations was to enhance analytical performance and the combination of integration of components and small size were seen as advantages. These chip devices are based mainly on chromatography[71], though electrophoretic separations play an increasing part. As the years go by, these on-chip separation devices have become more and more important as analytical instrumentation tools.

On-chip GC

Gas chromatography was the first type of separation to be miniaturised on chip in 1979 and has been used in many space projects. Due to its robustness and simplicity, GC has been realised in the micro domain for some time now. Sandia National Laboratories has produced the μChemLab for gas phase sample detection and analysis; it is a commercial handheld instrument based on chip technology. It is a gas phase system that concentrates, separates (by GC) and detects compounds at parts per billion levels. Typical analytes are chemicals, explosives and organic solvents. Electric fields are used to manipulate the microfluidics. To make the surface acoustic wave (SAW) detector work for chemicals, a sorbent polymer film is deposited onto the SAW propagation path. As each analyte exits the GC column and passes across the SAW sensor, the coating momentarily absorbs the analyte. A gravimetric response is registered by incorporating the SAW device as the frequency control element of an oscillator circuit, or by comparing the phase shift across coated and uncoated (reference) devices. The SAW detector gives a sensitive and unique response to each analyte.

Recent developments in chip-based GC have been significant[72], especially in terms of speed, rapid heating, stationary phases and sensitive and selective detection. For detection, the development of both a mini FID[73, 74] and a mini FPD[75] have been reported. Metal oxide semiconductor detectors[76, 77] and a plasma emission detector[78] have also been used for on-chip GC.

On-chip CE

Capillary electrophoresis has been on chip since the 1990s and recent advances in microfabrication technology have enabled fast development of CE microchips for a wide range of applications[79]. Because there is no need for a liquid pump, CE is relatively easy to miniaturise. The only issues can be Joule heating, which is due to the current generated by the high voltage power supply, and the size of the power supply unit itself (though they are getting smaller[80]). The correct choice of substrate material, e.g. PDMS, and the use of very narrow channels aid in the better dissipation of this heat when compared with traditional CE and hence higher voltages can be used on-chip. Etched channels in CE are available in the same size range ($<25\,\mu m$) as fused silica capillaries. Channels are not always at right angles to each other and curved (serpentine) channels can also be used. These curved channels enable longer separation lengths on the same chip.

Injection is easy to control due to the plug-like flow profile and the availability of cross-T, double-T and other microfluidic designs. The movement of analytes in microchip CE is based on both electro-osmotic flow (EOF), which occurs when aqueous buffer and sample in a capillary has a voltage applied across it and there is a bulk flow movement in one direction, and electrophoresis, which is the movement of ions in a fluid under an applied voltage. CZE and MEKC are the most common modes of CE employed. Both glass and plastic substrates are used. Chips can be stacked to allow flow through the channels in a vertical as well as a horizontal direction, leading to more complex and flexible designs (Figure 10.9). Also, both reaction and separation can be carried out on the one chip[81].

In CE, the principle detection schemes are spectrometric and electrochemical. Fluorescence is easy to implement (especially off-chip), is extremely sensitive, which is useful since the sample volumes are typically very small, and is well understood. However, some compounds may need to be fluorescently labelled[82, 83]. This can be done prior to, during or after separation. Renzi *et al.* from Sandia National Laboratories have reported a handheld microanalytical instrument for CE analysis of proteins using laser-induced fluorescence detection[84]. The fused silica chip is 2×2 cm and features on-chip sample introduction, inlet port filters and a 10 cm separation column. Nanomolar concentrations of fluorescamine-labelled proteins were detected.

Instead of fluorescence, absorbance can be used if the limits of detection are not too low and as long as the microchannels are transparent in the radiation region being exploited. Microfluidic CE devices have been fabricated from calcium fluoride (CaF_2) to allow optical detection in the ultraviolet, visible and infrared spectral regions[85]. Fast chiral CE separations have been carried out with UV detection[86].

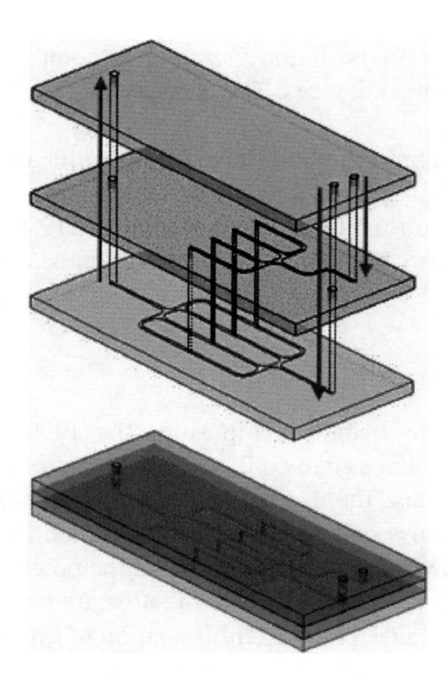

Figure 10.9 *A bonded stack of microfluidic chips allows vertical as well as horizontal flow. This approach is often used for increasing the number and/or density of fluid ports, whilst reducing the overall size of the final device. (Reproduced by permission of Epigem Ltd).*

Much effort has been invested in integrating mass spectrometry with on-chip CE. The flow rates typically used (nL to μL/min) are very suitable for electrospray ionisation (ESI) prior to MS. However, the buffers used in CE tend not to be compatible with ESI and there is also a need to decouple the two electric fields (one for the electrophoretic separation, one for the electrospray). One method that has been used for interfacing electrospray with chips is to bond electrospray nozzles/needles to the outlet of the microchannel[87]. Electrospray tips can also be incorporated onto the chip as part of the fabrication process[88]. Electrospray detection following separation by CE has worked well for proteins, carbohydrates[89] and many other compounds.

As electrodes are already small and compact, electrochemical detection works well for on-chip CE[90, 91]. Electrodes can be patterned onto the chip during fabrication by many methods, such as evaporation, sputtering and screen-printing. However, for amperometry, a challenge lies in the isolation of the detection electrodes from the high separation voltages. Ways around this include locating the electrodes outside the microchannel[92] or at the end of the microchannel[93], though some researchers have reported on the possibility of using electrodes inside the microchannel if they are well isolated[94]. Conductivity detection for on-chip CE has also been reported[95, 96]. In all cases of on-chip CE, poor detection sensitivity can be a limitation. Hence, various in-line preconcentration techniques have been developed, ranging in power and complexity, and there are now a number of well understood approaches capable of providing a 10^4 to 10^6 fold increase in sensitivity, even with complex samples[97].

Commercially, both Agilent Technologies and Caliper LifeSciences now cell CE microchips.

On-chip LC

Downsizing LC has many advantages but a very important one is that, by reducing the internal diameter of the column, separation performance is unaffected while sensitivity improves. Despite this, HPLC arrived on-chip later than GC due to difficulties with packing stationary phases and pumping. Regnier produced one of the earliest planar chip-based HPLC systems in 1998[98]. However, there were some disadvantages with his early design, such as the fact that the stationary phase was monolithic and nonporous, with a resultant small surface area. It was also made of quartz, which was labour intensive to prepare, and detection was difficult as the chip was small and the channels shallow. Researchers have since reported separations on C18 stationary phase in-chip channels, and other types of materials for packing into microchannels which have been investigated include continuous polymer beds[99], sol-gels[100, 101] and porous polymer monoliths. The pumping issue has already been discussed (Section 10.2.1) and solutions have varied from splitting the flow from conventional pressure-driven pumps (which is obviously not ideal) to using non pressure-driven micropumps. Glass is still the mainstay for LC separations due to compatibility with typical solvents used but more plastic chips are under development. Some research groups have focused on using chips to miniaturise the injector for multi-channel monolithic silica capillary HPLC[102].

Researchers at the Centre for Embedded Network Sensing have reported a number of LC chip developments using parylene microfluidics technology – an ion chromatography (IC) chip for separation of anions and an LC–ESI chip for protein and peptide separation. The LC–ESI chip incorporates three pumps (one for sample, two for solvent), a static mixer, a C18 column and an ESI nozzle for off-chip tandem MS[103]. Commercially, Agilent produces microfluidic LC chips that integrate a trapping column, separation column and electrospray source within a single structure[104, 105] (Figure 10.10). Crystal Vision Microsystems is developing and optimising a microscale HPLC LOC system for use in process analysis. A split flow is diverted from the main process and sent to the hybrid integrated sample pretreatment/HPLC cartridge. A H-filter performs the dilution,

Figure 10.10 *A commercially available HPLC chip (Copyright 2006, Agilent Technologies. Reproduced with permission.).*

filtration and/or pH conditioning required and the assay is carried out on the chip in under four minutes.

Microscale fluidic systems use small volumes so sensitivity of detection can be a challenge. Any detector for chip-based LC needs to be small and ideally have low power consumption. It is generally a problem of interfacing. Flow cell geometry is also a big factor, e.g. a U cell instead of linear flow cell can give a ten-fold increase in sensitivity for absorbance measurements. Electrochemical detection is very common, mainly amperometric and potentiometric, and very amenable to detection on chip. Fluorescence is more sensitive than UV–Vis absorbance and chemiluminescence is sensitive down to a single molecule, similar to LIF.

Potentiometric and refractive index detection are not affected by volume but are relatively insensitive in the nanolitre to picolitre range compared to amperometric detection (micro surface area) and fluorimetric detection (micro amount of material). At 1 pL, limits of detection are similar for potentiometry, amperometry and fluorescence. On-chip LC is very compatible with mass spectrometry due to the low volumes and flow rates required. Battery-operated ion trap MS has been reported but miniaturisation of MS offers no sensitivity or selectivity advantages. Electrospray ionisation (ESI) has been successfully integrated into a chip format allowing for many ESI nozzles on one chip. Arrays make pattern recognition possible.

As already mentioned, it is easier to miniaturise CE than pressure-based LC systems. This is because it is easier to apply a high voltage or heat (CE) to a chip than pressure from an engineering point of view. However, CE has some issues with limited selection of buffers, nonuniform (joule) heating and injection problems. Sample preparation can be easier in CE due to the use of mainly aqueous solvents but if organic solvents are required, LC might have an advantage subject to the tolerance of the chip. Capillary electrokinetic chromatography (CEC) may be a viable compromise between the two techniques.

10.3.4 Imaging Devices

Imaging instruments have certainly become much smaller in recent years and there are many examples of these devices in the literature. NASA in the USA is working on an optical scanning microscope based on microchannel filters and advanced electronic image sensors and that utilise X-ray illumination[106]. The FANSOM (fluorescence apertureless near-field scanning microscope) currently has a resolution of 4 nm, a limit that previously required the use of an electron microscope[107]. The focusing optics and lenses of a conventional optical microscope are replaced by a microchannel filter and a charge-coupled-device (CCD) image detector. There is a one-to-one mapping from a point on a specimen to a pixel in the image sensor, so that the output of the image sensor contains image information equivalent to that obtained from a microscope.

A new imaging device, called a two-photon microendoscope, that is less than 4 cm long and weighs only 3.9 g[108], has been developed by researchers at Stanford University and allows nerve cells and capillaries deep inside living subjects to be observed. It is a tiny portable handheld device with the full functionality of a microscope.

Nanosurf, which manufactures portable SPM instruments, has been selected by NASA to join a consortium to equip a Mars lander with an AFM in 2007[109]. The AFM weighs only 300 g and its power consumption is less than 10 W. The AFM scanner is

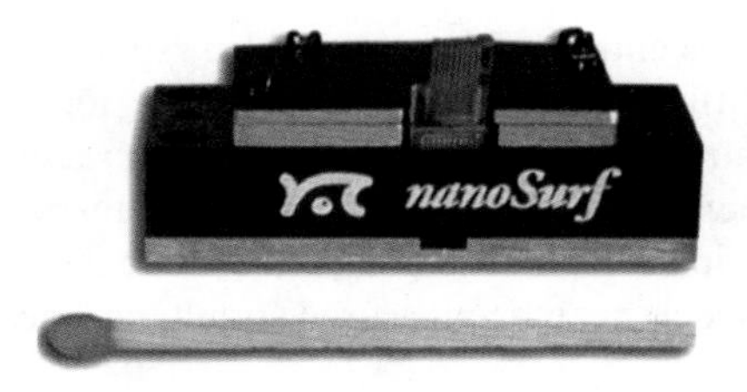

Figure 10.11 *The AFM scanner from Nanosurf will be part of the miniaturised suite of microscopic instrumentation on the Phoenix space mission to Mars (Reproduced by permission of Nanosurf).*

only 1.8 × 3.5 × 0.8 cm in size (Figure 10.11). It has eight sensors and cantilevers. Only one at a time is used for dynamic mode imaging and if one becomes worn out or dirty, the cantilever can be removed. The goal of the mission is to perform a scientific analysis of the Martian arctic soils for clues to its geologic history and potential for biology. One major challenge of the project is the extreme temperature conditions on Mars (−133 °C to 22 °C).

Given Imaging Inc. has developed PillCam, a tiny video camera in a capsule that can be swallowed. It is the same size as a multivitamin tablet. The device transmits colour digital pictures from the gastrointestinal tract to a receiver fitted against the patient's body. The images of the small intestine are transmitted by radio frequency to an array of sensors on a belt worn on the patient's waist. Approximately eight hours and 50 000 images later, the belt and recorder are returned to the clinic. Here, the images are downloaded onto a computer workstation and a doctor can then scan the video for abnormalities.

10.3.5 Electrochemical Devices

There are an increasing number of microelectrochemical devices being developed. A microfluidic chip for determining nucleic acids based on amperometric and coulometric detection has been reported[110]. This biosensor is composed of an ultramicroelectrode array on a glass chip and PDMS channels. In contrast to most microbiosensors, which are based on fluorescence for signal generation, only a simple potentiostat was required for both signal recording and interpretation. The assay took six minutes to complete, had a limit of detection for target DNA molecules of 1 fmol per assay and a dynamic range between 1 and 50 fmol.

Another research group has created an electrochemical microdevice which supports both amplification and detection of DNA[111]. The integrated chip houses an 8 μL reaction chamber, the associated temperature sensors and heaters and two electrochemical detection techniques, including metal complex intercalators and nanogold particles. The sensitivity of the device allows the detection of a few hundred copies of target DNA.

Zhu *et al.* have developed a chip for heavy metal ion electrochemical detection[112]. By controlling the microfluidics, a mercury droplet microelectrode is generated on the chip. Square wave stripping voltammetry is used for analysis.

Another electrochemical sensor for heavy metal detection in liquids has been fabricated by Miu *et al.*[113]. A three-electrode cell on silicon and voltammetric techniques for determining the metal concentrations are used. Also reported is an electrochemical chip which has a number of functions: as an oxygen sensor, for generating microfluidic pH gradients or for fluid control[114]. The device is based on a three-electrode (working, counter and reference) electrochemical cell design. Wang and Pumera have described a new dual electrochemical microchip detector based on both conductivity and amperometry[115]. Both ionic and electroactive species can be measured simultaneously.

10.3.6 Other Chip-based Devices

There are many other types of chip-based devices in development and, indeed, on the market. Many of these are for purpose point-of-care diagnostics[116]. A few of the main categories of chips are discussed here.

Protein Chips

A protein chip is a glass, plastic or silicon chip onto which different proteins have been attached at separate locations in an ordered manner to form a microscopic array. They are attached mainly by adsorption, absorption, covalent cross-linking or affinity binding. There are two main types of protein arrays: analytical arrays and functional protein arrays[117]. In the first type the capture molecules are antibodies or antibody mimics which are used to detect the presence and amount of protein in a sample. These are mainly used for diagnostics and expression profiling experiments. The second type of protein array involves immobilising proteins on a chip and using the chip to probe biochemical activity.

The most common type of capture molecules used on a protein chip are antibodies, though other proteins, such as peptides, nucleic acids, enzymes and receptors, have been spotted on to chips. Protein chips can be difficult to prepare, as the proteins need to be kept in an active state, in a high concentration and in a moist environment. Typical samples investigated are cell lysates for general research or blood and urine for diagnostic applications. Fluorescence detection methods are preferred since they are simple and sensitive. Other detection methods, where labelling is not required, can take the form of SELDI or AFM.

DNA Chips

A DNA chip (also known as a gene chip) is a glass, plastic or silicon chip on which different DNA segments have been attached at separate locations in an ordered manner to form a microscopic array. These are used for to monitor expression levels for thousands of genes simultaneously. For example, the chips can be used to identify disease genes by comparing gene expression in diseased and normal cells or to detect RNAs that are translated into active proteins. The DNA segments are oligonucleotides or fragments of polymerase chain reaction (PCR) products and are often referred to as probes or reporters.

Caliper/Agilent markets DNA chips which have become standard analytical tools for a number of life science laboratories. Commercial DNA chips are also available from companies such as Affymetrix and GE Healthcare. Microfluidic chips for carrying out purification processes for nucleic acids have been reported[118]. Cell isolation and lysis as

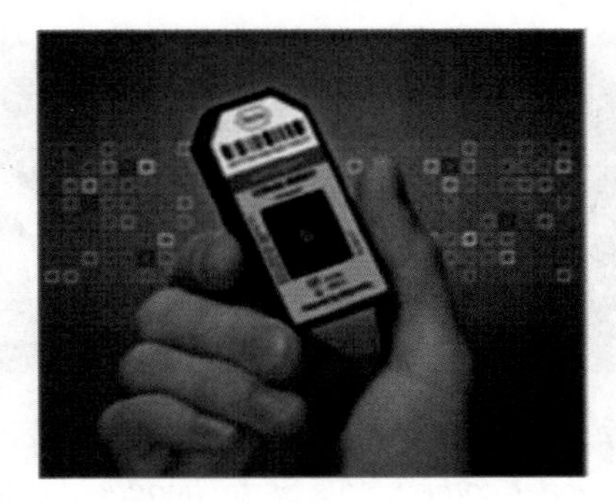

Figure 10.12 *The AmpliChip CYP450 (Reproduced by permission of Roche Ltd).*

well as the purification and recovery are done on one chip in nanolitre volumes. Different samples can be processed in parallel. A chip which can carry out PCR on 1 μl reaction samples has been reported[119]. Faster heating and cooling cycles reduce the amount of byproducts formed.

A single nucleotide polymorphism (SNP) is a variation at a single site in a strand of DNA. It is the most common type of variation in DNA and there are about 5–10 million SNPs in the human genome. SNP chips are a particular type of DNA chip that are used to identify the SNPs that cause genetic variation in a population. These genotyping experiments may be used for forensic applications, determining genetic predisposition to disease, or identifying DNA-based drug candidates.

Roche Diagnostics has recently released the AmpliChip CYP450 Test (Figure 10.12) which combines PCR technology and microarray technology in a single assay[120]. It analyses for 31 polymorphisms and mutations related to cytochrome P450 enzymes. The test reveals which variants of the genes for two of the most significant cytochrome P450 enzymes people have in their livers. Cytochrome P450 enzymes are produced in the liver and are involved in degrading the majority of prescribed drugs, hence they are a major focus of attention for researchers developing personalised therapeutics. For example, up to 10 % of the population do not experience any pain relief from codeine due to the specific version of one of the P450 enzymes found in their livers. The test has been approved for clinical use.

Chemical and Diagnostic Chips

A chemical chip is an array of chemical compounds spotted on a solid surface, such as glass, plastic or silicon. They are used to search for proteins that bind with specific chemical compounds and hence have applications in discovering new therapeutic targets. The chemical molecules can be immobilised onto the chip by covalent binding and the resultant chip is called a small molecule array. Alternatively the molecules can be simply spotted and dried onto the chip surface, in which case the experiments are referred to as micro arrayed compound screening experiments. The third method of immobilising the chemical compounds is to spot them on in a homogeneous solution, a method commercialised as DiscoveryDot™ technology.

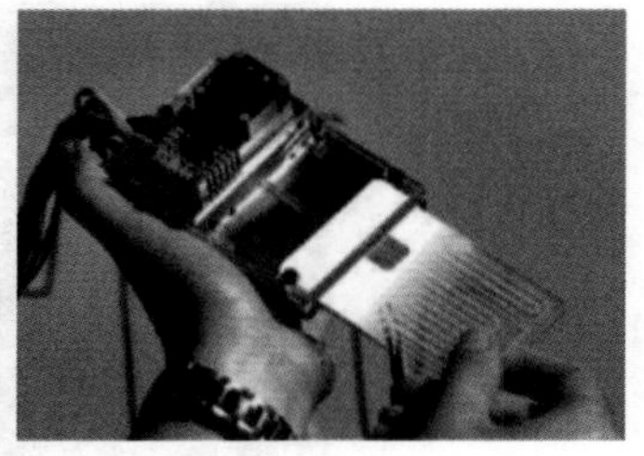

Figure 10.13 *The Quicklab device and diagnostic card; on the right, both the device and card are uncovered revealing the inner components of the handheld reader and the microfluidics on the chip (Photographs: Courtesy of Siemens AG).*

Diagnostic chips are widely reported in the literature. Ahn *et al.* have fabricated and developed a disposable chip for measuring oxygen, glucose and lactate in human blood[121]. The chip contains the biosensor array and integrated passive microfluidics and power source. There is also a small wrist-watch style reader for the chip. Goluch *et al.* have reported a microchip that can detect tiny amounts of a protein associated with breast and prostate cancers and that could, therefore, provide an early warning for those diseases[122]. The chip, by using gold nanoparticles and magnetic beads, was able to detect prostate specific antigen (PSA) at concentrations as low as 500 aM.

Siemens markets a device called Quicklab, which is used to carry out quick on-the-spot tests on blood or other biological samples in a clinical setting (Figure 10.13). It automatically extracts the genetic information of viruses, bacteria or body cells, so has wide-ranging medical diagnosis applications from detection of allergies to detection of hereditary diseases. The analytical system is based on an electrochemical microfluidic chip.

References

1. Manz, A., Graber, N. and Widmer, H.M. (1990) Miniaturized total chemical analysis systems: A novel concept for chemical sensing. *Sens Actuat B*, **1 (1–6)**, 244–248.
2. Terry, S.C., Jerman, J.H. and Angell, J.B. (1979) A gas chromatographic air analyzer fabricated on a silicon wafer. *IEEE Trans Electron Dev*, **26 (12)**, 1880–1886.
3. Koh, C.G., Tan, W., Zhao, M. *et al.* (2003) Integrating polymerase chain reaction, valving, and electrophoresis in a plastic device for bacterial detection. *Anal Chem*, **75 (17)**, 4591–4598.
4. Legendre, L.A., Bienvenue, J.M., (2006) Roper, M.G. *et al.* A simple, valveless microfluidic sample preparation device for extraction and amplification of DNA from nanoliter-volume samples. *Anal Chem*, **78 (5)**, 1444–1451.
5. Wang, H., Chen, J., Zhu, L. *et al.* (2006) Continuous flow thermal cycler microchip for DNA cycle sequencing. *Anal Chem*, **78 (17)**, 6223–6231.
6. Auroux, P.A., Koc, Y., deMello, A. *et al.* (2004) Miniaturised nucleic acid analysis. *Lab Chip*, **6**, 534–546.
7. Seong, G.H., Heo, J. and Crooks, R.M. (2003) Measurement of enzyme kinetics using a continuous-flow microfluidic system. *Anal Chem*, **75 (13)**, 3161–3167.
8. Roper, M.G., Shackman, J.G., Dahlgren, G.M. and Kennedy, R. (2003) Microfluidic chip for continuous monitoring of hormone secretion from live cells using an electrophoresis-based immunoassay. *T.Anal Chem*, **75 (18)**, 4711–4717.

9. Pihl, J., Sinclair, J., Karlsson, M. and Orwar, O. (2005) Microfluidics for cell-based assays. *Mater Today*, **8 (12)**, 46–51.
10. Yi, C., Li, C.-W., Ji, S. and Yang, M. (2006) Microfluidics technology for manipulation and analysis of biological cells. *Anal Chim Acta*, **560 (1–2)**, 1–23.
11. Kutter, J.P. and Fintschenko, Y. (Eds) *Separation Methods In Microanalytical Systems*. CRC Press (2005).
12. Tabeling, P. *Introduction to Microfluidics*. Oxford University Press (2006).
13. Juncker, D., Schmid, H., Drechsler, U. *et al.* (2002) Autonomous microfluidic capillary system. *Anal Chem*, **74 (24)**, 6139–6144.
14. Zoval, J.V. and Madou, M.J. (2004) Centrifuge-based fluidic platforms. Proc IEEE, **92 (1)**, 140–153.
15. Gustafsson, M., Hirschberg, D., Palmberg, C. *et al.* (2004) Integrated sample preparation and MALDI mass spectrometry on a microfluidic compact disk. *Anal Chem*, **76 (2)**, 345–350.
16. Xie, J., Miao, Y., Shih, J. *et al.* (2004) An electrochemical pumping system for on-chip gradient generation. *Anal Chem*, **76 (13)**, 3756–3763.
17. Munyan, J.W., Fuentes, H.V., Draper, M. *et al.* (2003) Electrically actuated, pressure-driven microfluidic pumps. *Lab Chip*, **4**, 217–220.
18. Tanaka, Y., Sato, K., Shimizu, T. *et al.* (2007) A micro-spherical heart pump powered by cultured cardiomyocytes. *Lab Chip*, **2**, 207–212.
19. A tiny pump promises big time performance; Binghamton University invention could ‘sweeten’ diabetes therapy within five years, *Science Daily*, 12 June 2003.
20. Portable ‘lab on a chip’ could speed blood tests, *Science Daily*, 19 October 2006.
21. Sudarsan, A.P. and Ugaz, V.M. (2006) Multivortex micromixing. *Proc Natl Acad Sci*, **103 (19)**, 7228–33.
22. Sudarsan, A.P. and Ugaz, V.M. (2006) Fluid mixing in planar spiral microchannels. *Lab Chip*, **6**, 74–82.
23. Xia, N., Hunt, T.P., Mayers, B.T. *et al.* (2006) Combined microfluidic-micromagnetic separation of living cells in continuous flow. *Biomed Microdev*, **8 (4)**, 299–308.
24. Herrmann, M., Veres, T. and Tabrizian, M. (2006) Enzymatically-generated fluorescent detection in micro-channels with internal magnetic mixing for the development of parallel microfluidic ELISA. *Lab Chip*, **6**, 555–560.
25. Sun, Y., Giebink, N.C., Kanno, H. *et al.* (2006) Management of singlet and triplet excitons for efficient white organic light-emitting devices. *Nature*, **440**, 908–912.
26. Liu, J., Tu, G., Zhou, O. *et al.* (2006) Highly efficient green light emitting polyfluorene incorporated with 4-diphenylamino-1,8-naphthalimide as green dopant. *J Mater Chem*, **15**, 1431–1438.
27. Chan, K.L., McKiernan, M.J., Towns, C.R. and Holmes, A.B. (2005) Poly(2,7-dibenzosilole): a blue light emitting polymer. *JACS*, **127 (21)**, 7662–7663.
28. Ashwell, G.J., Urasinska, B. and Tyrrell, W.D. (2006) Molecules that mimic Schottky diodes. *Phys Chem Chem Phys*, **8**, 3314–3319.
29. Vulto, P., Glade, N., Altomare, L. *et al.* (2005) Microfluidic channel fabrication in dry film resist for production and prototyping of hybrid chips. *Lab Chip*, **5**, 158–162.
30. Gross, G.M., Grate, J.W. and Synovec, R.E. (2004) Monolayer-protected gold nanoparticles as an efficient stationary phase for open tubular gas chromatography using a square capillary: Model for chip-based gas chromatography in square cornered microfabricated channels. *J Chrom A*, **1029 (1–2)**, 185–192.
31. www.newmaterials.com/news/9310.asp
32. de Mello, A.J. and Beard, N. Focus. (2003) Dealing with ‘real’ samples: sample pre-treatment in microfluidic systems. *Lab Chip*, **1**, 11N–20N.
33. Xu, Y., Zhang, C.X., Janasek, D. and Manz, A. (2003) Sub-second isoelectric focusing in free flow using a microfluidic device. *Lab Chip*, **3**, 224 –227.
34. Broyles, B.S., Jacobson, S.C. and Ramsey, J.M. (2003) Sample filtration, concentration, and separation integrated on microfluidic devices. *Anal Chem*, **75 (11)**, 2761–2767.
35. Jemere, A.B., Oleschuk, R.D., Ouchen, F. *et al.* (2002) An integrated solid-phase extraction system for sub-picomolar detection. *Electrophoresis*, **23 (20)**, 3537–3544.

36. Ekstrom, S., Malmstrom, J., Wallman, L. *et al.* (2002) On-chip microextraction for proteomic sample preparation of in-gel digests. *Proteomics*, 2002, **2 (4)**, 413–421.
37. Kuban, P., Dasgupta, P.K. and Morris, K.A. (2002) Microscale continuous ion exchanger. *Anal Chem*, **74 (21)**, 5667–5675.
38. Lion, N., Gellon, J.-O., Jensen, H. and Girault, H.H. (2003) On-chip protein sample desalting and preparation for direct coupling with electrospray ionization mass spectrometry. *J Chrom A*, **1003 (1–2)**, 11–19.
39. Yang, Z., Matsumoto, S. and Maeda, R. (2002) A prototype of ultrasonic micro-degassing device for portable dialysis system. *Sens Actuat A*, **95 (2–3)**, 274–280.
40. Jandik, P., Weigl, B.H., Kessler, N. *et al.* (2002) Initial study of using a laminar fluid diffusion interface for sample preparation in high-performance liquid chromatography. *J Chrom A*, **954 (1–2)**, 33–40.
41. Schilling, E.A., Kamholz, A.E. and Yager, P. (2002) Cell lysis and protein extraction in a microfluidic device with detection by a fluorogenic enzyme assay. *Anal Chem*, **74 (8)**, 1798–1804.
42. Crowley, T.A. and Pizziconi, V. (2005) Isolation of plasma from whole blood using planar microfilters for lab-on-a-chip applications. *Lab Chip*, **9**, 922–929.
43. VanDelinder, V. and Groisman, A. (2006) Separation of plasma from whole human blood in a continuous cross-flow in a molded microfluidic device. *Anal Chem*, **78 (11)**, 3765–3771.
44. Qin, J., Fung, Y., Zhu, D. and Lin, B. (2004) Native fluorescence detection of flavin derivatives by microchip capillary electrophoresis with laser-induced fluorescence intensified charge-coupled device detection. *J Chrom A*, **1027 (1–2)**, 223–229.
45. Namasivavam, V., Lin, R., Johnson, B. *et al.* (2004) Advances in on-chip photodetection for applications in miniaturized genetic analysis systems. *J Micromech Microeng*, **14 (1)**, 81–90.
46. Minas G, Wolffenbuttel, R.F. and Correia, J.H. (2005) A lab-on-a-chip for spectrophotometric analysis of biological fluids. *Lab Chip*, **5 (11)**, 1303–1309.
47. Benetton, S., Kameoka, J., Tan, A. *et al.* (2003) Chip-based P450 drug metabolism coupled to electrospray ionization-mass spectrometry detection. *Anal Chem*, **75 (23)**, 6430–6436.
48. Sillon, N. and Baptist, R. (2002) Micromachined mass spectrometer. *Sens Actuat B*, 2002, **83 (1–3)**, 129–137.
49. Stadnik, D. and Dybko, A. (2003) Fibre optic coupler as a detector for microfluidic applications. *Analyst*, **128**, 523–526.
50. Hüller J, Pham, M.T. and Howitz, S. (2003) Thin layer copper ISE for fluidic microsystem. *Sens Actuat B*, **91 (1–3)**, 17–20.
51. García, C.D. and Henry, C.S. (2003) Direct determination of carbohydrates, amino acids, and antibiotics by microchip electrophoresis with pulsed amperometric detection. *Anal Chem*, **75 (18)**, 4778–4783.
52. Wu C.-C., Wu, R.-G., Huang J.-G. *et al.* (2003) Three-electrode electrochemical detector and platinum film decoupler integrated with a capillary electrophoresis microchip for amperometric detection. *Anal Chem*, **75 (4)**, 947–952.
53. Hebert, N.E., Snyder, B., McCreery, R.L. *et al.* (2003) Performance of pyrolyzed photoresist carbon films in a microchip capillary electrophoresis device with sinusoidal voltammetric detection. *Anal Chem*, **75 (16)**, 4265–4271.
54. Galloway, M., Stryjewski, W., Henry, A. *et al.* (2003) Contact conductivity detection in poly(methylmethacylate)-based microfluidic devices for analysis of mono- and polyanionic molecules. *Anal Chem*, **75 (16)**, 2407–2415.
55. Pumera, M., Merkoci, A. and Alegret, S. (2006) New materials for electrochemical sensing. VII. Microfluidic chip platforms. *TRAC*, **25 (3)**, 219–235.
56. Hansen, C. and Quake, S.R. (2003) Microfluidics in structural biology: smaller, faster… better. *Curr Opin Struct Biol*, **13 (5)**, 538–544.
57. Dethy, J.M., Ackermann, B.L., Delatour, C. *et al.* (2003) Demonstration of direct bioanalysis of drugs in plasma using nanoelectrospray infusion from a silicon chip coupled with tandem mass spectrometry. *Anal Chem*, **75 (4)**, 805–811.
58. Vollmer, M., Hoerth, P., Rozing, G. *et al.* (2006) Multi-dimensional HPLC/MS of the nucleolar proteome using HPLC-chip/MS. *J Sep Sci*, **29 (4)**, 499–509.

59. Boernsen, K.O., Gatzek, S. and Imbert, G. Controlled protein precipitation in combination with chip-based nanospray infusion mass spectrometry. An approach for metabolomics profiling of plasma. *Anal Chem*, **77 (22)**, 7255–7264.
60. Zamfir, A.D., Bindila, L., Lion, N. *et al.* (2005) Chip electrospray mass spectrometry for carbohydrate analysis. *Electrophoresis*, **26 (19)**, 3650–3673.
61. Ostman, P., Jantti, S., Grigoras, K. *et al.* (2006) Capillary liquid chromatography–microchip atmospheric pressure chemical ionization–mass spectrometry. *Lab Chip*, **6 (7)**, 948–953.
62. Chang, Y.Z., Yang, M.W. and Wang, G.J. (2005) A new mass spectrometry electrospray tip obtained via precise mechanical micromachining. *Anal Bioanal Chem*, **383 (1)**, 76–82.
63. Muck, A. and Svatos, A. (2004) Atmospheric molded poly(methylmethacrylate) microchip emitters for sheathless electrospray. *Rap Comm Mass Spectrom*, **18 (13)**, 1459–1464.
64. Schilling, M., Nigge, W., Rudzinski, A. *et al.* (2004) A new on-chip ESI nozzle for coupling of MS with microfluidic devices. *Lab Chip*, **4 (3)**, 220–224.
65. Gustafsson, M., Hirschberg, D., Palmberg, C. *et al.* (2004) Integrated sample preparation and MALDI mass spectrometry on a microfluidic compact disk. *Anal Chem*, **76 (2)**, 345–350.
66. Guo, X., Chan-Park, M.B., Yoon, S.F. *et al.* (2006) UV embossed polymeric chip for protein separation and identification based on capillary isoelectric focusing and MALDI-TOF-MS. *Anal Chem*, **78 (10)**, 3249–3256.
67. Vollmer, M., Hörth, P., Rozing, G. *et al.* (2006) Multi-dimensional HPLC/MS of the nucleolar proteome using HPLC-chip/MS. *J Sep Sci*, **29 (4)**, 499–509.
68. Wensink, H., Benito Lopez, F., Hermes, D.C. *et al.* (2005) Measuring reaction kinetics in a lab-on-a-chip by microcoil NMR. *Lab Chip*, **5**, 280–284.
69. Sorli, B., Chateaux, J.F., Pitaval, M. *et al.* (2004) Micro-spectrometer for NMR: analysis of small quantities in vitro. *Meas Sci Technol*, **15 (5)**, 877–880.
70. Hilty, C., McDonnell, E.E., Granwehr, J. *et al.* (2005) Microfluidic gas-flow profiling using remote-detection NMR. *PNAS*, **102**, 14960–14963.
71. de Mello A. (2002) On-chip chromatography:the last twenty years. *Lab Chip*, **2**, 48N–54N.
72. Gross, G.M., Reid, V.R. and Synovec, R.E. (2005) Recent advances in instrumentation for gas chromatography. *Curr Anal Chem*, **1 (2)**, 135–147.
73. Zimmermann, S., Krippner, P., Vogel, A. and Müller J. (2002) Miniaturized flame ionization detector for gas chromatography. *Sens Actuat B*, **83 (1–3)**, 285–289.
74. Zimmermann, S., Wischhusen and Müller J. (2002) Micro flame ionization detector and micro flame spectrometer. *Sens Actuat B*, **63 (3)**, 159–166.
75. Thurbide, K.B. and Anderson, C.D. (2003) Flame photometric detection inside of a capillary gas chromatography column. *Analyst*, **128 (6)**, 616–622.
76. Lorenzelli, L., Benvenuto, A., Adami, A. *et al.* (2005) Development of a gas chromatography silicon-based microsystem in clinical diagnostics. *Biosens Bioelectron*, **20 (10)**, 1968–1976.
77. Zampolli, S., Elmi, I., Stürmann J. *et al.* (2005) Selectivity enhancement of metal oxide gas sensors using a micromachined gas chromatographic column. *Sens Actuat B*, **105 (2)**, 400–406.
78. Bessoth, F.G., Naji, O.P., Eijkel, J.C.T. and Manz, A. (2002) Towards an on-chip gas chromatograph: the development of a gas injector and a dc plasma emission detector. *J Anal Atom Spectrom*, **17 (8)**, 794–799.
79. Dolnik, V. and Liu, S.R. (2005) Applications of capillary electrophoresis on microchip. *J Sep Sci*, **28 (15)**, 1994–2009.
80. Erickson, D., Sinton, D. and Li, D. (2004) A miniaturized high-voltage integrated power supply for portable microfluidic applications. *Lab Chip*, **2**, 87–90.
81. Belder, D., Ludwig, M., Wang, L.-W. and Reetz, M.T. (2006) Enantioselective catalysis and analysis on a chip. *Angew Chemie*, **45 (15)**, 2463–2466.
82. Götz, S., Revermann, T. and Karst, U. (2007) Quantitative on-chip determination of taurine in energy and sports drinks. *Lab Chip*, **1**, 93–97.
83. Skelley, A.M., Scherer, J.R., Aubrey, A.D. *et al.* (2005) Development and evaluation of a microdevice for amino acid biomarker detection and analysis on Mars. *PNAS*, **102**, 1041–1046.
84. Renzi, R.F., Stamps, J., Horn, B.A. *et al.* (2005) Hand-held microanalytical instrument for chip-based electrophoretic separations of proteins. *Anal Chem*, **77**, 435–441.

85. Pan, T., Kelly, R.T., Asplund, M.C. and Woolley, A.T. (2004) Fabrication of calcium fluoride capillary electrophoresis microdevices for on-chip infrared detection. *J Chrom A*, **1027 (1–2)**, 231–235.
86. Ludwig, M., Kohler, F. and Belder, D. (2003) Chiral CE – High-speed chiral separations on a microchip with UV detection. *Electrophoresis*, **24 (18)**, 3233–3238.
87. Lion, N., Rohner, T.C., Dayon, L. *et al.* (2003) Microfluidic systems in proteomics. *Electrophoresis*, **24 (21)**, 3533–3562.
88. Huikko, K., Kostiainen, R. and Kotiaho, T. (2003) Introduction to micro-analytical systems: bioanalytical and pharmaceutical applications. *Eur J Pharm Sci*, **20 (2)**, 149–171.
89. Zamfir, A.D., Bindila, L., Lion, N. *et al.* (2005) Chip electrospray mass spectrometry for carbohydrate analysis. *Electrophoresis*, **26 (19)**, 3650–3673.
90. Wang, J. (2005) Electrochemical detection for capillary electrophoresis microchips: a review. *Electroanalysis*, **17 (13)**, 1133–1140.
91. Jiang, L., Lu, Y., Dai, Z. *et al.* (2005) Mini-electrochemical detector for microchip electrophoresis. *Lab Chip*, **9**, 930–934.
92. Martin, S., Gawron, A.J., Lunte, S.M. and Henry, C.S. (2000) Dual-electrode electrochemical detection for poly(dimethylsiloxane)-fabricated capillary electrophoresis microchips. *Anal Chem*, **72 (14)**, 3196–3202.
93. Wang, J., Tian, B. and Sahlin, E. (1999) Integrated electrophoresis chips/amperometric detection with sputtered gold working electrodes. *Anal Chem*, **71 (17)**, 3901–3904.
94. Martin, R.S., Ratzlaff, K.L., Huynh, B.H. and Lunte, S.M. (2002) In-channel electrochemical detection for microchip capillary electrophoresis using an electrically isolated potentiostat. *Anal Chem*, **74 (5)**, 1136–1143.
95. Lichtenberg, J., de Rooij, N.F. and Verpoorte, E. (2002) A microchip electrophoresis system with integrated in-plane electrodes for contactless conductivity detection. *Electrophoresis*, **23 (21)**, 3769–3780.
96. Pumera, M., Wang, J., Opekar, F. *et al.* (2002) Contactless conductivity detector for microchip capillary electrophoresis. *Anal Chem*, **74 (9)**, 1968–1971.
97. Breadmore, M.C. (2007) Recent advances in enhancing the sensitivity of electrophoresis and electrochromatography in capillaries and microchips. *Electrophoresis*, **28 (1–2)**, 254–281.
98. He, B. and Regnier, F. (1998) Microfabricated liquid chromatography columns based on collocated monolith support structures. *J Pharm Biomed Anal*, **17 (6–7)**, 925–932.
99. Ericson, C., Holm, J., Ericson, T. and Hjertén, S. (2000) Electro-osmosis and pressure-driven chromatography in chips using continuous beds. *Anal Chem*, **72 (1)**, 81–87.
100. Breadmore, M.C., Shrinivasan, S., Wolfe, K.A. *et al.* (2002) Towards a microchip-based chromatographic platform. Part 1: Evaluation of sol-gel phases for capillary electrochromatography. *Electrophoresis*, **23 (20)**, 3487–3495.
101. Breadmore, M.C., Shrinivasan, S., Karlinsey, J. *et al.* (2003) Towards a microchip-based chromatographic platform. Part 2: Sol-gel phases modified with polyelectrolyte multilayers for capillary electrochromatography. *Electrophoresis*, **24 (7–8)**, 1261–1270.
102. Shintani, Y., Hirako, K., Motokawa, M. *et al.* Development of miniaturized multi-channel high-performance liquid chromatography for high-throughput analysis. *J Chrom A*, **1073 (1–2)**, 17–23.
103. Xie, J., Miao, Y., Shih, J. *et al.* (2005) Microfluidic platform for liquid chromatography-tandem mass spectrometry analyses of complex peptide mixtures. *Anal Chem*, **77 (21)**, 6947–6953.
104. Yin, H., Killeen, K., Brennen, R. *et al.* (2005) Microfluidic chip for peptide analysis with an integrated hplc column, sample enrichment column, and nanoelectrospray tip. *Anal Chem*, **77**, 527–533.
105. Fortier, M.H., Bonneil, E., Goodley, P. and Thibault, P. (2005) Integrated microfluidic device for mass spectrometry-based proteomics and its application to biomarker discovery programs. *Anal Chem*, **77**, 1631–1640.
106. Scanning microscopes using x-rays and microchannels. *NASA Tech Briefs*, 2003, **27 (12)**, NPO-20873.
107. http://solarsystem.nasa.gov/scitech/display.cfm?ST_ID=869

108. Flusberg, B.A., Jung, J.C., Cocker, E.D. *et al.* (2005) *In vivo* brain imaging using a portable 3.9 gram two-photon fluorescence microendoscope. *Optics Letts*, **30 (17)**, 2272–2274.
109. www.mars-afm.ch, http://monet.physik.unibas.ch/famars/afm_prin.htm, http://phoenix.lpl.arizona.edu/science_meca.php
110. Goral, V.J., Zaytseva, N.V. and Baeumner A.J. (2006) Electrochemical microfluidic biosensor for the detection of nucleic acid sequences. *Lab Chip*, **3**, 414–421.
111. Lee, TM.-H., Carles, M.C. and Hsing, I.-M. (2003) Microfabricated PCR-electrochemical device for simultaneous DNA amplification and detection. *Lab Chip*, **2**, 100–105.
112. Zhu X.-S., Gao, C., Choi, J.-W., Bishop, P.L. and Ahn, C.H. (2005) On-chip generated mercury microelectrode for heavy metal ion detection. *Lab Chip*, **2**, 212–217.
113. Miu, M., Angelescu, A., Kleps, I. and Simion, M. (2005) Electrochemical sensors for heavy metals detection in liquid media. *Int J Environ Anal Chem*, **85 (9–11)**, 675–679.
114. Mitrovski, S.M. and Nuzzo, R.G. (2005) An electrochemically driven poly(dimethylsiloxane) microfluidic actuator: oxygen sensing and programmable flows and pH gradients. *Lab Chip*, **6**, 634–645.
115. Wang, J. and Pumera, M. (2002) Dual conductivity/amperometric detection system for microchip capillary electrophoresis. *Anal Chem*, **74 (23)**, 5919–5923.
116. Chin, C.D., Linder, V. and Sia, S.K. (2007) Lab-on-a-chip devices for global health: past studies and future opportunities. *Lab Chip*, **7**, 41–57.
117. Zhu, H. and Snyder, M. (2003) Protein chip technology. *Curr Opin Chem Biol*, **7**, 55–63.
118. Hong, J.W., Studer, V., Hang, G. *et al.* (2004) A nanoliter-scale nucleic acid processor with parallel architecture. *Nature*, **22**, 435–439.
119. Neuzil, P., Pipper, J. and Hsieh, T.M. (2006) Disposable real-time microPCR device: lab-on-a-chip at a low cost. *Mol BioSyst*, **2**, 292–298.
120. Medicine gets personalised. *Chemistry World*, 2005, **2 (7)**, 36.
121. Ahn, C.H., Choi, J.-W., Beaucage, G. *et al.* (2004) Disposable smart lab on a chip for point of care clinical diagnostics. *Proc IEEE*, **92 (1)**, 154–173.
122. Goluch, E.D., Nam, J.-M., Georganopoulou, G.D. *et al.* (2006) A bio-barcode assay for on-chip attomolar-sensitivity protein detection. *Lab Chip*, **6**, 1293–1299.

Index

Page numbers referring to illustrations are indicated in italic type, tables in bold type.

Analytical Instrumentation: A Guide to Laboratory, Portable and Miniaturized Instruments G. McMahon